BECOMING FEARLESS

FINDING COURAGE IN THE AFRICAN WILDERNESS

BRENDA E SMITH

The names of certain characters have been changed to protect their privacy.

Crocodile illustration sourced from Vecteezy.com

Author photo credit - JLBowler

Publishing Services provided by Paper Raven Books LLC

Cover Design by SusansArt

Printed in the United States of America

First Printing, 2023

Paperback ISBN: 979-8-988-3878-1-7
Hardback ISBN: 979-8-988-3878-0-0

savor the taste of honey-drenched flapjacks and warm, crusty bread, baked in the coals of a campfire, and you'll marvel at a dazzling sunset halfway up Mt. Kilimanjaro. Smith's life-altering journey encourages readers to reach past fear and discover the wonders of the unfamiliar."
—Robin Clifford Wood, author of *The Field House*, 2022 Maine Literary Award winner for non-fiction

"In *Becoming Fearless*, Brenda Smith writes an adventure story from a valuable and unique point of view, that of someone resisting the adventure, rather than embracing or fetishizing it. Throughout the rather amazing journey she documents, she constantly questions her decision to be in this position, in this place, yet carries on. Ironically, it is through recognizing her limits while on Mt. Kilimanjaro that Smith discovers a true courage, and it is this courage which serves as the motivation for her later, very different types of adventures, which we learn about in an epilogue."
—Michael D. Burke, author of *The Same River Twice*, and Emeritus Professor of English and Creative Writing at Colby College

"If you can learn to face adversity in some manner, particularly in the wilderness, it will be something that becomes an ingrained lifelong process for dealing with issues through the rest of your life outside the wilderness."

—Richard Bangs, *Leading Steep* Podcast, 2021

MAP OF TANZANIA

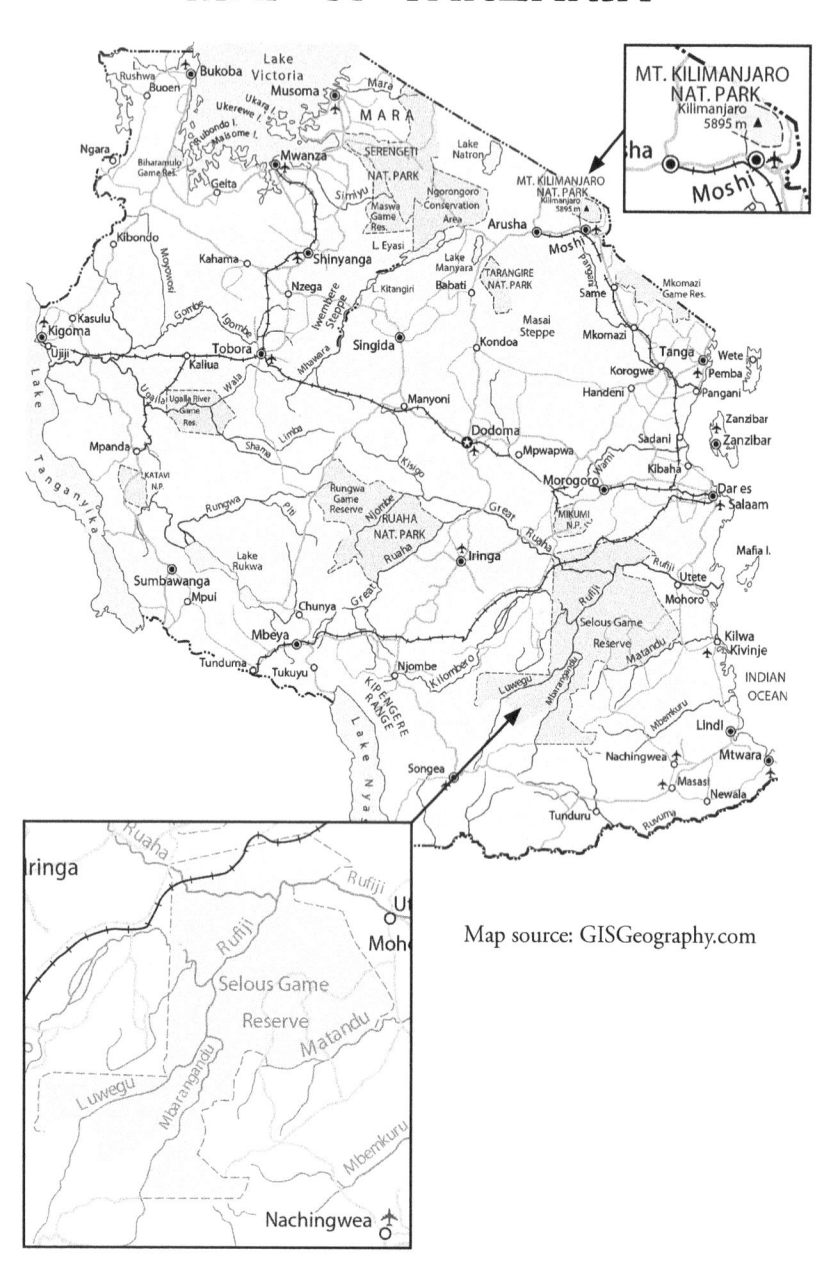

Map source: GISGeography.com

SELOUS GAME RESERVE

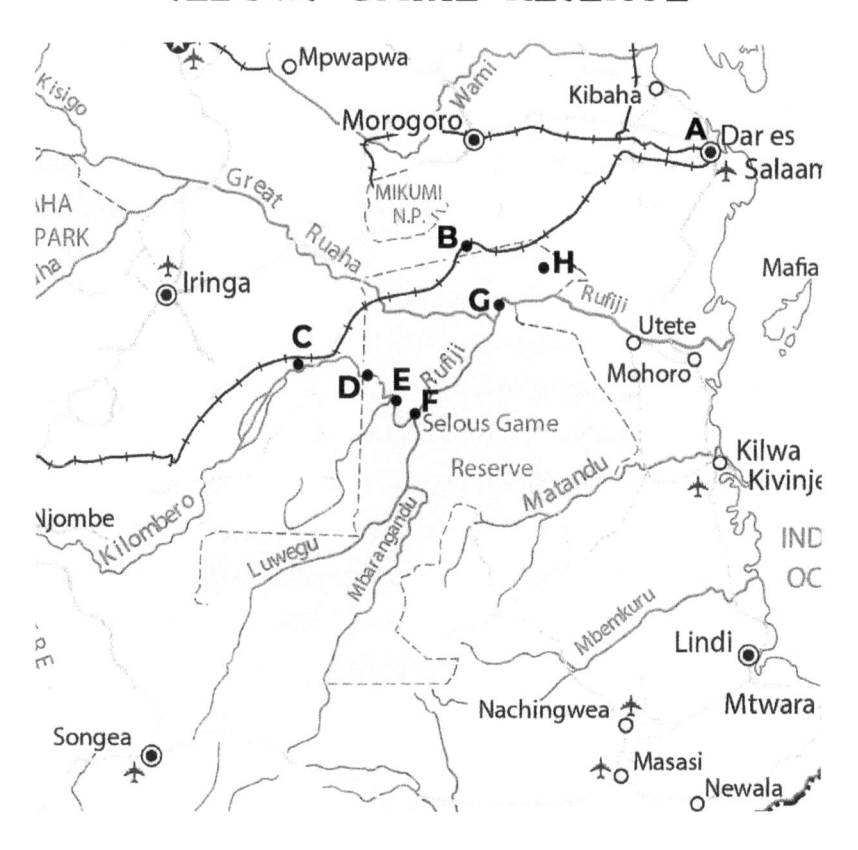

A : Starting Point

B : Kerosene Mango - Tanzam Railway

C : Ifakara

D : Hippo Bites Richard's Boat

E : Ranger Shoots Hippo

F : Shiguri Falls

G : Stiegler's Gorge

H : Beho Beho Camp

Map source: GISGeography.com

ROUTE ON MT KILIMANJARO

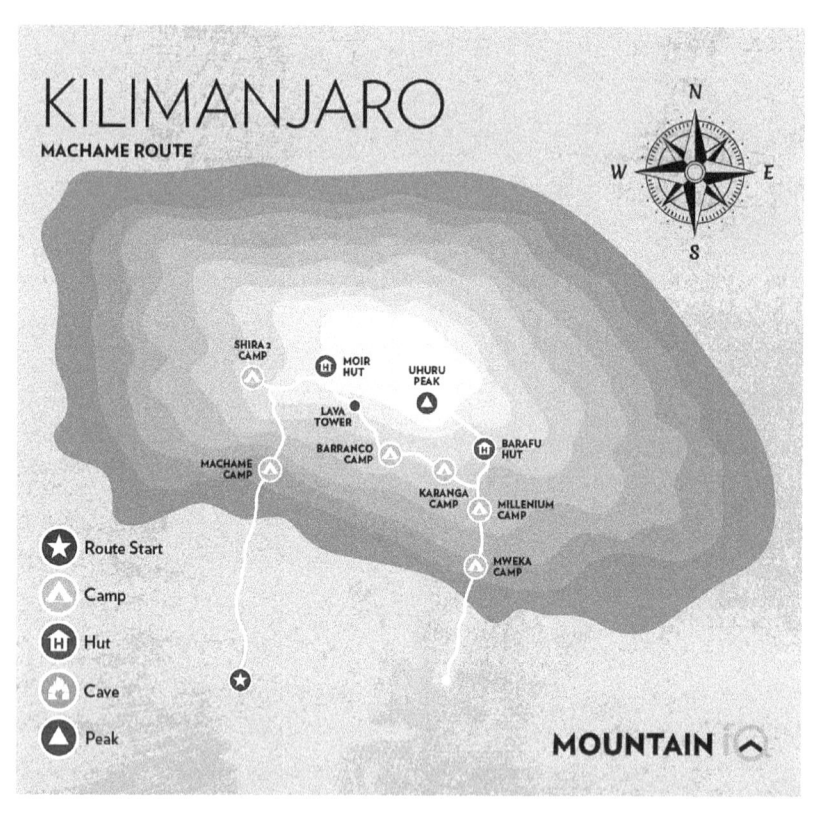

Map courtesy of Mountain IQ

1

IN OVER MY HEAD

I BLAME JIMMY BUFFETT. If I'd never heard the catchy lyrics of his 1977 hit song "Changes in Latitude, Changes in Attitude," I might never have gone on a rafting trip. And if I'd never gone rafting, I wouldn't be in Tanzania risking my life navigating a river teeming with hippos and crocodiles in an inflatable raft. That song had pierced my soul and excited the playful spirit within me. Jimmy lived in a world where nothing remained the same, and as enticing as that sounded, I never expected I'd have the courage to follow in his footsteps.

Everything about the first 25 years of my life had revolved in a familiar loop of sameness. Same Boston suburbs. Same group of friends. Same ambition to excel in schools and jobs. I'd had no desire to break out of my sensibly molded life. But Jimmy had a different message. His song encouraged us to escape from our routine: go traveling, meet people, and laugh. Buffett had found the key to unlock the good life, and I wanted what he had. It turned out that getting it wasn't as easy as he promised.

Just three years earlier, I worked as a staff auditor for Arthur Andersen, one of the country's most prestigious public accounting firms. With the generous salary I earned, I lived comfortably. I could easily afford the stylishly conservative vested business suits required by our dress code. In this job I traveled to corporate client offices spread across New England, where I put in long hours of tedious devil-in-the-details work demanded by our audits. But as a diehard homebody, I especially dreaded the Monday morning rat race to catch planes to out-of-town clients where I'd live out of hotels, sometimes for weeks.

When not reviewing audit work papers in client offices, I worked in an open bullpen on the 26th floor of a financial district skyscraper with a bird's eye view of Boston. The daily grind of reconciling account balances and documenting variances lacked excitement, but my training had prepared me for auditing's repetitiveness. I felt in control knowing what to expect.

One day, while poring over work papers with a colleague in the bullpen, we discussed our plans for our summer vacations. As an afterthought, I said, "I wish I had more courage, because someday I'd like to go rafting on the Colorado River through the Grand Canyon. I'd never go by myself, though. If I could ever find a brave soul who would go with me…" As I spoke, it occurred to me that even if I found a fellow traveler, such a rigorous trip would require a boldness I just didn't have.

My fellow auditors tended not to be outdoorsy sorts, so it astonished me when my colleague replied, "I've always wanted to do that trip too."

"No way! Really?" Out of the blue, in the least likely of places, I'd discovered a daring accomplice. Now I needed to dig deep to find the courage to make a reservation. After a few days of weighing options, we booked a 14-day river trip with an outfitter called OARS, Inc. Jimmy Buffett would have cheered us on.

We both imagined this trip as the greatest adventure of our lifetimes. But the more I read to prepare for the river journey, the more fearful I became of the infamous Lava Falls, mightiest of all rapids on the Colorado. This tumultuous cascade of bank-to-bank thundering, frothing whitewater had the reputation of swallowing rafts whole. Even destroying them.

Months of mounting anxiety about drowning followed, including a few sweat-drenched nightmares that jolted me wide-eyed awake. Three times I picked up the phone to cancel my reservation. But I never actually dialed because backing out on my colleague would have been reprehensible. I tried to console myself with logic. Boatloads of people ran Lava Falls every day during the summer. With the raft's navigation in the hands of a professionally trained guide, I figured the odds of being a victim had to be extremely low. Still, I had to face it; I was *not* prepared to die at 25.

ON OUR 10TH DAY, 179.2 miles into our trip, I finally faced my dreaded nemesis. Our raft had floated to within 10 yards of the gargantuan thundering pour-over at the top of Lava Falls, before our guide, Tom, realized our raft had drifted off course. At least five yards separated our raft from the telltale line of bubbles dancing on the water's surface that guides used to locate the safest entry point into the rapid. Panicking, he screamed at us to hyperventilate. For a few seconds, I sucked hard to fill my lungs with oxygen. Then our raft plunged into a monstrous, crashing abyss. The force of the water flung our raft backward, flipping it end over end. Instantly, the deafening foaming current ingested me, gulping me helplessly downward into its black depths.

The river clutched and violently flushed me ever deeper. My feet dragged over submerged boulders. Was I scraping along the bottom of the river? Helpless to escape the turbulence holding me under, I felt terrified that I might have only seconds left to live. My lungs burned with the need to inhale. "Don't breathe!" my brain screamed.

An eerily silent blackness surrounded me, in stark contrast to the sparkling brilliance and deafening roar of seconds before. I realized the darkness meant I'd become trapped under the boat. I frantically tried to recall what Tom's last-second safety instructions had been. Raising my hands above my head, I felt for the rubbery smoothness of the side tube. Eons passed before I found it.

I shoved hard to drive myself away from it, deeper into the river. Meanwhile, the raft's side tube floated past over my head. A patch of light appeared far above me. During the eternity it took for my life jacket to buoy me upward toward that light, I struggled against a frantic urge to inhale. The instant I broke the surface, I gulped greedily, sucking in a mixture of air infused with aqueous froth.

Less than a second later, a massive wave curled over me. It pounded my body back below the surface. I coughed violently, trying to rid my lungs of the water I'd just inhaled. The swift current dragged me through the monster wave. My head popped above water again as I choked and gasped for precious oxygen between wave after wave. Our raft bobbed through the waves just an arm's length in front of me. Suddenly, I felt my shoulder squeezed from behind. "Are you okay?" Tom shouted. I nodded. With the superhuman strength of Clark Kent, he launched himself up out of the river onto the smooth neoprene floor of our overturned raft. Then he hauled me up out of the rapid onto the raft.

Another raft waiting downstream raced to our rescue. It grabbed our raft's bowline. Tom, another client and I clung to each other for dear life, on the slippery surface, as it towed us through the next rapid, then finally to shore. Back on land safely, I blurted a foolish comment about wanting to go back upstream to run it all over again, this time the right way. Clearly, I must have been in shock, because no way in hell would I have gone back through that rapid again, now that I knew the real odds. I'd learned the hard way no one should ever take Lava Falls for granted.

Aside from my ongoing phobia of that one disastrous rapid, I'd fallen in love with the majestic Grand Canyon's solitude, the kaleidoscopic color changes on its walls painted by the rising and setting sun, the hidden waterfalls tucked up in flood-carved side canyons that rained cool massaging showers on our heads and shoulders, and the playful rapids that dared us to "drop" in. Beyond all that, I was utterly in awe of how our multitalented guides not only survived, but thrived in this remote wilderness paradise.

Impressively, saddled with a group of city slickers totally separated from all their life-easing amenities, these boatmen kept us safe and comfortable. They lectured about the ancient geologic formations, some nearly two billion years old, steered us away from rattlesnakes, taught us to rappel down cliff walls, and fed us some of the tastiest food I'd ever eaten, prepared solely on camp stoves and over campfires. Honestly, I idolized these guides.

So it floored me when, the day after Lava Falls, our tanned, muscular, and charming head guide, Dave Shore, impressed by my contrived resilience, proposed that I come to California to work for OARS, both to manage their finances and to become a guide myself. While Dave's pitch tantalized me, I had to admit these guides belonged to an exclusive league of their own. I felt flattered that he believed there could be a place for me among their ranks, but I knew beyond any doubt I did not have the right stuff to be a river guide. I'd had a delightful detour through Jimmy Buffett's world, but I was ready to return to my safe, comfortable life, bringing with me enough memories of adventures on the wild side to last a lifetime.

BACK AT MY JOB, I threw myself into work. Deadlines and reports took priority. The solitude I'd experienced in the canyon faded behind the everyday noises of emergency vehicle sirens, taxi cabs' blaring horns, and the constant chatter of my colleagues. The panoramic vista from our floor mostly offered views of adjacent office buildings.

Historic brick structures mingled with newer architecture flaunting entire exteriors of dark tinted glass. I missed the colorful canyon walls and its landscape of tamarisk, cottonwood, and willows. I'd had my eyes opened to a vibrant natural world I didn't know existed.

In the distance, the rubber-streaked asphalt runway system of Logan Airport buzzed with an endless stream of planes arriving and departing. The only green thing in our field of vision was an occasional Aer Lingus jet. As time passed, I settled comfortably back into everyday, routine urban life.

Then, five months after returning home, I received an unexpected piece of mail from George Wendt, the owner of OARS, Inc. His letter encouraged me to visit their office in Angels Camp, California. He wanted to meet me and gauge my interest in working for his company. This unexpected opportunity dumbfounded me. Over the next few days, thoughts of George's invitation repeatedly sabotaged my work. I'd have to be crazy to contemplate such a radical departure from the job I held and the career path I'd chosen. Or would I? Could I really imagine myself working for decades alongside "cream of the crop" CPAs, brilliant but lackluster Brooks-Brothers-suited men to whom money and balance sheets meant everything? After just a taste of adventure in the Grand Canyon wilderness, I wasn't sure.

THE MORE SERIOUSLY I considered accepting the nontraditional position with OARS, the more my mind drifted westward. I could already hear the patronizing scolding from my parents. "After we've spent our hard-earned money on your college education, how dare you throw away your wonderful job… just to paddle a raft down some river out in the boondocks?"

2

DISCOVERING SOBEK

THE FLIGHT FROM BOSTON to San Francisco for my interview with OARS took six hours. Then a three-hour drive east into the Sierra Nevada foothills transported me back a hundred years in time to the sleepy historic gold mining town of Angels Camp. OARS's yellow, wood-shingled office sat beside Highway 49, in front of the Calaveras County fairgrounds, site of the famous Frog Jump competition that celebrates Mark Twain's classic tale every May.

I came prepared with a stash of questions for George. First, I asked him why and how he'd started his business. The former Los Angeles-area middle-school math teacher told me he'd fallen in love with wild rivers in 1962 during his first trip on the Colorado River through Glen Canyon. He and his friends floated downstream on a Huck Finn-style raft they constructed from inner tubes and planks.

Just two years later, a new hydroelectric dam drowned Glen Canyon at the bottom of Lake Powell. The loss of that glorious natural paradise devastated George. He made it his mission to use summer breaks to introduce wild rivers to his friends and boys in his scout troop. Before long, more friends, the scouts' parents, and even strangers begged to come on his river trips. By 1969, the summer fun had evolved into a full-time business.

"You know, Brenda, what I've discovered over the years," George said, "is that we save what we love, and we love what we know. Our job is to show people how truly special these rivers are, so they will engage with us to preserve them forever."

His pitch hooked me. Like a poker player confident in the hand she held, I wanted to go all in. I wanted to help him achieve his mission. Humble and a bit of a nerd, George excelled as a teacher, but his answers to my questions revealed he knew little about how to run a business. We both realized that OARS sorely needed a business professional with my financial management skills. He sweetened the deal by promising me I could take the OARS intensive river guide training program for free. I just needed to show up during the spring runoff, when the rivers ran high.

Though I doubted I'd ever be capable of the requisite skills for guide certification, George promised I *would* learn to navigate white water. My brain argued, "This is nuts!" But my heart pleaded, "Go for it!" Somehow, I just couldn't say no.

MY FIRST MONTH at OARS flew by. I set up accounting ledgers and payroll journals in the office, while two or three days a week, with eight other rookies, I headed to the river for guide training. One day, a tall blond stranger wearing the typical t-shirt, shorts, and Teva sandals of a river guide bounded into my office. Despite his casual attire, his outgoing, self-confident air hinted of leadership qualities. His blue eyes twinkled with excitement, as if seeing an old friend as he extended his hand.

"Hey, I'm Richard. Welcome to the Sobek gang. It's great to have a financial genius on board. John and I need all the help we can get."

"Sobek gang?" I asked, puzzled.

"Sorry I haven't had the chance to meet you until now. John and I were off on expeditions when you interviewed with George. I just got back to Angels Camp today."

As I struggled to put the pieces together, I recalled George briefly mentioning he'd invested as a partner in a second company with two unconventional 29-year-old daredevil adventurers, Richard Bangs and John Yost. Friends since childhood, the two spent most of their time overseas, leading trips on unexplored rivers. I faintly remembered George saying this second company operated out of a dinky room in the back of OARS headquarters.

"It's nice to meet you, but… well, I'm confused. George hired me to work for OARS. He didn't say anything about Sobek."

"What? He told me you were going to do our books, too."

Baffled, we stared at each other, wondering if George had blindsided us.

"Sobek Expeditions is way smaller than OARS, but we run the coolest rivers all over the world. And we have the coolest guides on the planet. It won't take very much of your time, I promise. Besides, we really need you."

Richard puffed out his lower lip in a show of faux disappointment. Obviously, he'd already mastered the skills of persuasion. His goofy, outgoing personality charmed and disarmed me. How could I possibly say no to this man who'd just called me a genius and had pleaded with me to work for him? So, with that brief exchange, Richard became my second boss.

He didn't have quite the deeply tanned, chiseled physique of many other guides, but I learned that what he lacked in brawn, he made up for with brains and an unbridled fearlessness. I found his quick wit refreshing and marveled at his extensive command of the English language. When not exploring vanishing wildernesses, his degree in journalism had prepared him to write in vivid detail about every hair-raising aspect of his latest expedition.

George kept his promise. At the end of May, I finished training and qualified to work as a whitewater river guide on a few of California's wild rivers. During the summer and fall, I escaped from my office chores to work as a guide two or three days a week. I found joy in the camaraderie of working with my fellow guides as we navigated boatloads of appreciative passengers through the exciting but easily maneuverable rapids in the gorgeous canyon of the nearby Stanislaus River.

While OARS operated popular rafting trips on accessible white-water rivers of the Western United States, Sobek's groundbreaking exploratory rafting expeditions on wild rivers around the globe cemented its reputation as the premier rafting company in the world. Any new lead about a scenic wild river in a remote location drove Richard to giddiness. He'd research the river and its surroundings, analyze whether an accessible "put in" and "take out" even existed, entice adventure sports gear shops, airlines, and travel magazines to fork over funds and freebies to sponsor the exploration, and then handpick trustworthy experienced guides to sign on with the expedition.

Sobek's crews included the most accomplished and self-reliant river guides in the world. These river warriors regularly traveled to far-flung places like Pakistan, Chile, Papua New Guinea, and Ethiopia to challenge obscure rivers whose names I could barely pronounce. It was no exaggeration to say that just about every river guide in America dreamed of being invited to train with the crew on a Sobek Expedition. All the young guides I worked with in California yearned for a long-shot chance to brave the turbulent currents in those exotic locales. They dreamed of the coveted prize—bragging rights earned as a crew member of a successful Sobek conquest.

Each time a Sobek crew returned from a new exploration, the OARS California guides and staff gathered like school kids eager to hear the latest exploits of their fearless heroes. This brotherly band of Sobek adventurers hungered for daunting challenges to defeat and craved the addict's high, elicited by the discovery of unique places we'd likely never see. They recounted backbreaking portages around 100-foot-high cascades, tales of epic end-over-end raft capsizes in raging currents, even a near-death escape when a crew member's sneaker, with his foot still in it, wedged in a crack between two boulders after he'd been tossed from his raft. A breathtaking slideshow documented each adventure, supplying visual proof of their death-defying claims.

Ironically, I'd seen and heard enough to know that I wanted *no* part of *those* types of death-defying trips. Count me out. Even the Colorado's mighty Lava Falls equated to child's play compared to the labyrinth of boulder-choked stretches of river and mini-waterfalls that congested some of Sobek's trips.

Becoming certified as a whitewater guide had heightened my appreciation for how quickly a simple mistake could turn disastrous. Fortunately, I guided on rivers where the worst consequence of screwing up resulted in a harmless surprise dunking. But a miscalculation on the rivers Sobek ran could be deadly. I wanted no part in a trip with that degree of risk. Unlike Richard and John, I entered this world with the daredevil gene missing from my DNA.

Nope. I belonged safely put within the United States borders. I'd traveled abroad only once before. On my high school senior year trip to France, I'd suffered terribly from homesickness. 10 years had passed since that fiasco, yet still I lacked any inkling of a desire to travel abroad again.

ON A LOVELY INDIAN SUMMER afternoon in September 1980, Richard strode into my office like a man with a mission. "You've been working here for more than a year. You've earned yourself a *real* adventure. So, which expedition do you want to take?"

I had a specific trip in mind, but wasn't sure it would qualify under Richard's definition of a real adventure. For him, that meant an exciting, never-attempted undertaking involving unusual hazards in the middle of nowhere. Whereas my dream envisioned floating down an Alaskan glacier-fed waterway past translucent blue icebergs under snow-capped mountains. I imagined evenings gathered around the dancing flames of a driftwood bonfire, sipping river-chilled wine and dining on freshly caught grilled salmon.

Before calling it a day, we'd crowd into a teepee sauna for an authentic sweat lodge experience. Once salty beads of perspiration dripped from our bodies, we'd take a quick rinse in the icy river. Since Sobek marketed their Alaska trips as safe for older children, I figured my chances of survival in our northernmost state, despite occasional sightings of grizzly bears, ranked much higher than on a crazy desolate river in a place like Papua New Guinea.

"Richard, I would love to do our Copper River trip in Alaska."

He stared at me in disbelief, as if I'd just said something blasphemous. "We give you the chance to go anywhere in the world! To exotic, magical, wondrous places where few humans have stepped foot! And you pick Alaska?"

His reprimand stung, leaving me speechless. After a long silence, he shook his head and said, "You can't go to Alaska. Pick somewhere else. Outside the United States. We'll talk about this later." He turned and left before I could object. I had no intention of venturing beyond US borders. Forget that.

Unfortunately, I learned that my Copper River choice would be impossible. The last Alaska trip of the year had departed two weeks earlier. Even waiting until next summer wouldn't work because the warm weather of summer fueled the busiest financial season for OARS. Were I to leave the office for a month or more, it would have to be during winter. I made a mental note to search our trip catalogue for a "safe" trip I could take during winter. For the next few days, I purposely steered clear of Richard. If I could avoid crossing paths with him, perhaps I could avoid being sent into harm's way.

3

TANZANIA? REALLY?

EVENTUALLY, RICHARD CAUGHT UP with me. When he called me into his office, the words he spoke shocked me.

"You and I are going to do the Kilombero and Rufiji Rivers in Tanzania!" Richard's head of sun-bleached curls bobbed in self-approval at his startling proposal that I accompany him on Sobek's newest white-water adventure. His declaration caught me off guard. I'd never heard of Tanzania. It sure as hell wasn't on my travel wish list. I cringed just hearing the syllables: Tan-zan-i-a.

My cheeks blushed hotly as I struggled to hide how flustered I felt. I couldn't let Richard know I had no clue where on Earth one would even find a Tan-zan-i-a. He explained, "Two years ago, I asked Conrad Hirsh to scout these two remote rivers."

I knew Conrad served as our senior guide in Africa. That offered a clue that Tanzania must be somewhere in Africa. Richard's voice bubbled with enthusiasm as he shared Conrad's finding that, "there are no other rivers on earth quite like these two, which flow through a vast uninhabited tract teeming with elephants, lions, hippos, and crocs."

Did he have a death wish? Was the man out of his mind? Wild whitewater rivers were dangerous enough without throwing in predatory beasts. No way would I go where the animals could kill me.

"This will be the adventure of a lifetime," Richard continued. "We'll be part of the first commercial trip to journey across the massive Selous Game Reserve by raft, with a full crew and paying clients. I'll only do the river trip, but you can climb Mount Kilimanjaro afterward with the rest of the crew."

I stood speechless in front of him. Surely, he must have expected some sort of positive reaction from me. I wanted to remind him it was supposed to be *my* choice of which trip to take, not his. And *my* choice would definitely be a river in the United States.

As Richard continued his sales pitch about the Tanzanian adventure, I realized he'd already developed a rock-solid plan for our trip. Not even the smallest detail remained open for debate.

I attempted to fake a smile. Inside, I was freaking out. He could risk his own life, but not mine. No way. I was *not* going to Tanzania. And I would *not* get into an inflatable raft surrounded by thousands of hippos. What an insane idea! He couldn't make me. Could he?

My brain revved into overdrive, searching for an escape route. But how could I admit to this icon of adventure travel the thought of tiptoeing around ravenous lions and dallying with hippopotami terrified me, when these were the very thrills he sought to experience? I didn't want to disappoint him; more importantly, I didn't want to lose his respect. With all my heart and soul, I wanted politely to decline his invitation. But I simply lacked the courage to say no to my boss.

Simple logic demanded I stand my ground, but I didn't have time to devise any valid excuse to avoid the dreadful fate assigned to me. Unable to mount a spontaneous counterargument, my shocked silence signaled acceptance by default.

Leaving his office, I raced to our travel agent and asked for the new brochure on our trip in Tanzania. She handed me a 12-page booklet describing our *'Kilimanjaro and River Safari in Tanzania.'* "Are you going on this trip?" I sensed jealousy in her voice.

Still doubtful this might actually happen, I hedged. "I'm thinking about it."

That night in the sanctuary of my bedroom, I pulled out my world atlas and located Tanzania in the index. Holding my breath as the pages ruffled through my trembling fingers, I found the correct map plate. There it was. Right between Kenya and Zambia. Right there in the heart of tropical Africa. I regretted opening the atlas. Impulsively, I shut it and shoved it back on the shelf in my closet. Out of sight, it didn't seem so real.

Flopping back down on my bed, I closed my eyes and searched the recesses of my memory for anything I knew about Africa. My education in accounting, economics, taxes, and business law had done nothing to prepare me to be thrust into the world of Joseph Conrad's terrifying novel, *Heart of Darkness*, the only book I'd ever read about Africa. The images that raced through my mind were anything but comforting: stalking lions circling for a kill; an undulating spitting cobra poised to strike; droning mosquitos laden with incurable diseases; and hostile cannibals with bones through their noses, stirring a monstrous cooking pot of steaming *Homo sapiens* stew. They all shared the common theme of death. I shuddered.

To force my thoughts back to a safer reality, I glanced through Sobek's brochure. The overview stated, "The 170-mile float down the Kilombero and Rufiji Rivers is a superb introduction to Tanzania's Selous Game Reserve, the largest uninhabited game reserve in Africa. Selous's 18,000 square miles of untouched wilderness are home to more than a million animals, including the world's greatest elephant population, thousands of hippos, countless crocodiles, and the rare black rhino.

One portage is necessitated by Shiguri Falls, an awesome cascade where the river drops 250 feet in a short distance."

Another section dealt with trip preparations: passport, visas, and vaccination requirements. I worried that getting five vaccinations against cholera, typhoid, tetanus, hepatitis, and yellow fever might do me in on their own. A prescription for chloroquine would serve as a prophylaxis against malaria. I couldn't imagine a sane person purposely putting themselves into an environment with so much disease, deadly wild animals, and god knows what other obstacles.

I lay in bed, tossing and turning. Anxiety wrenched my gut until the early hours of the morning. Two people had already accidentally died on trips during the seven years since Sobek launched its first trip down the Awash River in Ethiopia. I didn't want to be the third. But that night I could do nothing to rescue myself from Richard's madness.

He'd been infected with a serious case of wanderlust for years, I reminded myself. No known cure existed. Treating its symptoms could get expensive. Richard grappled unsuccessfully with the concept that wanderlust didn't course through everyone's veins as it did through his and those of his closest circle of friends. It seemed to be highly contagious. It was not something I wanted to catch. At last, fear-producing hormones overwhelmed my body with fatigue, and I drifted into a shallow, restless sleep.

The next morning, the reality of my reluctant acquiescence to Richard's decision hit me like a boxer's gut punch. I wanted to throw up. If I revealed my true feelings, Richard, George, and every guide I worked with would see me as a lily-livered coward. Bravery and quick decision-making in the face of danger topped the list of traits required for safely transporting our rafting clients and ourselves. No head guide would ever accept a self-confessed coward on their crew. But why weren't *they* afraid of dying in a god-forsaken wilderness, distant days of trekking on foot across trackless terrain away from medical help? Why was *I* the only terrified one?

I'd backed myself into a corner. Panic fueled my doubt of finding *any* graceful way to extricate myself from this mess. Later that day, a coworker stopped by my office to give me a thumbs up. "Richard says you're going to Africa. Lucky you!"

My head drooped forward in a show of defeat.

"Luck has nothing to do with it. I'm not even sure I want to go. I *wanted* to go to Alaska."

"Are you kidding me? I'd switch places with you in a heartbeat," he said, rolling his eyes in disbelief. "You really don't know how lucky you are."

For a moment, I thought about proposing that switch to Richard, but I knew it would never fly. Days slipped by without revealing any legitimate exit ramp to get me off this highway headed straight to hell.

A WEEK LATER, I called my parents on the east coast. We'd barely said our hellos when I blurted out my news. "I'm going to Tanzania in January."

"*Where* are you going?" my mother asked.

"Tanzania," I repeated.

"And where is Tanzania?" she demanded suspiciously.

"In Africa."

"Why?"

"To raft a river and climb a mountain there," I said without enthusiasm.

"That sounds dangerous. Aren't those rivers full of crocodiles?"

Don't forget the hippos, I wanted to tell her, but her questions just kept coming. Her voice edged toward hysteria.

"Why in the world would you ever want to do something like that? Do you have a death wish? You just need to tell them you aren't going to do this."

My mother had always held a powerful grip over my life. She considered it her motherly prerogative to find fault with the food I ate, how I dressed, or whatever I did that didn't measure up to her strict standards. She'd pointed out the flaws in every boyfriend I'd had and lectured me about whether my friends were worthy of my time. Activities I planned with friends required her approval. If she didn't like our plans, she'd shut them down by simply saying, "I'm your mother, and I'm saying no."

Once I finalized my decision to move to California, she did everything in her power to prevent me from leaving. Guilt, anger, tears; she used every tool in her toolbox. Once settled in Angels Camp, if I mentioned I missed something on the East Coast, she'd pertly remind me, "You made your choice. Now you've just got to live with it."

As I expected, she objected to my news about going to Tanzania. In a million years, *she* would never take a trip like this, so why would she consent to her daughter doing such a foolish thing? It had been painful to discover that being physically 3,000 miles away barely deterred her from trying to orchestrate my life.

However, during that conversation, I felt a radical shift inside me. Instead of deferring to my mother's wishes yet again, I rebelled. Why not try something I'd never dreamed I would do? Why not travel to a far-flung destination and learn something about the world?

The next thing I knew, I heard myself defending *my* "choice" of traveling to Tanzania, to raft down a wild isolated river filled with crocodiles and hippopotami, followed by climbing the highest mountain in Africa.

Unlike my mother, on the verge of tears, my father listened quietly to our conversation. Toward the end of the phone call, he interrupted to wish me good luck. "Follow your heart and have fun. I'll be waiting to hear all about your trip when you get back."

From that day forward, my attitude changed. I cautiously began to look forward to the adventure. That Richard, already elected into the elite membership of the prestigious Explorer's Club, and the leader of successful first descents of at least a dozen rivers all over the world, would be my personal escort on my first trip abroad presented a rare opportunity. I trusted he wouldn't let anything terrible happen to me.

He nudged me to apply for my passport and Tanzanian visa. We arranged for all my immunizations, most locally, but the live yellow fever vaccine required a trip to San Francisco. Getting all those shots wasn't much fun, but I gladly offered my arm, knowing they'd prevent me from contracting killer diseases that came with excruciating pain or even death.

I bought a new camera, highly recommended by other guides, and a state-of-the-art waterproof case to protect it from rain or splashes from a boisterous rapid. I calculated that 20 spools of film would be adequate if I used one roll of film per day. I could hardly wait to capture close-up shots of wildlife like the photos I'd seen taken by other Sobek guides, a naïve ambition given our goal would be to *avoid* close contacts with wild animals!

4

THE JOURNEY BEGINS

FOUR MONTHS LATER, on January 21st, fearless world explorer Richard Bangs and I, petrified neophyte, boarded a plane at San Francisco International Airport to begin our 10,000-mile journey. For the next 48 hours, our flights would carry us to New York City, London, and Cairo before landing on the tarmac in Dar es Salaam, Tanzania.

We had a 13-hour layover in London. Since I'd never been there, Richard proposed a visit to the city center. I welcomed his suggestion even though my body's clock told me I ought to be sleeping, not sightseeing. We took the Tube into Trafalgar Square, a busy crossroads featuring several monuments, a spouting water fountain, and about 20,000 pigeons.

Richard bought a bag of popcorn from a street vendor and stuffed a handful of the popped kernels in his mouth. He beamed as we walked toward the fountain.

"Only half of this is for me. Watch this!" With the grace and speed of a magician, he raised his arm to shoulder height, his palm cupped full of the pigeons' favorite treat. Instantly, eight pigeons swooped down to line up along his arm like dutiful toy soldiers. Their antics tickled me.

He handed me two fistfuls of popcorn and dared me, "Go ahead, you try." Cautiously, I raised my arms, hiding the kernels in my fists.

A different feathered regiment glided in to perch on my arms. As I uncurled my fingers, a shoving match ensued, with each pigeon fighting for a position closer to the prize. Small talons suddenly gripped my scalp. A pigeon had hunkered down to oversee the mayhem from on top of my head.

Richard roared with laughter. "Don't move. I've got to get a picture!"

And so, the first photo of my first international expedition showed a still naïve, bewildered me, covered from head to fingertips with the royal not-so-wildlife of Trafalgar Square. My world would never be the same. The first thrill of wanderlust had infected me. Helplessly, I suspected nothing could prevent this incurable infection from growing.

We checked out a cozy pub where we consumed a cheap meal of some sort of tasteless, glutinous mush. By then, exhaustion and jet lag tempted us to rest our heads on the pub's checkered tablecloth. But instead of risking the pub owner's wrath for tying up his table while we took a quick nap, we walked down the street and caught a movie where, covertly under the cover of darkness, we could take a 90-minute snooze. Despite the matinee's cheap price, only a handful of other movie-goers occupied the theatre. We hoped our snoring wouldn't annoy them. When the lights came up, we gathered our backpacks and boarded the Tube back to the airport.

Two hours remained until our departure. Both of us desperately needed more shut-eye, but I knew if *I* fell asleep in the boarding lounge, I wouldn't wake up for hours. If we missed this flight, the next flight to Dar es Salaam wouldn't depart for two days. Richard promised he would hear the boarding call even if he fell asleep. I seriously doubted him. He instantly conked out while I struggled to keep myself awake. One of us needed to hear that call.

The sound of his rhythmic snoring irritated me. It also aggravated me that Richard had taken for granted I'd stay awake to guarantee we'd catch our flight. Finally, when the first boarding call for our flight blared from the departure lounge speakers with enough volume to wake Rip Van Winkle, I waited to see if Richard would wake up. Nope. Sleep had snatched him off to another realm.

10 minutes later, a second announcement warned of our flight's impending departure. Still, Richard didn't stir. Tired and grumpy, I considered leaving him behind. If I'd had the courage to get up and check myself onto the plane without him, I'd have done it. Truthfully, though, the idea of setting off on the last leg of this journey alone petrified me. Of course, Richard had known this all along. It didn't matter how dead tired I felt. He knew I'd be his trustworthy alarm clock.

"Hey, Richard, our flight is about to leave," I said loudly enough to pierce whatever dream had him entranced. I stood and walked toward the gate. Startled, he sat up, rubbing his eyes.

"Geez, thanks for standing watch," he stammered. He grabbed his stuff and penitently followed me onto the plane. Score one for the neophyte.

10 HOURS LATER, our Ethiopian Airlines jet thumped onto the macadam of Tanzania's international airport. The single runway it taxied down bisected an overgrown field of grasses and vines. It came to a stop in front of a small gray cinder block building with empty spaces in the walls where windows should have been. Between the building and the plane stood a row of soldiers dressed in full camouflage, each wearing a beret and carrying an AK-47.

A few middle-aged laborers toted an oversized wagon up to the cargo hold and began unloading luggage. Two others pushed a wheeled stairway up to the plane's door. After hours of sitting confined in my seat, standing and stretching my legs felt freeing. As soon as the flight crew opened the hatch, the crush of passengers eager to get off swept us forward along the center aisle. I clutched the railing as we spilled down the stairway. With Richard in the lead, we stepped onto foreign soil in this land of exotic foliage and warm equatorial breezes.

I didn't know quite what to expect. The men in charge of the one and only luggage cart already had tossed their first load of suitcases through the opening in the cinder-block wall next to the building's doorway. The soldiers scowled as we walked past. Not much of a friendly welcome for a tourist.

Other soldiers stood guard inside the luggage retrieval room, presumably to quell baggage claim chaos if it got out of hand. The impatient throng of passengers thinned with the delivery of each cart load of rope-reinforced, overstuffed suitcases, wooden crates, and scuffed duffel bags. At last, we spotted our luggage as it came sailing through the opening.

Just one uniformed immigrations inspector had the authority to inspect passports and vaccination booklets. He occupied a desk near the exit where his duties also included choosing which pieces of luggage to search for contraband. On an adjacent table, the inspector's underlings pawed through the disheveled contents of three open suitcases for illegal valuables to seize. I prayed he wouldn't pick my duffel. The mortifying thought of these men handling my panties and bras in plain sight caused my cheeks to flush with embarrassment.

The dour-faced inspector demanded our documents. My hand trembled slightly as I handed him mine. He eyed me, then the photo in my passport. Satisfied, he officially stamped and dated my arrival in Tanzania, then waved me on to the exit. Thank God! Richard got the same treatment.

"We made it. I can't believe I'm truly in Africa," I said. A swirling jumble of excitement, fear, curiosity, and wonder made me dizzy while I took in my strange new surroundings.

Our head guide in Tanzania, Conrad Hirsh, wearing a tan t-shirt and khaki shorts, stood outside waiting for us. The deeply tanned

40-year-old, topped with an unruly mop of black curls, greeted us with hearty hugs, then hailed a taxi to transport us into the heart of Dar Es Salaam, Tanzania's largest city. I'd met Conrad two years earlier on my fateful Grand Canyon rafting trip. He'd flown over from Africa to work the trip as a trainee boatman. Since then, we'd kept in touch over money matters. As Sobek's controller, I wire-transferred money to Conrad about a month before each African trip departed so he could buy food and supplies, arrange for transportation to and from the river, and pay for reservations at local hotels for our clients. As the trip leader, I expected him to provide me with a complete accounting of how he spent that money, along with the receipts to prove it.

Several years before I met Conrad, he'd grown disillusioned with life in the United States. A graduate of Stanford University, he had a brilliant mind and taught college-level math in Texas, but felt smothered by academia's inflexible rules for advancement. Shortly after an acrimonious divorce, he sold everything he owned and moved to Ethiopia, where he'd previously served as a Peace Corps volunteer from 1964 to 1966.

On his return, Haile Selassie University in Addis Ababa hired him to teach math. While he still occasionally taught at the university, he'd settled into a more relaxed expat life, completely free of other people's expectations, preferring to spend most of his time exploring vast parcels of East African wilderness. One of his friends described him as a literate, perceptive, witty, but often cranky adventurer, a skinny math nerd with the soul of an artist.

During our 20-minute drive into the city, Conrad caught us up on his plans. Seven crew members would work on this expedition. The other six were men, so Conrad had booked sleeping accommodations at the YMCA. He had wrangled three rooms with double cots and a shared bathroom down the hall. As a female, the YMCA would not permit me to stay at their guest house. Conrad hadn't yet figured out what to do with me. This freaked me out, but I stayed silent.

The lush green vegetation that surrounded the airport gave way to thatched-roof, one-room mud houses and a hodgepodge of small shops built from sheets of corrugated steel and gray cinder blocks. Outside, shirtless men toiled at everything from bicycle repairs to welding.

I'd always imagined Africa to be one enormous, dense jungle, brimming with salivating wild animals, waiting to pounce on and devour whoever entered their domain. I never envisioned that cities could sprout up in this untamed wilderness. The cities I knew, like Boston and San Francisco, boasted gleaming steel and glass multistory buildings, smooth paved roads edged with concrete sidewalks, and streetlights at intersections to control traffic. I certainly didn't expect to find any fancy cities like those here in Africa. Truthfully, I didn't know what to expect.

Yet over 900,000 residents, nearly double Boston's population, lived in Dar es Salaam. As the taxi penetrated deeper into its labyrinth of crumbling buildings, dilapidated automobiles, and pothole-pocked roads reminiscent of the ruins of a war zone, I struggled with the shocking sight of the city's decrepit façade. Trucks spewed clouds of black exhaust, overcrowded buses honked, and barely functional bicycles tottered. People, cows, dogs, chickens, and children crammed into the rutted, muddy, garbage-strewn thoroughfares. The city air stunk of a blend of rotting garbage and pungently spiced food from street vendors' carts. I'd never experienced a sensory overload like this. Nothing seemed familiar.

Conrad laid out a few options for my lodging. "I don't know, Richard. We could rent a room in a cheap hotel, but I don't think it's a good idea for her to be on her own. She doesn't know the language, and the crime around here is pretty bad. We could rent a room in one of the tourist-class hotels, but that will cost 10 times what I have budgeted."

I pleaded with Richard, my eyes growing misty, "There's no way I'm leaving you guys. I don't want to stay alone anywhere in this city."

"Calm down. We aren't going to leave you alone," Richard assured me before asking Conrad, "Do you think we can get another room at the Y?"

"For her?" Conrad turned around from the front seat and shot Richard a dubious look. "They aren't going to let a woman stay in the YMCA. Besides, they're full up. I had to grovel to get the rooms we have."

Determined not to be separated from them, I pleaded again, "I can sleep on the floor in one of your rooms."

"The only way you're getting into the Y is if they think you're a man." Conrad's remark on the impossibility of that ploy stung like a slap on my face.

"That's no problem," said Richard. "We'll just find somewhere to buy a Mao cap. She can stick her hair up inside it. It'll be dark when we get back to the Y, and the rest of us can just surround her. We'll make it work." His confidence convinced neither Conrad nor me.

Conrad shrugged, not wanting to argue against his boss's hairbrained idea. I didn't have the foggiest idea what a Mao cap was, but I had to trust that the two wouldn't leave me out on the street to fend for myself.

When the taxi stopped in front of the YMCA, Conrad took my duffel bag and motioned for Richard to follow. At least Richard would get to see *his* room and a tour of the Y. Meanwhile, I fidgeted on a bench in the Y's front yard, protected from the heat of the midday sun by the canopy of a gorgeous purple jacaranda tree in full bloom. All I wanted to know was where I'd be sleeping that night.

5

NAVIGATING DAR

A FEW MINUTES LATER, Conrad, Richard and the four other crew members who had arrived earlier to prep for the trip sauntered out of the building. On the drive into Dar, Conrad had briefed Richard and me on the progress of the preparations. They'd spent the last two days searching for and buying buckets, rope, tarps, water jugs, an axe, paint, paper products, and other necessities.

Earlier that morning, Conrad had dispatched two of the crew to buy solid blocks of ice half the size of the coolers we would later fill with frozen meat and perishable dairy items. Only in Dar could we buy ice blocks of this size. We'd need to consume all the perishable foods during our first few days on the river before the ice blocks melted. After that, we'd have no refrigeration.

Conrad doled out our next assignments. "This afternoon, we're going food shopping. I've made a couple of lists of the things we need. We'll split into two groups. Michael, you and Bart will go to the open-air market and the butcher's shop. The rest of us will hit up the pharmacy and grocery shops. Use taxis to get the stuff back to the Y. If you see anything I've forgotten to put on the list, buy it. Once we're in the game reserve, we'll be days away from food, supplies, medicine, and doctors, so we've got to pack everything we think we'll possibly need. See you in a couple of hours."

As a frequent traveler throughout East Africa, Conrad spoke Swahili proficiently. So did our guide, Michael Ghiglieri, a professional wildlife biologist, who'd picked up the language from his years of doing chimpanzee research in Uganda. Both also possessed well-honed skills in the art of bartering, the local practice used in Tanzanian markets. Conrad handed one list and a fistful of shillings to Michael, who, with Bart, headed toward the open market. The rest of us followed Conrad, lugging baskets woven from dried banana leaves into which we'd load our purchases.

Watching Conrad skillfully bargain for the best deal with vendor after vendor, I discovered how speaking this language proved a tremendous advantage when shopping. I didn't know a single word, yet I could sense the drama of the negotiation as it unfolded. Conrad would ask a shopkeeper the price of an item. The shopkeeper would reply with an exorbitant price. Conrad would act shocked, shaking his head before countering with an equally outrageous lowball offer.

Then the shopkeeper, acting insulted, would wave Conrad off. But after a few seconds, he'd come back with another lower price. The standoff volleyed back and forth for a few rounds before they finally reached a mutually acceptable price. It reminded me of the back-and-forth deal making with car salespeople that I detested back home. It would exhaust me to go through this process for every single thing I bought. I happily assumed the role of spectator watching the animated bartering. I'd do my share by carrying baskets.

Going from one small grocery shop to the next, we filled our baskets with the precious food products that would keep us alive on our journey. Our list consisted largely of sacks of dried pasta (lots of it), rice, dried beans, cans of tomato sauce, tuna, sardines, SPAM, jars of peanut butter and jelly, crackers, flour, sugar, and tins of coffee and tea. As our heap of purchases expanded, the items on Conrad's shopping list dwindled to just the perishable fruits and vegetables he planned to buy in the local village from where we'd launch.

By the time we finished shopping and transporting our purchases back to the Y in the trunks and back seats of two taxis, sweat soaked my clothes. I felt weary, ravenous, and thirsty. My male colleagues stored our purchases inside their rooms in the Y while I again waited outside.

When they reappeared, Conrad led the way along noisy streets still busy from the day's-end rush of locals returning to their homes. Along the route, shops selling goods from clothes to counterfeit Rolex watches cozied together with only an occasional alley providing access to the next parallel street over. Goods sold inside the shops overflowed out onto racks or into bins lining the street in front, tempting those passing by to stop for a look.

We stopped at a shop sporting a rack of Mao caps just outside its entrance. These visored cotton caps worn by Mao Tse Tung and other Chinese leaders had become popular in Tanzania because of its close political and economic ties with China. Each cap had a one-inch-wide band at its base that ballooned into a three-inch-high roomy compartment above.

As I tried on the caps, I noticed that most of the younger Tanzanian men and a few older ones passing by wore these caps. I found a navy blue one that rested tightly enough on my head that my hair would stay tucked up and in place. It cost a dollar. Adorned in my new cap, perhaps just maybe, a big maybe, I might pass as a guy. Conrad and two other guides also bought caps to sport the fashionista look. With four of us wearing the caps, and with them surrounding me, I prayed I'd be less noticeable.

A short time later, we entered a hole-in-the-wall café where the aroma of exotic spices wafting from the steaming kettles on the stove reminded me I'd not had a decent meal in two days. My empty stomach grumbled its need for food. We shoved two of the café's six tables together to accommodate our group. In Swahili, Conrad placed our order with the cook. Soon, a server delivered plates heaping with basmati rice and topped with beans stewed in tomatoes and coconut milk for each of us. Though I'd never eaten this type of food before, the slightly sweet flavor of the fluffy rice melded well with the tangy sauce of the beans. The simple meal quickly satisfied my hunger and quieted my noisy stomach.

The crew (except me) all had years of experience rafting rivers in exotic locations, but Conrad alone had seen this river. Like an electric current, anticipation charged our imaginations as we asked Conrad critical questions about how we were going to pull off this trip, particularly getting around the gigantic waterfall midway that marked the end of the Kilombero River and at its base the origin of the Rufiji River.

Jim Slade, one of the first guides hired by Sobek, who had rafted dangerous whitewater rivers and climbed grueling mountains around the globe, voiced a question that was on all of our minds. "Was your contact able to get anything on the weather forecast for the next two weeks?"

Conrad nodded. "Mostly dry and hot."

"You got the hot part right," said Mike Wynn, who occasionally joined our expeditions, renting his personal raft to Sobek in exchange for a spot on the crew.

"We purposely scheduled this trip in the window between the short rains that come in November and December and the long rains that start in March and go into May," Conrad said, glancing toward Richard.

"Yeah," Richard explained. "We have to do the trip when the water levels run high in these rivers. We already have a bunch of other trips that go out in the summer, so logistically, January into February works best for this trip."

The only reason Bart Henderson sat across the table from me owed to this timing. Bart, another of our most experienced guides, also held the area manager position for our warm season Alaskan river trips.

Bart sported the ruggedly handsome look of a mountain man with the confidence and skills to live comfortably in the harsh wilderness. While our northernmost state hibernated in a frozen winter wonderland, Bart had leaped at the opportunity to swap glacial icebergs and salmon-fishing grizzly bears for tropical close encounters with hippos and lions. I wouldn't have made that trade, but then again, I had nothing to trade.

"So, what about the water level? Any idea how many cubic feet per second the Kilombero is currently running?" Bart asked.

"My friend said the short rains ended early in December and brought less rain than usual. So I think we can expect lower river levels than when I scouted it two years ago. How much lower, I'm not sure."

We all shared concern about the water level. A strong current pushing us downstream meant we'd spend less time on the water and more in camp. It would also make it easier to stick to our itinerary's schedule. A lower water level meant less space for us to coexist with hippopotami and crocodiles. As well, a slow meandering current would demand longer hours of back-straining paddling, with the risk of missing our planned completion date. That would throw a monkey wrench into everyone's departure plans.

"Do you think two days is enough time for us to carry the boats and all our gear around Shiguri Falls?" Richard's brow raised as he posed this question.

"I think we can do it. It won't be a cakewalk, that's for sure. Depends on whether we can keep everyone healthy and energy levels up. Everyone will have to at least carry their own personal gear, and perhaps we can convince some passengers to carry items of food or kitchen gear for us. It'll be a slog, but a doable slog." Conrad's answer contained a lot of ifs that surprised me. I didn't recall our trip brochure mentioning anything about turning our paying clients into pack mules for this portage.

"Speaking of keeping people healthy, do we have a plan for what to do if someone gets injured or is really sick?" Michael asked. This question loomed foremost in my mind. Worried I might get scolded for asking, I had kept my mouth shut. Now I was all ears to hear Conrad's answer.

"Yup, we're going to make sure no one gets hurt or sick. Got it?"

It wasn't the answer I expected. Shit happens all the time. If you trip and break a leg at home, it's easy to call 911. If a wild animal chomps on your leg, miles and miles away from any form of medical help, and you've got no way to communicate with the outside world, what *does* one do? All of us were first aid and CPR trained to take care of basic injuries. But what if…? Suddenly, this expedition seemed like sheer madness to me.

With my hunger satisfied and knowing I'd processed as much of Conrad's orientation information as my weary state would allow, all I wanted was a good night's rest. Conrad herded us back to the Y. On the way back I noticed that even my colleagues who didn't speak Swahili appeared comfortable in these strange environs. It amused me how adept they were at using body language to flirt innocently with timid young females hurrying home from the market carrying woven baskets filled with bread and vegetables perched atop their heads.

These girls dressed in traditional *kangas*, a matching set of two large rectangular pieces of cotton fabric printed with wood-blocked designs in vibrant primary colors. One rectangle served as a skirt, wrapped around their lower bodies, the top edge rolled and tucked in at their waist. They wrapped the matching piece around their torsos, secured by a knot above one shoulder. The colorful patterns of their *kangas* brightened the otherwise drab gray and brown tones of the city. We'd run out of time for touristic pursuits, but I vowed to purchase one of the stunning *kangas* as a keepsake when we returned to Dar after our trip.

Just before arriving at the Y, Richard reached down and pinched some dirt between his fingers.

"Hold still," he said as he smeared the dirt on my cheeks.

I recoiled. "Stop! What are you doing?"

The other guides chuckled as they watched Richard's antics.

Grinning, he answered, "Now you really look like a street ruffian. It's perfect."

The men circled around me, and we made a perfectly choreographed entrance past the clerk at the registration desk. He glanced up barely long enough to recognize a few of the white foreigners he'd registered earlier and to wave them in. We retired early to recharge for the next day's task of transferring all of our equipment and supplies to our final staging area.

I slept with two other guides in an already-cramped room. The two cots alone took up 70 percent of the space. A single bare lightbulb hung from the ceiling. A small rectangular screened window high on the outside wall permitted only the slightest amount of air exchange. The sleeping bags already laid out on the two cots marked claims previously staked by my roommates.

In the hierarchy of guiding experience, my place was at the bottom of the totem pole. The concrete floor between the two cots had my name on it. So, with all our gear tucked under the two cots, I laid out my foam mat and sleeping bag in the two-foot space between the cots and crawled in. My sleeping bag's sack stuffed with a sweatshirt and four t-shirts served as a lumpy pillow.

While my sleeping space bordered on claustrophobic, I felt safe wedged between my two broad-shouldered bunk mates. I only hoped they wouldn't stomp on me in the middle of the night in a rush to relieve themselves of the local beer they'd downed with dinner.

Sleep rushed in as quickly as if I'd been injected with anesthesia and told to count backward. 100, 99, 98, 97… out. The room in the YMCA swirled, then went black.

6

THE DREAM

OUT OF THE BLUE, I found myself dressed uncomfortably in the garb of a Victorian adventurer: long-sleeved, belted, khaki bush jacket; matching Jodhpur pants; knee-high, stiff leather boots. A safari pith helmet perched atop my head, and a pair of binoculars looped around my neck. Wearing this bizarre costume, I felt like an actor standing in the spotlight, center stage of an otherwise blacked-out theatre. Horrified, I realized I knew nothing about this play and not even a single line of its script. What was I doing here?

As I readied to flee into the wings, Richard, wearing similar vintage clothing of a mid-1800s explorer, strolled onto the stage to join me in the spotlight. He spun me around to inspect my outfit, then nodded with approval.

"You've got the perfect outfit for our exploration of the great and mysterious terra incognita."

His intended compliment missed its mark. Irritated by the tight fit of the strange costume, and angered by Richard's expectation of a mystifying but imminent journey, I demanded, "What and where the hell is terra incognita?"

Taken aback by my ignorance, he explained, "Why, it's a place where no human being has ever been. It's a starkly blank space on a map. A wilderness full of never-seen marvels. Don't you ever dream of being the first person in the world to see an exotic plant or creature no human has ever set eyes on?"

He sounded like a con man promising visions of grandeur as he passed me a hand-drawn map. The date 1860 appeared in an upper corner. Much of the center of the African continent it depicted stood devoid of detail, though one small notation caught my eye.

"Richard, there's a note here in the middle that says, *'Here be dragons and lions.'* Why on earth would you want to go where there are dragons and lions?"

"Because we can. That's the whole point! The cartographer that drew this map doesn't have any idea what's actually there. He just invented that description to tell people that anything could be out there. We can be the first to find out what is really there."

Richard's rationale sounded suspicious to me.

He insisted, "No one will truly know what's there until a bona fide exploration can survey the area. Don't you want to see an actual dragon?"

"No! Nor do I have any desire to see a wild lion. Freaking lions can kill you in an instant," I fired back.

"True, but we won't know if there are lions or dragons in terra incognita unless we go look for ourselves."

"Frankly, Richard, surprises freak me out. I don't think I have the stomach for this. Why do you feel so compelled to find out what's in terra incognita, anyway?"

"Simple, because I'm curious. There is only *one* first time for seeing and knowing something previously unknown. Only *one* chance to experience discovery in its purest form. The more that is known about the world, the more limited and obscure opportunities become for unprecedented encounters. We are running out of time for unearthing those rare novel prizes."

"Curiosity killed the cat, you realize," I reminded him.

Richard grinned, then jabbed back in triumph, "Yes, but don't forget that satisfaction brought him back! Time is of the essence. As we speak, other great explorers are rushing into the African unknown. Dr. David Livingstone is right now searching for the source of the Nile River. Henry Morton Stanley is attempting to map the Congo wilderness, and don't forget Mungo Park, who, after exploring almost 1,000 miles of the Niger River, met an untimely death by drowning. We have to go now while there is yet some terra incognita to explore!"

I counted the many ways that one might meet their maker in this terra incognita place, while Richard seemed ecstatic to face them head-on. I asked, "Doesn't the possibility of an untimely death bother you? Personally, I'd choose a less thrilling long life over a premature death any day."

"Look, everyone dies eventually." His outstretched, upward-facing palms dared me to argue the inevitability of the obvious. I said nothing. "For me, it's the quality of the life you lead that matters more. Imagine experiencing awe, excitement, and exhilaration every day you're alive. Beats a long life of tedious, repetitive, mind-numbing work."

Richard's words bore more truth than I cared to admit. As I contemplated our opposing philosophies, a wisp of white, odorless smoke swirled up from the floor, obscuring everything. As the fog receded, my mother and father stood before us. Eyeing our unusual garb, my mother's face contorted in horror. Behind her, my bemused dad teased, "I never thought I'd see you dressed up as an explorer!"

"Dear child," my mother begged with tears in her eyes, "don't let this buffoon fool you. He shows no fear and would lead you beyond safety to a certain death. You are utterly unprepared for this type of journey. Look at how scared you are of snakes, and thunderstorms, and noises in the dark. Let alone lions and dragons!"

Immediately, Richard countered, "Which is exactly *why* you must accompany me. You must face your fears. Only then can you escape the paralyzing power they hold over you. Leading a life cloaked in fear weighs on you, sapping your energy, destroying your natural curiosity, leaving you with nothing but a dull existence."

Angered by Richard's petulance, my mother glared at him. She chose her words carefully. "You, more than anyone, should understand that fear serves a critical function in keeping us from danger, thus increasing the prospect of our survival. If you persevere in your plan to lead my daughter into peril, I will pray for God to strike you down to spare her life!"

This emotionally charged moment between Richard and my mother left me speechless. Each spoke words of truth, but their motives couldn't have been more contrary. Richard wanted me to experience all the things I'd never seen or done in my life. My mother wanted to protect me from danger. I felt more confused than ever.

My father stepped forward and addressed all of us. "Brenda is a wise young woman, but naturally fears the danger in what she does not know. We love her and want to protect her." Turning to my mother, he continued, "But our own fears have held us back from letting her make her own decisions. We've been afraid to let her fail, to let her learn from her own mistakes. God knows how much we've learned from making our own."

I searched his face and saw a trace of remorse as he faced Richard. "She needs confidence to find her own path in this world. You aren't much older than her, yet you seem to have a wiser soul. Can you teach her how to survive and thrive in this world that is constantly changing?"

Richard looked surprised. After being excoriated by my mother's blistering tirade, he had not expected this sincere request from my father.

"I can only promise to show your daughter many curious things unknown to most humans. We will no doubt encounter dangers in terra incognita, but, sir, I do not have a death wish. We will proceed with great caution."

Dad still appeared skeptical. But he seemed to know the time had arrived to set me free. Even if I wasn't convinced that I wanted to be set free.

"Brenda, your happiness means more than anything in the world to me. Follow your heart. Open your eyes. Heed this man's advice and learn what he has to teach you. Most of all, have fun."

"What are you doing?" my mother demanded, dumbfounded by the swiftness with which my father had granted me permission. I was free to accompany Richard on his travel in search of untold treasures.

"One last thing." From behind his back, my father withdrew a gleaming gold saber. He offered me its handle. "Just in case you *do* happen across any dragons or lions…" Laughing at my father's protective instinct, I graciously accepted the formidable weapon. I experimented with waving it about in the air.

"Whoa there!" Richard dodged the swish of my blade. "I'd like to keep my head on top of my shoulders a while longer."

Richard nodded toward my father. "I give you my word that Brenda will return to you with amazing stories to entertain you for days."

My father winked at me. "I'll certainly want to hear about all those lions and dragons you find."

As my mother sobbed, Richard announced, "We must be on our way."

Instantly, the scene went blank.

My eyes fluttered open. I was back on the floor in the Y. Yet for hours after waking, the memory of that vivid dream remained. So did my fears.

7

THE TRAIN TO IFAKARA

LIKE A DRILL SERGEANT, Conrad barked out our marching orders when he woke us just before sunrise. He'd reserved seats on the train departing later that morning, which would take us to the village from where we'd start our river trip. Only a few hours remained to finish tightly packing all our supplies into coolers, wooden crates, buckets, and baskets before transporting them to the train station. Conrad needed more Tanzanian shillings to settle our bills with a few vendors in Dar, so Richard and I set off to exchange the extra dollars we'd brought for the trip.

Under Tanzanian law, the exchange of foreign currency for shillings could *only* be done at the National Bank. Despite stiff penalties for breaking that law, including prison time, a thriving underground black market existed for US dollars, always in high demand. Using the black market, we could get up to 15 times more shillings than the extortionate official exchange rate the government offered. Illegally exchanging our American currency would drastically lower the cost of running the expedition, but at the risk, if caught, of being subjected to the torturous hospitality afforded an inmate in a Tanzanian prison.

Of course, Richard, the risk-taker, had no qualms about going for the better rate. He assured me he knew of a black market contact who could safely change half of our dollars. The other half we would change legally at the bank. We set off on foot with a small city map to find the shop of the unconventional money changer. After a few wrong turns, we found the right street. Wandering down it, we scrutinized the names painted on the retail signs and banners. Crammed together like canned sardines, the small shops sold everything from hardware to clothing to junk. We stopped in front of a dingy storefront Richard suspected was "our place."

He told me to stay outside on the opposite side of the street to watch for any sign that someone had followed us. As Richard crossed the street, an older man of East Indian descent rose from a stool just inside its open door. They exchanged a few words; then the merchant invited Richard to follow him into the shop's dim interior. Richard glanced back over his shoulder, gave a slight nod, and disappeared into the secretive darkness.

Time stalled as I pretended to look interested in the goods displayed in nearby shop windows. I paced back and forth along the storefronts, wondering what could take so damned long. How was I, a naïve tourist, supposed to know if someone was tailing us? That sort of thing happened in spy movies, usually with dire consequences. But… what if I *did* notice a suspicious stalker? Richard hadn't given me any instructions about what I should do. This uncertainty raised goosebumps on my arms.

Tanzanians traveled up and down this street on foot, on bicycles, on scooters, and in autos. It bustled with customers entering and exiting the shops. Women loaded down with baskets hurried on their way to the market. I alone loitered awkwardly, waiting for the "deal" to conclude. Many passersby stared at me. Most seemed merely curious, but a few creeped me out. I decided Richard deserved a scolding for leaving me standing conspicuously alone outside. But I doubted I'd summon the courage to chide him.

Finally, both men emerged from the gloom. They hugged each other like long-lost friends. Richard shook his hand and wished him well. Then he crossed the street to where I stood waiting.

"Do you know that man, Richard?" I asked, thinking they acted a bit too friendly for merely having done one money exchange.

"I do now." He coyly dismissed my question. I let it go. If some secretive intelligence operation connected Richard with this East Indian, the less I knew about it, the better.

"Let's go. The National Bank is down this way." He pointed down the street in the opposite direction that we'd come.

At the bank, Richard exchanged the rest of our trip money while I exchanged $200 of my own. With shillings in hand, along with the legal exchange receipts we'd need to show the customs control officer when we left the country, we headed back to the Y. Conrad took the extra shillings and immediately left to square up accounts with two of the Dar shop owners who'd allowed him to run a tab.

Soon afterward, the four taxis Conrad had hired pulled up in front of the YMCA. We stuffed our personal gear, rafts, equipment, food, and supplies into their trunks, interior spaces, and rooftops. Using ropes and straps, we secured the overflowing loads firmly in place. Then, like a wandering band of gypsies, we rattled our way to the Tazara Railway station.

Tanzanians beamed with pride at their new Chinese-built single-track railway connecting Dar with Kapiri Mposhi in Zambia, 1,106 miles to the southwest. The Dar es Salaam station, inaugurated in 1975, appeared to be the cleanest and most modern building in the city.

Conrad signaled for a couple of porters to assist with hefting our mountain of luggage onto the train. I wanted to take a photo of our gear being loaded, but Conrad hollered at me to put my camera away just as one of the military police guarding the station noticed I was about to break the law. He pointed at a sign I hadn't seen hanging on the wall of the station. It strictly prohibited photography at any official government facility. His face looked stern as he marched toward me. I hurriedly stashed my camera and mumbled a quick, "Sorry."

It didn't take long for the porters to load our gear into one of the freight compartments. A conductor then directed us to board the passenger car where our ticketed seats were located. Since seven of us were making the journey, Conrad had purchased tickets in two separate six-passenger compartments so no one would be stuck sitting alone. Richard and I shared the polished wood bench seats in one compartment with four Tanzanian men. Amused by their two foreign compartment-mates, they giggled and stole glances at us as they conversed in Swahili. Meanwhile, Richard and I speculated on what they might be saying about us. Soon we were all laughing.

Timidly, the youngest man tried his English on us. "You are Americans?"

We nodded.

"My name is Solomon." He extended his hand to shake ours as we introduced ourselves. "Where you are going?" he asked.

"We're traveling to Ifakara, 224 miles down the tracks. It should take eight hours *if* all goes well," Richard replied, then paused while Solomon translated for the other Tanzanians. "We are going to launch four rubber rafts on the Kilombero River, following it through the Selous Game Reserve for 12 days." Solomon's eyes widened. I could hear disbelief in his voice as he translated Richard's tale of putting rubber rafts into a river full of crocodiles and hippopotami, intending to float out of touch with the outside world for 170 miles. I still hadn't completely convinced myself about the wisdom of undertaking this feat. The alarmed look on all the Tanzanians' faces seemed ominous.

With each passing hour, Solomon gained more confidence in his English. We traded stories to pass the time. He'd translate our side of the conversation for the others, who occasionally raised an eyebrow or shook their heads at the craziness of what they were hearing. Our young translator might have added a few embellishments while he described our plans for what I suspected he deemed a terribly misguided adventure.

Solomon was traveling with his older brother, who had recently enlisted in the army. For his first deployment, he'd be helping to secure the western border from attacks by guerilla rebels. Since Solomon had made a last-minute decision to accompany his brother to the border, they ended up seated in different compartments. The brothers took turns "visiting" each other by standing in the open hallway that allowed access to all the compartments. Safety regulations, however, strictly prohibited standing in this hallway while the train was moving. So, whenever a passing conductor caught a "visiting" brother, he issued a stern reprimand and ordered the offender back to his own seat.

Solomon told us how Tanzania's 20-year experiment with socialism had failed under President Julius Nyerere's political manifesto. Paupers refused to leave the cities for newly constructed rural villages built to relieve urban overcrowding; farmers opposed his agricultural collectives; corruption in the government had exploded. More recently, significant shortages of consumer goods, including cotton, had led to rampant black-market activity. Most Tanzanians were suffering.

His mention of cotton made me think of the women's *kangas*. "Solomon, when I get back to Dar, where's the best place to buy a *kanga*?"

"Oh, not possible." Seeing my surprise, he explained, "Government owns only *kanga* factory in Tanzania. Not much cotton, so not make many *kangas*. Only Tanzania ladies buy. Maybe one, maybe two new *kangas* in year. Only factory store sells, but ladies need show government ID card."

"But I thought I saw some for sale at one of the outdoor markets in Dar?" I said.

"Fakes from China. Bad work. Are not real *kanga*. Real *kanga* need Swahili proverb printed on side."

Solomon shrugged his shoulders in sympathy. I leaned back in my seat, sorely disappointed.

Every half hour or so, the train arrived at a clearing in the jungle. Local vendors would race along next to the train as it slowed to a stop. It took roughly 10 minutes to offload the freight destined for tiny villages hidden behind the tangled vegetation. At each station, some passengers left the train, while others boarded. Squawking vendors held their merchandise outstretched up to the open windows, tempting passengers with bananas, tropical fruits, pastries, and deep-fried sweets.

Richard had advised me that eating anything from these vendors would invite a nasty case of gastric distress. My canteen was nearly empty. I took only tiny sips when my mouth felt parched. The intensifying humid midday heat drained us. My thirst grew as sweat soaked my t-shirt. At least my new Mao cap held my hair coiled up inside and off the back of my neck where beads of sweat trickled south. With the train stopped, the sultry breeze ceased, and the air instantly reeked of the damp stink of overheated bodies.

By three o'clock, my parched mouth felt as dry as a desert, and my stomach grumbled. I dreamed about biting into a juicy fruit. At our next stop, a woman sporting a bright orange-and-red *kanga* with a basket of brightly colored fruit balanced on her head paused beneath our window. Our eyes met. In Swahili, she coaxed me to buy her fruit. I'd never seen this plump green-and-orange-skinned oval fruit. Whatever they were, they looked juicy.

"Richard, look at the fruit that lady has. It looks pretty fresh. Do you think it would be alright to eat?" The woman lifted her basket higher.

"Those are mangos. Haven't you ever had a mango?"

I shook my head. "First time I've seen one. I don't think Angel's Food Market back home carries these."

"Yeah, they look pretty good." He eyed the basket with the confidence of a mango connoisseur. "Let's get some."

We bought two of her best-looking mangos and settled back in our seats as the train chugged down the tracks.

"So, Richard, how do we eat this thing?" I assumed I could just follow his lead.

Sheepishly, he replied, "I'm not really sure. I just know you don't eat the skin. I've had plenty of mango shakes in cafes, and they're delicious."

Regrettably, Solomon hadn't returned from checking on his brother at the last stop, so we couldn't ask him. The other three Tanzanians watched with anticipation as it became apparent that we had no clue how to eat a mango. I felt silly not even being able to ask them for help.

I pulled out my buck knife. Richard and I figured a single cut straight through the center might work, as with an apple or orange. My knife struck a very solid pit in the center and stopped dead.

Suppressed giggles sounded from the bench across from us, the first sign of a flawed technique. I sliced around the pit as if slicing an avocado. I hoped I didn't look like a fool. Next, I tried to twist the two halves apart. They failed to budge, but they oozed a sticky liquid.

I appealed to Richard. "What do you think?"

"Let me try it." Gladly, I handed him the mango. He gripped both halves and twisted until sticky-sweet mango juice ran down both his arms and his fingers left deep indentations in its skin. The halves were no looser.

Unable to contain their composure as we mangled the mango, the Tanzanians roared with laughter. Deeming the first butchered mango unfit for consumption, I tossed it out through the open window. Maybe some hungry jungle beast would have a feast.

Our comical massacre of the first mango had so riveted our compartment mates they sat ready for Act Two. I thought maybe cutting the fruit into wedges might work. After wrestling with the fruit, I pried a piece somewhat resembling a wedge loose. I offered it to Richard, then cut another for myself. My thirst was so great; I didn't care what it tasted like as long as it had juice.

I bit into the center of the slice, sucking its liquid into my mouth. It tasted peculiar, both sweet and tart. I swallowed and started to cut another wedge. I noticed Richard looked confused. Unexpectedly, a foul aftertaste permeated my mouth. I suddenly regretted ingesting any of this strange fruit.

"This tastes horrible, Richard!"

"It doesn't taste like any mango shake I've ever had, that's for sure. It tastes like kerosene!"

"Kerosene, exactly! Do you think it's not ripe?" I asked, desperately wishing for something to cleanse the vile taste from my mouth. Richard shrugged. He, too, had been stunned by the nasty flavor. I took a few precious sips from my canteen, but the kerosene taste persisted.

Impishly, I extended the fruit toward him. "Want some more?"

He pushed the mango away in disgust. "Get rid of it."

The departure of the second mango through the window set off another round of hysterical laughter from our Tanzanian comrades. I could only imagine what they were thinking about the two ignorant Americans, faces and arms sticky with kerosene-laced mango juice, who'd just butchered two seemingly perfect mangos.

When Solomon returned from visiting his brother, we explained our predicament to him. Graciously, he pulled a mango from his bag to show us the proper way to eat it. He sliced two thick slabs from each side of the elliptically shaped fruit, cutting close to but avoiding the pit. Then he crosshatched the deep yellow flesh of one slab and inverted its curve to separate the flesh into easily bitable chunks. The simplicity of his technique wowed me.

He generously offered us some of his mango, but with the flavor of kerosene still pungent in our mouths, we emphatically declined the offer. I vowed right then never to eat another mango.

THAT AFTERNOON, as the train skirted the northwestern boundary of the Selous Game Reserve, breaks appeared in the dense jungle as we temporarily entered the savannah grasslands of the Kilombero River floodplain. The river itself meandered too far away in the distance to glimpse from the train, but this region offered a sense of the vast wilderness we eventually would enter. I gazed out with wonderment at the swaying tall grass that stretched forever.

We finally reached the village of Ifakara as the sun's rays tinted the early evening sky with pastel hues. My heart ached to bid farewell to Solomon, my first Tanzanian "friend," knowing I would never see him again. His stories, sharing so many facts, practical and trivial, about his country, had made the long, sweltering journey pass more quickly. Concerned for our safety, he begged us to be extra careful of wildlife in the game reserve. He must have thought our insane plan would take us on a one-way journey to meet our maker. That possibility kept resurfacing in my mind.

Once we had off-loaded our two tons of gear, the train chugged westward. In the gathering dusk, I could just make out Solomon's distant silhouette, still waving through an open window until the train veered around a bend half a mile down the track.

The cooler air of evening reenergized our team. The driver of Conrad's prearranged flatbed truck backed up to our small mountain of gear. We loaded all our equipment, then climbed atop the heap and held on for dear life.

Darkness engulfed us as we set off toward the village. Nightfall served as the alarm on nature's clock, waking the jungle critters who napped during the day's heat. The jet blackness through which we traveled seemed to amplify the screeching, groaning, cackling, and croaking of unseen predators. Judging by the surprising intensity of these sounds, I worried the animals might be only a few feet from our truck. Despite my mind-numbing exhaustion, I wondered if I would ever fall asleep with all this commotion.

8

DOUBLE STANDARDS

OUR LATE ARRIVAL made the idea of trying to find and set up a campsite in the village impractical. So Conrad splurged on indoor accommodations at the Lutheran Mission for our first night in Ifakara. When our truck rumbled into their courtyard, the matron and a few of her staff welcomed us warmly in front of their office as we hopped down from the truck. Conrad had negotiated a spacious, empty, one-room building at a bargain rate rather than reserving their pricier traditional guest rooms. The matron led us to our quarters and unlocked the door.

The building appeared recently constructed with cinder-block walls, a concrete floor, a tin roof, and screened windows. The bare lightbulb hanging from the ceiling revealed the room looked clean and safe. Instantly, my mood improved; in here there would be no close encounters with the critters making all that racket outside. This building offered plenty of space for all of us to spread out sleeping bags.

The transport truck backed up to the door of the building. With the help of Mission staff working under the watchful eye of the stout but strong-willed matron, we unloaded the truck and stashed our gear along the inside wall closest to the doorway. With our trip gear secured, we collected our personal duffel bags and backpacks and toted them into the building.

As soon as I entered "our" room, the matron rushed in behind me, grabbing my duffel bag from my hand. "No, no, no!" she objected. Her eyes glowered with accusation as she carried on with her protest in Swahili. Conrad tried to calm her, to no avail. While they argued, she kept a death grip on the handles of my bag. Her scowl gave me a bad feeling that I wouldn't be getting my bag back until she got her way. Glances my way by both Conrad and the matron as their debate raged only heightened my angst.

Conrad finally nodded at her and explained to the rest of us, "The Lutheran Mission has very strict rules. Unless people of the opposite sex are married, they may not sleep in the same room." His eyes fixed on mine. "She says that you can't stay in this building with us tonight."

"No, Conrad!" I could hear the panic rising in my voice. "I have to stay with you guys. There's six of you. What does she think we're going to do? Have an orgy?"

I was determined not to let my fellow guides out of my sight, not here, not in the middle of the jungle in the middle of nowhere. Conrad remained silent. I turned to Richard. "Please, Richard, don't let her split us up." He could see how terrified I was, but said nothing.

I turned to the matron, still clutching my duffel. "Please, ma'am, I've never been to Africa before. I'm scared. Please, just let me stay in here, where it's safe. Please?" I pleaded, staring into her big brown eyes so she could see the fear in mine. She tilted her scarf-covered head toward Conrad, indicating the answer would come from him.

He put his arm around me and gently led me outside.

"Let's go look at where she wants to put you. It's just over there."

He pointed to a thatched roof structure with mud-plastered walls about 30 yards away. Conrad, Richard, and I followed the matron, while Conrad calmly explained that these villagers held strong religious beliefs, so it would be best if we didn't insist that I sleep in the same room with the men.

The smaller building housed six tiny, undecorated, partitioned cubicles opening off a narrow central hallway. Each room was barely large enough to hold a single-sized cot and a gauzy mosquito net hanging from a rafter over it, along with a small table and chair. Between the top of the six-foot-high wooden wall partitions and the thatched roof was a sizable open gap.

Besides the locking door at the entrance of the building, each cubicle had its own locking door. This added level of security wasn't worth much, since a swift kick would easily topple the door and the partition wall that framed it. Conrad tried his best to mollify me.

"Look, you have the luxury of sleeping on a cot instead of a concrete floor. There are no other guests expected, so you've got the entire building to yourself. You won't have to put up with our snoring. And we're maybe 30 yards away. All you have to do is scream and we'll hear you if you need help."

Already I felt like screaming.

Richard added, "You know, you're getting the best deal. You should consider yourself lucky. Compared to our room, these are high-class digs!"

None of this was high class by any stretch of any imagination.

"I just feel safer staying with you guys. Please?" They both shook their heads to emphasize that wasn't going to happen.

"Okay, but if I get scared, I'm gonna scream." With monster-sized misgivings, I agreed to stay by myself that night.

I couldn't remember when I'd felt this famished. The "special" offered in the Mission's dining room that night featured boiled white rice and some sort of mystery meat stew. At that point, I would have been happy with anything except a mango. They offered three beverage options: room temperature (about 80 degrees Fahrenheit) Tanzanian beer, Coca-Cola, or chai, the bitter black Tanzanian tea diluted with unpasteurized, unhomogenized milk and sugar. None appealed to me. I just wanted a glass of clean water. But that wasn't on the menu. I chose the last bottle of Coca-Cola they had in stock. The guys drank local beer while we waited for the food to be served.

"Pretty rude stuff." Jim Slade smirked as he hoisted his bottle in a sarcastic salute.

Soon, the kitchen delivered steaming bowls of stew to us, which, to my surprise, proved tasty and filling. I hadn't expected the meat to be so tender. Cooked with it were pieces of onion and potato in a sauce flavored with curry, cinnamon, and other spices I couldn't identify. Within minutes only empty bowls remained.

With our stomachs full and our bodies weary from loading and unloading equipment on both ends of that tedious train ride, we headed off to our sleeping quarters. The matron, smug with her success in arranging proper sleeping arrangements, led me back to "my" building. She unlocked the door and twisted a bare lightbulb in its socket to illuminate the hallway. She had placed my duffel bag on the cot inside the cubicle nearest to the building's entrance.

As I stepped inside, she launched into a lecture in Swahili, punctuated by hand gestures. I didn't have a clue what she said, but toward the end, she handed me a key to my cubicle door. She pantomimed the motion of locking my door with the key. She obviously wanted me to keep my cubicle door locked.

The matron had promised I'd be alone in the building, so why did I have to lock my door? Who did she think might prey on me? I was just too tired to pursue answers. I nodded obediently. She lit a partially melted candle and jammed it into the top of an empty soda bottle, standing on the small table. Satisfied, she exited to the hallway. She waited to hear the click of my key. Then she unscrewed the hallway light bulb and locked the outer door as she left.

I groped through my duffel for my headlamp since the meager light cast by the candle only illuminated a tiny portion of the cubicle. Once I could see again, I spread out my sleeping bag on the rickety cot and unraveled the loose knot holding the gathered mosquito netting suspended above the bed. As gravity unfurled it, the netting revealed two gaping holes the size of grapefruits.

Those holes needed to be sealed somehow. I remembered I'd packed a miniature sewing repair kit. I dug it out of my backpack. To my delight, it included two safety pins. I carefully folded over the netting around the holes twice, then fastened the four layers of netting shut with the pins.

Carefully, I tucked the bottom edges of the netting between my sleeping bag and the cot. I zipped up my duffel and my backpack and slid them under the cot. Then I blew out the candle and dove inside the canopy through a small space I'd left untucked. After I'd tucked myself inside the netting, I settled into the trench formed by the sagging cot, and turned off my headlamp.

As exhausted as I felt, I should have quickly drifted into dreamland, but sleep refused to come. I hadn't heard them earlier, but now, in the blackness surrounding me, the thunderous whining of jungle mosquitos roared in my ears. They were so loud I fretted that a section of the netting had come untucked and the hungry bloodsuckers had snuck into my protective cocoon.

Instantly, my mind entered crisis mode. "Malaria! Beware!" The debilitating symptoms of the mosquito-borne parasitic illness, so prevalent throughout Africa, included high fevers, headaches, vomiting, and diarrhea. If not treated promptly, it could be fatal. It killed a million people worldwide every year, 90 percent of them in Africa.

Following my doctor's instructions, I'd begun taking a prophylactic dose of chloroquine two weeks before the trip. One of our guides had told me about a friend who died after being bitten by a mosquito carrying a chloroquine-resistant form of the parasite. His story took center stage in my thoughts. It strengthened my resolve that these nasty whining buggers would *not* feast on my blood.

I turned on my headlamp, searching for the noisy culprits. My horror grew when I saw outside the netting a veritable air force of mosquitos attacking the fine mesh, driven into a frenzy by the irresistible scent of my unreachable blood. My repairs to the netting were holding fast. I did not find a single mosquito inside my canopy. Between the unending din of the starving mosquitos and the ruckus of jungle animals outside on the hunt, I wished I'd brought some industrial-strength earplugs to help me fall asleep.

Minutes later, some small creature scurried across the floor on the far side of the building. Could it be a mouse or, worse, a rat? I preferred not to know. In a moment of honest self-reflection, I conceded how painfully ill-prepared I was to be in such an unfamiliar environment. Doubts about whether I could survive this trip again crept into my thoughts. Would *any* of us survive it? This was the first commercial trip ever to raft through the Selous. So many things could go wrong. Fortunately, I didn't realize the true extent of how deadly the encounters could be.

At this point, I had to accept there was no escape. I had to endure whatever we faced. To divert my thoughts from mental mayhem, I tried to imagine the things I enjoyed most in the world: a spectacular multicolored sunset over the ocean; my fluffy-furred baby kitten; coffee ice cream slathered with hot fudge, whipped cream, and nuts. Mmm. As I recalled the joy brought by these favorites, my body relaxed. I welcomed the escape that sleep promised.

A key clicked, turning in a lock. I bolted upright with terror. It wasn't my imagination. The door into the building slowly creaked open. I trembled and readied a scream that would be so earsplitting that every creature within a mile would stop dead in its tracks. Who the hell was sneaking into my sleeping quarters?

From just outside my cubicle door came a female's hushed giggle. Then the sound of a big wet slurpy kiss, while a male murmured deep tones of pleasure. Another giggle and the illicit couple, obviously familiar with the building layout, snuck down the hallway and into a cubicle to consummate their clandestine intentions.

I had no intention of being entertained by their midnight tryst. To let them know they weren't alone, I coughed loudly. Twice. I heard no further noise. Nor did I hear them sneak back out of the building.

Ironically, I felt a twinge of reassurance knowing I wasn't alone. But the hypocrisy of separating me from the male guides because of "strict" Mission rules banning men and women from sleeping together in the same building? Really? That royally pissed me off.

If it hadn't been the middle of the night, I would have found that holier-than-thou, rule-spouting matron to reveal the carnal activities happening right under her nose by someone with access to her building's keys. As my adrenaline surge waned, I finally fell into a deep sleep.

A pair of roosters raucously crowing in the yard outside woke me several hours later. Refreshed and hungry, I emerged from my cubicle. I glanced down the hallway. All the doors were open. No trace remained of my midnight visitors.

9
STAGING CHORES

TWO 60-FOOT TALL, magnificent mango trees stood as guardians over an open square in the village center. These leafy giants provided a welcoming canopy of shade from the fierce tropical sun. We set up our staging area that morning beneath their massive branches.

Conrad had convinced an expat friend living in Tanzania to store the Sobek equipment he'd used on his exploratory run in a shed behind his bungalow for the last two years. Everything needed to be inspected, cleaned, and, if needed, repaired. To begin with, we inflated our rafts, meticulously checking them for leaks by pouring soapy water over the bloated tubes, searching for any tiny telltale bubbles escaping from leaks in seams or worn areas.

Word of our arrival swiftly spread through the village. Like the mysterious intrusive Coke bottle falling out of the sky in the middle of the Kalahari Desert in the recent popular movie *The Gods Must Be Crazy*, here in Ifakara, foreign white people messing with an assortment of strange objects had landed without warning in the middle of their village.

Curious children interrupted their games to squat in a circle around our work area. A few adults kept a wary eye on them and us from doorways of huts opening onto the square. They, too, must have been wondering what in the world we were doing.

Our chores became a spellbinding event in these youngster's lives. They jostled for the best view, worried they might miss a critical moment of the "action." By midmorning, 30 barefoot boys and girls dressed in worn hand-me-downs had parked themselves along our perimeter, inspecting our every move.

"We need someone to paint our name on these rafts." Richard knew that having the name of our company, "Sobek," in large bright yellow letters on the unmarked gray side tubes of our rafts would stand out in photographs taken during the trip. This brand could prove valuable were he to be successful in convincing magazines to publish his account of this history-making adventure—provided we all survived.

"Who's good at freehand lettering?" he asked.

"You definitely don't want me doing it if you expect it to be legible," said Bart.

"How about you, Brenda? You've got beautiful handwriting. I know you can do this."

"I can try," I said. "But I can't promise perfection."

"You're hired!" Without hesitation, Richard pulled a couple of paint cans and brushes from the pile of supplies purchased in Dar.

As I carefully outlined the letters with marine-quality waterproof paint, it occurred to me I had landed the cushiest job of all. I got to sit in the shade on an overturned bailing bucket and, with the precision of an artist, spell "SOBEK" in foot-tall letters on each side of our four rafts.

While I painted, others on the crew constructed support frames from sturdy bamboo poles, measured to fit snugly against the inside curve of the boat's side tubes. Others used more bamboo to build fruit and vegetable baskets that would hang suspended from the frames inside the boats. Bart triple-checked the supplies in the first aid kit and the raft patch kit to verify that Conrad had included everything we might need.

As the blazing sun climbed higher, the heat and humidity intensified, turning the surface of my skin sticky. Our work pace slowed. The adult villagers moved to the cooler interiors of their huts. To stay within the shrinking circle of shade as the sun moved directly overhead, we moved our projects closer to the tree trunks. Being the height of mango season, hundreds of ripe mangos hung ready to drop from the limbs above us. Staying in the shade practically guaranteed a few crushing hits by these trees' natural warheads.

All morning, we listened for the whishing of plummeting mangos rustling through the leaves on their earthward plunge. When the leaves right above us rustled, we dipped our heads, protecting them with our hands to soften a bruising blow should one hit us. Once we heard the mango's terminal thud, we'd resume our chores, safe until the leaves rustled again. Each thud signaled the start of another race. The children closest to a fallen mango sprang to their feet in a mad dash to be the one to retrieve the prize.

Occasionally, a breeze would shake four or five mangos loose at once. Chaos ensued as kids sprinted in hot pursuit of the sweet treasures. When two emerged from a chase, each with their hands wrapped around the same mango, a scuffle would ensue, with shouting and shoving until one reluctantly relinquished his or her half of the trophy. The kids' dogged quests to collect every fallen mango amused us as much as we entertained them with our basket-making and boat-painting.

For lunch, we walked back to the Mission for another simple meal of boiled white rice and mystery stew. Given the lack of refrigeration and the blistering midday heat, I hoped the lighter color of the noontime stew was proof we were eating a freshly made batch. Less spicy than the previous night's stew, this new sauce tasted as though the cook had added coconut milk and fried onions.

As usual, the guys downed warm beer, figuring the alcohol would help ward off gastric upsets. Not being a fan of a foam-topped cold beer, let alone beer warmed to a steamy 90 degrees, I opted for the excessively sweet and anything but thirst-quenching tea. After one sip, I uttered, "Pretty rude stuff."

"Shoulda had a beer with us," Slade teased.

SWEAT SOAKED MY T-SHIRT in the stifling afternoon heat. Richard helped me to paint the insides of the letters that I had meticulously outlined that morning. We noticed a scrawny, barefoot boy, maybe five years old, inching his way closer to us. When Richard beckoned to him, a jubilant smile spread across his face. He scampered the last few steps to join us, then sank into a flat-footed squat between the two bailing buckets where Richard and I sat.

Richard spoke to the boy as if they were buddies. The lack of a common language didn't stop them. Richard alternated speaking in English and a gibberish imitation of Swahili. To my surprise, the boy responded in Swahili, which neither of us understood. I couldn't tell whether he was answering Richard's question or asking a question of his own.

"So, kid, do you like Mick Jagger and the Rolling Stones?"

The boy immediately gave a detailed response, using gestures and spirited facial expressions.

"Okay, you don't like the Stones. What do you want to be when you grow up? An astronaut? A race car driver? An international money launderer?"

The boy's brown eyes grew wide. He danced about as he answered.

"Oh, you want to be Tanzania's Chubby Checker!" As the child nodded his head, Richard continued, "Well then, you're going to need a costume. Here, let me make you one."

He reached for the boy's hand, pulled him forward, and then rotated him 180 degrees so the boy's back faced him. Confused, the kid fidgeted, clueless about what this white foreigner had planned. Richard dipped his brush into the syrupy yellow paint.

I couldn't believe that Richard intended to deface the boy's threadbare shirt. I waited one more second, praying I was wrong. He lifted the paintbrush even with the boy's shoulder blades.

"No, Richard! Please don't!"

Paintbrush in midair, he protested, "Why not? I'm gonna make this kid famous."

"You're gonna get this kid killed."

Richard looked puzzled.

"That's probably the only shirt the kid owns. His mother is going to kill him when she sees it's ruined."

"Nah, she'll love it."

While I challenged Richard, the boy posed stoically in front of him.

"I bet his mother's gonna ask you to pay for a new shirt." That's what my mother would have done.

Richard ignored me. He smeared a glob of yellow paint across the back of the boy's shirt. It took him less than a minute. After his last stroke, he whirled the boy around to expose his masterpiece. I expected a symbol or pattern, but no, the bright yellow paint against the dark blue fabric of the shirt spelled SOBEK.

"Oh no," I groaned. "How could you do that?"

Sporting a goofy grin, he reckoned, "Free advertising."

Proud he'd been chosen as the model for Richard's artistic expression, the boy gleefully ran off to show his friends his new logo. In no time, he returned, leading a few of his friends. They eagerly lined up to have their own threadbare shirts painted by the visiting American artist. News of Richard's marketing effort blew through the village like a raging nor'easter. Several of the children's parents gathered to watch the "*en plein air*" art show. Rather than being angry, they seemed curious. Amazingly, some even laughed and smiled watching Richard's creative endeavor. His quirky, innate ability to bask in the good graces of everyone everywhere fascinated me.

When I realized we wouldn't have to flee into the jungle for destroying every piece of children's clothing in Ifakara, I relaxed. Richard happily spent the rest of the afternoon creating goodwill by painting the shirt of nearly every child in the village.

"Hey, Richard," I said. "Maybe the elders will decide to change the name of the village to Sobek, now that half the town is sporting clothing with our company's name on it."

Richard laughed. "Ralph Lauren, eat your heart out, buddy."

AT MIDAFTERNOON, two Tanzanian men uniformed in official-looking moss-green pants and khaki-colored shirts arrived at our staging area. They asked to speak with Conrad. The older man introduced himself as Magigi, then his associate as Masengela. Conrad greeted them with a hearty handshake. Knowing how dangerous it would be to spend two weeks in the Selous wilderness without some form of protection, Conrad had arranged for knowledgeable armed escorts from the Tanzanian Game Department.

The puzzled look on Conrad's face turned to exasperation as he ranted, "These are the game rangers I hired to keep us safe. But I requested four of them, one per raft. I'm not sure why there are only two here."

Conrad asked to inspect their weapons. Only Magigi produced a rifle, which resembled a relic from World War I. Masengela was unarmed. His anger growing, Conrad demanded to see Magigi's stash of ammunition. From his pocket, Magigi dug out the three bullets the Game Department had issued to him.

Their conversation, initially cordial, deteriorated into a full-fledged argument. Although I couldn't understand a word of the increasingly nasty quarrel in Swahili, I watched Conrad seethe with unmistakable fury. He pointed in the distance and gave what sounded like an order, then shook his head in disgust and walked back toward us. The disgruntled rangers turned in unison and angrily marched off in the direction they'd come.

"Idiots!" Conrad shouted. "Frigging idiots! I told the goddamned Commissioner of the Game Department that we'd be taking four boats and 17 people for 12 days through the Selous. I asked him to provide us with adequate security for the trip." Mimicking the Commissioner's patronizing British dialect, Conrad continued. "He tells me, 'No problem at all. You'll have everything you'll need to be safe. Don't worry. No problem.' So, what does he send me? These ragtag lackeys. One of them has a rifle and only three goddamned bullets."

That sure didn't sound like adequate protection to me. Not when we were entering the realm of seriously wild animals with nothing but their natural fear of unfamiliar beings to deter them from approaching us. And that caveat only held true if we didn't accidentally provoke them. If a hungry pride of lions attacked us on our third day out, we'd easily need all three bullets (or God forbid more) during just one encounter with the hungry man-eaters. How then could we even think of continuing for nine additional days with no ammunition? Coming from America, the land of plenty, I didn't understand why they couldn't just bring 100 bullets to be on the safe side. If we didn't need them all, the rangers could return them to the Game Department's inventory after the trip.

"What did you tell them?" Richard asked.

"I told them to take their sorry asses back to the Commissioner and tell him we aren't going into the Selous without more rangers, weapons, and much more ammunition. And I told them, 'Don't dare come back here unless you all have rifles and a full supply of ammunition, and you'd better be back here by tomorrow night.'"

The rest of us got back to our trip preparation tasks. I wondered if a Plan B existed if the rangers didn't return, but I sure as hell wasn't going to be the one to risk upsetting Conrad even more by asking.

10

MAKING A RAFIKI

BY FOUR O'CLOCK that afternoon, my energy reserve was running on empty. The breeze that swirled through the tree canopies above us should have provided some relief, but the scorching heat relentlessly sapped the moisture from our bodies.

"Hey, guys, let's take a break." Conrad pointed at a group of local men he'd just been talking with. "Those guys say there's a stream nearby where we can get washed up and cool off. Finish up what you're doing, and we'll head down there in a couple minutes."

This revelation filled me with joy. Grungy, tired, and with the heat stressing my body in a way I'd never experienced, I quickly stuffed my towel and soap in my backpack. All of us followed Conrad on a narrow dirt footpath through a grassy field toward the stream. When we reached a fork in the path, Conrad glanced in both directions, then chose the trail curving to the right and sloping gently downhill into a grove of trees.

We hadn't gone far when a village man, wildly waving his arms, ran up the path blocking our way forward. He gestured directly at me as he sternly lectured Conrad. Something was wrong. A moment of déjà vu raced through my mind. Again, fear surged in my heart. Conrad translated the Cliff Notes version of the man's warning.

"Ifakaran men and women don't bathe together. They use two separate spots in the stream. We men can continue on this way."

Of course *they* could. Ready to explode in frustration, I knew what was coming next.

"Brenda, you need to go to the place where the women bathe. Go back to the fork and take the trail that goes left."

"You want me to go there all by myself?" Tears welled in my eyes.

"You have to, if you want to get cleaned up." Conrad shrugged his shoulders. "Or you can go back to camp and wait for us."

His ultimatum struck a punishing blow, forcing me yet again to abandon the security of my colleagues. The jolt of adrenaline coursing through my body made me even more uncomfortably hot. Droplets of sweat beaded on my forehead and stung my eyes as gravity blended them with salty tears.

"I'm really nervous. Can't you just walk with me down to the stream?" I pleaded.

"No, I can't. We need to respect their rules. You can do this. He says a group of women and young children are down by the stream now, so you won't be alone. Go ahead. You'll be fine."

Continuing to protest would be fruitless. I had to do this on my own. Because I was so damn hot, the idea of finding cool water partially vanquished my fear of searching for the bathing spot alone, then mingling with female strangers, whose language I couldn't speak.

I turned and began my solo journey. With trepidation, I retraced my steps to the fork. I paused to muster courage before heading into unknown territory. The dusty path descended gradually through a field of chest-high stalks of ripening maize. The rippling of rushing water grew louder.

Above the sound of water lapping against rocks along the shore, I heard the gleeful shouts of children playing and splashing. Edging closer, I wondered how furious these village women and children would be at my startling intrusion. I readied myself for a quick retreat if it appeared I was unwelcome.

Summoning all my courage, I stepped into their clearing. Five bare-breasted women bantered with each other as they beat and kneaded bright gold-and-maroon *kanga* cloths on smooth river rocks while keeping one eye on their young ones frolicking in the water. One child spotted me and screamed in alarm. The water roiled with fury as it emptied tiny naked bodies. The children raced to safety, hiding behind their mothers. Instantly, all work and play froze.

I stood still, focusing on the slippery brown clay bank next to the swift 25-foot-wide current. 26 eyes fixed on me, waiting for my next move. Never had I felt so clueless about what I should do. I tried using a friendly smile and a little wave at them. Since I didn't know a single word of Swahili, I meekly added, "Hi everybody."

Smiles bloomed on the faces of both young and old. The children cautiously emerged from hiding behind their moms. The women resumed washing clothes. Relief! It seemed they'd accepted me.

I would need to remove my clothes to scrub the sweat and dirt from my body. Coming from an uptight, prim, and proper New England upbringing, I wondered how I could ever do that in front of complete strangers. Exposing my naked body to anyone outside in daylight was unthinkable. Heck, it even bothered me that some guides back in California regularly went skinny-dipping in rivers to cool off on hot summer days.

As I stood there pondering my dilemma, I laughed when I realized I was the only one in the group that wasn't naked! It didn't seem to bother the women at all that I had a clear view of every inch of their plump, bare bodies. They were obviously comfortable being *au naturel* together to wash clothes. We were all women, so what was the big deal?

My puritanical values demanded, "Don't do it!" But I desperately craved being clean and cool. I took off my sneakers and socks and placed them on the ground with my backpack well away from the stream. Somehow, I found just enough bravery to pull my T-shirt up over my head. My cheeks flushed red with embarrassment as I pulled off my shorts, then my underwear. I stood stark naked on the banks of that stream, feeling the most vulnerable I'd been in my life. Thankfully, God did not strike me dead for committing the crime of indecent exposure.

The women never stared directly at me, but they all checked me out with sly glances while they worked. Was I the only one feeling self-conscious? I was pretty sure I was the subject of their jovial chatter. My light tan skin drastically contrasted with their dark cocoa skin. I was an oddity with blue eyes and long, straight, light brown hair.

I longed to plunge under the protective cover of the rust-colored water. These mothers showed no signs of worry that crocodiles might lurk nearby. I didn't know the depth of the stream or the power of the current. So, I decided against an impulsive dive. But I still needed help to figure out a safe way to enter the water. None of the boys or girls had returned to the water. How could I get one of them to show me the way? I cautiously walked over to the edge of the slick bank, where I motioned the curious children to come near.

I engaged them in a game of charades. First, I pointed to the river, then placed my flat palms sideways against the middle of my thighs, then my waist, then my chest, and finally in front of my neck. How deep was this stream? Amused blank stares were all I got. I tried my gestures again. This time a petite girl with closely braided cornrows and twinkling brown eyes nodded emphatically.

She wrapped her small fingers around my right hand and pulled me a few steps closer to the edge of the bank. Then she swung our arms forward together, encouraging me to jump in with her. I resisted her second attempt to swing our arms. She turned to face me, her eyes questioning my reluctance. I pointed at myself and shook my head no. She seemed puzzled. Then I pointed at her and swung my arms forward, conveying that she should jump first without me.

Understanding flashed in her eyes. She backed up two steps, then charged forward, leaping in the air. She hit the water in a cannonball tuck. Spluttering, she surfaced and stood proudly, with her arms held high above her head. The current concealed her thin body up to the middle of her chest, close to waist deep on me. She could hold her own against the current, and I saw no sign of crocodiles. I beckoned for her to come back. She paddled to the bank and hauled herself onto shore.

I held out my hand. She grasped it enthusiastically. This time we backed up together, then rushed straight at the river. We hit the water in unison. Concentric waves swelled outward. Within seconds, a tidal wave of the children's bodies hurtled back into the water behind us. No one under three feet tall wanted to miss out on the fun. The instant coolness of the water on my sunburned skin felt heavenly. I submerged completely under the water again and again, feeling the sweat and grime being washed away by the current.

We climbed out and repeated the dive several times. Some of the other children grew impatient, demanding their turns to jump with me. For the first time on this trip, I felt happy. Truly happy, totally on my own, without a shield of bronzed, muscular boatmen to protect me. Though I didn't know a darned word of these women's and children's spoken language, our universal body language proved a perfect substitute.

The children and I leaped together from the riverbank repeatedly, splashing and frolicking for half an hour. They must have thought it strange to see a grown woman splashing playfully in the water like a child. Meanwhile, the mothers continued with their work, ever mindful of our watery shenanigans. Using the sunbaked surfaces of the largest rocks along the stream's edge, the women laid out their clean *kangas* to dry in the sun.

Then I remembered the reason I had come here in the first place—to bathe. I climbed out of the stream to fetch a brand new white oval of Dove soap from my backpack. Dipping back into the water, I lathered up, creating a creamy foam of white bubbles that coated my body. Some bubbles floated away on the water's current with the kids in hot pursuit.

My charming little "guide" waded up to me with her palms together as if praying. She eagerly pointed at my bar of soap. I handed it to her. Just as she had watched me do, she rubbed the bar over the top half of her thin muscular body and along her neatly braided cornrows.

Less than a minute later, I realized the folly of relinquishing control of my soap. The children swarmed toward the magical foaming bar in my little buddy's hands. As they passed the soap from hand to hand, I watched their tiny upper bodies disappear beneath a foamy layer of pearlescent bubbles. With each body washed, my bar of soap diminished in size.

At one point, the bar slid out of the grasp of one little boy's hand, plopping into the stream. For an uneasy moment, I figured I'd seen the last of my soap, but I was wrong. Several kids nearby quickly dove for the sinking oval. One child victoriously raised his arm. His hand tightly clutched the slippery bar. When every child had taken a turn soaping up, I retrieved my Dove and climbed out of the stream.

Before I could stash away the shrunken bar of soap, the women surrounded me, curious to learn for themselves more about this bubbling white block. One woman held out her hand for my Dove. The women touched it, smelled it, and rubbed it on themselves while chattering animatedly. Before I understood her intentions, one daring woman headed toward the stream with the soap eagerly followed by the others.

The woman with my soap laughed heartily as she rubbed the bar over her bare skin, clearly thrilled with the foam of silky bubbles. Again, the bar traveled from one to another, transforming the stream into a sudsy human washing machine. Watching from shore, hearing their delighted squeals made me smile. It finally dawned on me that soap might have been one of the many items the government couldn't afford to import. It gave me satisfaction to know that all these women got to enjoy a luxurious bath simply because I shared my 50-cent bar of soap with them.

The woman who originally took my soap finally emerged from the stream. The oval she returned had shrunk to one-third its original size. Tentatively, she reached out and touched my wet hair and stroked the bare skin of my arm. The others gathered around for a closer inspection, gently poking me, twisting my hair and giggling at the oddness of my body.

Over an hour had flown by. Much longer, I supposed, than my male colleagues would have spent getting clean. I toweled dry and put on my clothes and sneakers. Now, instead of rushing back to safety with the other crew, I felt a twinge of sadness leaving behind my bathing buddies. Hoisting my backpack, I gave them a little wave, then began my trek back to our staging area.

Halfway through the field of maize, I heard running steps behind me. I stopped and turned to see the young girl who dove into the stream with me racing to catch up. Now wrapped in a bright gold-and-green patterned *kanga*, she stopped in front of me. Speaking slowly and with great conviction, she tried to tell me something. But what? Had I left something behind? No, I thought, because she would have brought it to me. I shrugged my shoulders and shook my head, baffled.

Again, she slowly repeated the same incomprehensible words to me. Her pleading eyes showed me she was trying her best to make me understand her. I smiled at her efforts, but I had no inkling of what she was saying. Uncertain how to respond, I spoke to her in English, "I'm sorry. I don't know what you're saying. I don't speak Swahili."

My English words to her proved equally meaningless. But her eyes twinkled excitedly. At least I had tried to communicate with her. One more time she repeated her message, enunciating every word and conducting with her hands like a first-grade teacher reviewing basic vocabulary. This time, I heard some distinct words that sounded like "jumbora feekee."

I committed those two words to memory so I could ask Conrad what they meant. I motioned I had to go and began to walk away from her. But she ran along beside me, still chatting up a storm. When we arrived at our staging area, she took a place among the other children on the fringes of our work area.

A chorus of "Where have you been?" greeted me. My colleagues looked relieved they wouldn't have to mount a search party.

"I was hanging out with a bunch of the women and children. I let them try out my Dove bar and let's just say the millions of soap bubbles we made were a big hit. I might need to borrow soap from you guys before the end of the trip."

"We wondered if a crocodile had made a meal of you!" Richard said, partly in jest, yet clearly concerned. "We've been waiting for you to get back so we could go for dinner. We're all starving."

I said, "I'm sorry." But I wasn't because every minute I'd spent with the women and children had been delightful.

Before I forgot the little girl's strange foreign words, I told Conrad about our struggle to communicate. I repeated the two words I had memorized and asked if he knew what she was trying to say. Smiling, he said, "Well, well. I guess you've made yourself a friend. She was saying *jambo, rafiki*. It means 'hello, friend.'"

Touched beyond words by this girl's simple sentiment, her offering of friendship, I searched for her face in the throng of children intently observing us. When I found her, I went to where she squatted. I looked into her inquisitive eyes and carefully repeated her words, "*Jambo, rafiki.*"

Her face lit up with an enormous smile. That moment bonded us forever as rafikis.

Despite craving a fresh green salad and a plate full of vegetables, I settled once again for the Mission's "daily special," stew with white rice.

A positive aspect of starting on our river voyage involved the variety of identifiable fresh foods we'd be preparing each day. No more mystery stews for us! After dinner, we returned to our staging area to set up our tents. Michael worried that his lightweight bivy sack, a waterproof cocoon that surrounds a sleeping bag, would be too flimsy to fend off the inevitable mango meteor storm that night. He asked if any of the crew would consider sharing their structurally stronger aluminum-framed tent with him. All five of the other male crew members guffawed and declined his request.

Michael's plea triggered my nurturing instincts. I certainly didn't want anyone to get hurt. But honestly, *I* wanted a bodyguard. I still feared being by myself in these strange environs, especially on my first night sleeping outside under the stars in Africa. I figured my tent would protect *Michael*, and Michael would protect *me*. So, I offered him half the space in my two-person tent. He quickly took me up on my offer.

We arranged ourselves on opposite sides of the tent. Just before we turned off our headlamps, I noticed my *rafiki* squatting just outside the zipped-up mesh doorway of the tent. She might have been just curious, but I had the feeling she wanted to know that her new friend was safely tucked in for the night.

The blackness of night seemed to amplify the jungle animal noises. "How is anyone supposed to sleep with all that racket?" I grumbled.

"Are you kidding? This racket is nothing compared to the noises I put up with for two years studying wild chimpanzees in the Kibale Forest of Uganda," Michael said. I hadn't known this tidbit about him. I knew he had a PhD in Biological Ecology but didn't know what and where he had studied for it.

"Have you ever met Jane Goodall?" I asked.

"Of course I've met her. I greatly admire the woman. I almost worked with her at one point."

"What's she like?" I couldn't believe he knew this legendary champion of chimpanzees.

My question set Michael off on a mini-lecture. While he used a lot of her research findings in his own work, their research methodologies differed. Goodall favored "moving in" with the chimpanzees by feeding them, and she often interacted directly with them.

Michael opposed direct interactions. For his research, he relied purely on observation of natural behavior with as few direct interactions as possible, allowing the animals to maintain their wild intrinsic behaviors. His research findings about our closest animal cousins fascinated me. I kept asking questions, and he kept providing more information. We talked for hours until we realized we'd better get some sleep or we'd pay for the lack of it the next day.

JUST AFTER DAWN, a falling mango crash-landed on the side of my tent before sliding onto the ground. The impact, sounding like a small bomb explosion, startled us awake. A mad dash of tiny feet raced toward the tent to capture the fruit. Already the kids were on the job collecting mangos. Sunlight had brightened and warmed the inside of the tent. Peeking through the mesh netting at the tent's entrance, I saw an astonishing sight—my little *rafiki* squatting in the same place she'd been before we retired for the night. Had she ever returned home, or did she sleep all night just outside the door of the tent?

Conrad pounded on the side of my tent to alert us the crew was ready to make a breakfast run to the Mission. Still groggy from not enough sleep, Michael and I crawled out of the tent. Conrad muttered something about keeping the entire village awake until the wee hours, but I didn't believe him because there had been plenty of loud snoring coming from the other tents.

11

A NEW HAIRDO

FOR TWO DAYS, my new *rafiki* became my constant companion, staying by my side wherever I went. She sat next to me while I worked on our pre-trip tasks, and in the late afternoon, we went down to the stream together to wash up. After lunch on our second day, she took my hand and led me to the far side of the village square.

Outside one small hut, a plump, middle-aged woman occupied a single rickety wooden chair placed on a level spot of dirt. Six colorfully dressed local women comfortably squatted flatfooted around the chair. A younger woman, thin with enviably smooth high cheekbones, wearing a red, white, and black *kanga*, worked at braiding the woman's hair into an intricate geometric pattern of cornrows.

It appeared my little *rafiki* had led me to the open-air salon of the village hairdresser. One of the squatting women rose and ducked into a nearby hut. She emerged with another rickety chair for me. I assumed my *rafiki* had brought me to watch the stunning hair art being created right there in her village. Curiously, I studied how the young woman fervently tugged one three-inch-long strand of hair and laid it over another. The strands, as if glued, stayed perfectly in place as she

79

separated another strand and laid it across the last strand. Each strand of hair stayed put as the design grew.

While the patron's coarse curly mop of hair transformed into a tightly interwoven pattern of tiny flat braids, the squatting women did what women around the world do in hair salons. They gossiped. From the glances cast in my direction during their lively chatter, I guessed they were discussing Sobek's invasion of their village.

At one point, my young companion spoke directly to the hairdresser, who nodded in agreement. When the plump woman's stylish new "do" was complete, the hairdresser motioned to me to sit in her styling chair. Hesitantly, I held back. But my *rafiki* boldly grabbed my hand, dragging me to the chair.

The local women shuffled closer to get a better view of what their stylist would do with my foot-long, straight hair. The absence of scissors in her work area somewhat comforted me. She rubbed a strand of my hair between her thumb and fingers. Once she sensed the thinness and smoothness of my hair, I felt gentle tugging as she began creating tiny braids.

She kept up a constant commentary with her audience as she worked. Initially, she seemed frustrated because my strands refused to stay in place, slipping to one side or the other. After a while, she got the hang of working with my uncooperative hair, and the cornrows grew faster. She covered my entire head with cornrows, then pulled the ends of the 20 long braids together at the nape of my neck.

Reusing the hair elastic that had secured my ponytail, she tightly wrapped it around the cluster of new cornrows already unraveling at the ends. I gestured an offer to pay for my new hairstyle, but she firmly refused, forcefully shaking her head no. She must have felt a great sense of relief that she would never have to work with my exasperating type of hair again.

Back in camp, Richard immediately noticed my new "do." "Hey, everyone, look at Brenda. She's gone rogue. She thinks she's a native now."

"Well, I think my stylist did a great job," I said, defending my native hairstyle. "The cornrows are cooler, and the breezes can't snarl my hair into a bird's nest."

Conrad winked at me, hinting, "You might even get a marriage proposal from one of these village men now that you've got those stylish cornrows."

I spurned their teasing comments, adding, "If any of you want to get your own cornrows, I'll take you to get them done." I didn't get any takers.

LATER THAT AFTERNOON, three game rangers trotted back into camp. Only two of them carried rifles. Conrad demanded something of Magigi in Swahili. He opened his ammunition pouch and withdrew five bullets. Conrad threw up his arms in disbelief. I could see the fire in his eyes. The sharpness of his voice conveyed how pissed he was.

Unfortunately, time had run out to send them back for more weapons and ammunition. Our 10 clients on this expensive and unique expedition would arrive in just a few hours. We'd be floating down the Rumemo tributary by noon the next day.

As we walked to our last dinner at the Mission, the biblical analogy of the last supper came to mind. After all, wasn't our crew a bunch of devoted believers willing to follow Richard, at enormous risk to each of us, on the path he had chosen for us? I prayed that our last meal in Ifakara would be a magnificent feast.

Instead, we got an instant replay of the same white rice and mystery meat stew we'd eaten at every meal since we arrived. While we ate, Conrad updated us on the plans for the next 24 hours. He still seethed as he described his exchange with the game rangers.

"I can't fuckin' believe the Game Department expects us to spend 12 days in the Selous with only five bullets. Who knows if those goddamned Winchester bolt action rifles still even work? When we get back, that motherfuckin' Commissioner is going to be screwed when I'm through with him."

"Hey, man. Take it easy." Michael attempted to quell Conrad's rant. "With Tanzania's economic crisis, the Game Department is lucky to have *any* funding! It wouldn't surprise me if these rangers got half the entire inventory of the department's ammunition. It is what it is. We've got five bullets. We'll make it work."

Conrad glared at Michael, refusing to admit Michael's assessment was likely accurate. Michael shrugged his shoulders, daring Conrad to refute him. Tense seconds passed while Conrad regained his composure before he spoke.

"The third ranger is nothing but dead weight. No gun. No ammunition. We'll put him in one of the middle two rafts. Magigi, with his rifle, can ride up front in the first paddle raft, and Masengela, with his rifle, can ride in the last boat. Hopefully, one of the armed ones will have a shot if anything attacks."

None of this was reassuring. It only reinforced my doubts about going on this ridiculous trip. We finished our meal just as the sun sprinted toward the horizon, coloring the sky with deep glowing gold. All of us, except Richard and Conrad, headed back to camp for our last night beneath the murderous mango trees. They took off in two of the mission's dilapidated vans to meet the train carrying our 10 paying clients.

When the train chugged into the station, they welcomed our weary and hungry clients as they staggered off the train. After loading them and their gear into the vans, they drove back to the Lutheran Mission. The matron doled out room keys for the Mission's best guest rooms in a modern wing, fronted with landscaped gardens next to the office.

Later, after our clients had eaten dinner, Conrad and Richard held an impromptu orientation. To protect their sleeping bags and clothing, Conrad had left a black rubber dry bag stenciled with "Sobek" and a unique number for each passenger at the hotel in Dar. The instructions accompanying each bag specified to bring only what fit inside. Any extra possessions, not listed as one of the 12 essential items Sobek recommended clients bring on the trip, needed to be left in safekeeping at the hotel. This was of crucial importance because each client would have to carry their own possessions on the two-and-a-half mile jungle hike around the massive Shiguri Falls.

Conrad demonstrated the proper technique for packing the bags, including how to create a waterproof seal by kneeling on the packed bag to force out any extra air. He then triple-folded the top edges before tightly cinching the side straps to external buckles on each side. While we expected only a few rapids on the sections we'd be floating, mastering this technique would protect the contents from drenching rain showers or, god forbid, damage to a raft from an encounter with an angry hippo or hungry croc.

Our trip brochure instructed clients to purchase a military surplus 12" x 6" x 7" steel ammunition can. These top-opening boxes used a rubber gasket to create a waterproof seal. On the river, nothing offered better protection for cameras, toiletries, medicines, passports, and money than an "ammo" can. After checking that everyone had complied with their packing instructions, Richard and Conrad fielded last-minute questions.

"Eat a hearty breakfast tomorrow. Our midday launch means that your next meal will be dinner under the stars at our first river-side campsite." Conrad's description sounded glamorous.

Richard added, "Enjoy the last night you'll have in a comfortable bed for the next 12 days."

THAT NIGHT, my faithful *rafiki* resumed her position guarding the entrance to my tent. Michael and I agreed to forgo late-night conversation. My mind wandered as I waited for sleep to erase conscious thought. The crew and passengers, with one notable exception (me), couldn't wait for our river journey to start. Were we insane to enter the equivalent of an enormous zoo with no cages or barriers to keep tens of thousands of wild animals away from us? How could just five bullets be enough to protect us from a pride of stalking lions?

The shortage of rifles, combined with the scant supply of ammunition, worried me. Did these game rangers, entrusted with protecting our lives, even know how to shoot these rifles? Conrad's expectation of one expert marksman in each of our four rafts wasn't going to happen. So I took it upon myself to request a bit of divine help to watch over and protect us.

12

LAUNCH DAY

CROWING ROOSTERS ANNOUNCED the dawn. My last chance to "chicken out" of heading into the great unknown was now or never. I honestly felt paralyzed, as the terror of the dangers we might encounter competed with the excitement of the adventure. This same feeling had almost caused me to back out of my first Colorado River trip. Only now the intensity of my internal conflict felt 10,000 times greater. With the wisdom of hindsight, it occurred to me that had I backed out of the Colorado River trip, I wouldn't be here struggling to resist a panic-induced meltdown.

Getting pitched into a huge rapid, where the worst consequence had been getting drenched and a few minutes of heart-pounding panic, hadn't exactly been fun, but floating into a massive uninhabited game reserve with millions of wild killer beasts would be enormously more perilous. No sane person did that. Thoughts of the extreme risks we'd be taking prompted an urgent warning from the rational part of me. *"Run! Save yourself while you still can!"*

But if I couldn't dredge up the courage to step into a raft and push off on this Selous adventure, I'd be labeled a coward by the very people I so admired and respected. If I left now, how would I find my way out of Tanzania and back home on my own? I'd gotten myself stuck in the middle of the biggest lose-lose situation of my life. Beads of sweat formed on my forehead, not from the already uncomfortably warm air, but from inescapable dread.

Do-or-die time had arrived. I prayed it wouldn't be the latter. Once again, the path of least resistance drew me down it with the power of a magnet. The only way forward required my taking a giant leap of faith. I had to trust that my Sobek colleagues, collectively, possessed the experience and skills to deliver us safely to the far side of this most unusual escapade.

"Just do it. Don't be a wimp," my heart commanded.

AFTER BREAKFAST, I reluctantly crammed all my personal gear into my black-rubber, waterproof dry bag, then broke down and packed my tent. We still needed to finish packing food and supplies. Then we had to move our rafts and equipment from our staging area under the mangos to our "put-in," a small clearing on the banks of the Rumemo tributary, the same waterway we had bathed in, but further downstream where it widened.

My *rafiki* seemed to understand that my final preparation assignments would keep me busy, so she politely stayed out of my way as I hustled to complete my tasks. I labeled the contents of every cooler or container using a Sharpie pen on duct tape. We'd need all the Sobek crew, our game rangers, and a few local volunteer men in order to move the mountain of gear and supplies we'd depend on for the next 12 days down to the put-in.

We used a high-volume hand pump to top off all 24 inflatable chambers of our four rafts, vigorously forcing additional air into the tubes until our biceps rebelled. Six men then hoisted each inflated raft onto their shoulders to carry it down the path to the river. There, they carefully lowered each raft down the slick clay river bank into the

Rumemo. Soon all four rafts floated snug against the shoreline with their bowlines secured to small trees.

Next, we and our local helpers lugged the rest of our gear to the launch site. Even the strongest crew members groaned, straining their muscles to carry the sturdy coolers, packed with the enormous blocks of ice, perishable meats, cheese, and dairy items. Those items we'd consume before the ice melted in four or five days. Emptying these coolers early in the trip also meant less weight would need to be portaged.

The bamboo-pole baskets constructed by the crew to hold canned goods, along with fresh fruits and vegetables, fit snugly between the side tubes inside the rafts. Once filled, we covered the baskets with tarps to protect their contents from the scorching sun. These tarps would do double duty, keeping our camp kitchen dry in the event of rain. To minimize weight, most items we packed served at least two purposes.

I grabbed two, 10-foot-long wooden oars and centered them on each of my shoulders to serve as a makeshift clothesline, while Conrad hung 13 life jackets through their armholes onto the round shafts. Then I joined the procession, headed for the river. Sweat dripped from my head and streamed down my back. The sooner we got everything moved, the sooner we could launch. Once on the river, we'd get relief from the cooling breezes sweeping over the water.

On my second trip down to our launch site, I carried our fully provisioned first aid kit and a tightly covered, thick plastic bucket packed with toilet supplies. We worked nonstop to meet the goal of a noon launch. I'm sure we looked like worker ants trekking single file, dragging morsels of food to our anthill. This careful packing and moving took hours. As our gear disappeared from the center of the village, local adults wandered down to the launch site to observe the loading of the rafts.

Conrad, meanwhile, went to the Mission to escort our paying clientele, a married couple in their fifties and eight solo travelers, to our launch site. Two of the men had traveled with Sobek previously. All of them appeared to be financially well off based on their fancy Columbia and Orvis-type safari wear, aviator-style sunglasses, and expensive binoculars strung around their necks. Each of them lugged their numbered dry bag and ammo can. All had filled their canteens with the purified water freshly boiled that morning for us by the Mission.

Conrad led the introductions. Though we were a diverse group, young and old, experienced travelers and first-timers, the one thing shared by all (except me) was an eagerness to depart on this once-in-a-lifetime adventure. The entire group brimmed with enthusiasm. Could it be contagious? I wondered if I'd ever feel that excited about entering the dangerous unknown. While they chattered fearlessly about the wild animal encounters they hoped to have, I felt a twinge of envy.

The married couple, Martha and Jay, bragged that they frequently traveled to exotic locales, giving the impression that, for them, entering the wild terrain of the Selous wouldn't be a serious challenge. Martha, brimming with confidence, explained her eagerness to check off another trip on their travel bucket list. On the Jeep safaris they'd already done, they'd seen many of the wild animals we hoped to see on this trip, but the different perspective of traveling by raft appealed to them.

One of the solo travelers, Kate, in her late thirties, had chosen this trip as a vacation from her high-pressure management job. Despite her slight build, she appeared physically fit, more so than two of the older men standing nearby. She wore an extra wide-brimmed canvas hat with stampede strings to prevent the wind from catching it and sending it airborne. The hat's circumference cast a protective circle of shade over the fair skin on her face and neck. Kate had always dreamed of encountering wildlife in their natural habitat.

They seemed as comfortable embarking on this journey as they would if attending the premiere of an Indiana Jones movie. Why did they not feel the fear I felt? Didn't they understand? Genuine terra incognita loomed downstream, yet they were embracing every unknown challenge they imagined it held. They wanted to be explorers. They wanted to experience what others might never have the chance to do. For the privilege of feeling adrenaline jolt through their bodies, each had paid Sobek a small fortune to join this expedition.

Now, as I listened to our passengers, I sensed I was missing out on something important. Something that lifted the human spirit. Had my sheltered life shackled by fear been the reason Richard so adamantly insisted that I needed to go on a *real* adventure? Could I find courage deep inside me to cope with the unknown obstacles the Selous would expose?

Tanzania now held me firmly in its grip. My safe, middle-class American thinking had me tied in knots of dread. To escape them, I needed to change my mindset. Worrying every minute of the day would keep me from embracing the newness and strangeness of this wild land. I prayed I'd be able to savor this experience like my fellow travelers. I would do it, I told myself. I *had* to do it. To get me through this wild adventure, I would try channeling a bit of Harrison Ford's Indiana Jones bravado. Was the kid within me ready to take a gigantic leap of faith? Was I ready for fame and glory? Maybe. Only here there would be no retakes.

Conrad asked me to hurry back to our staging area for one last check that we'd left nothing behind. Waiting under one of the mango trees sat my rafiki. She looked pitifully forlorn. After a quick survey of the area, finding no trace of anything belonging to Sobek, I approached my little friend, crouching against the tree trunk with her spindly arms holding her bent legs close to her chest. The moment felt heartbreaking for both of us as we realized our time together had run out.

I held out my hand. She jumped up and put hers in mine. Bravely, she matched her stride to mine as we headed back down the path. My eyes became teary as I thought about having to say goodbye to her. We both marched along without looking at each other, our eyes trained directly on the path ahead.

Suddenly, she broke stride, running a few steps ahead of me. She stopped and turned to face me, blocking my passage. Her wide brown eyes, moist and glassy, held back tears too. She looked straight into my eyes as she voiced a last-minute appeal. The only word I understood was the word she had taught me, rafiki. She wove that word generously throughout her speech.

The longer she spoke, the faster she spoke. She waved her arms and pointed her finger. Her body language conveyed the gist of her plea. I listened attentively, realizing that the "words" were not as important as the love and concern she was expressing. Though only a child, I sensed she might already know of the danger that lay ahead. She wasn't merely saying she would miss me. She was warning me, as a mother would beg her child, to heed that there be dragons and lions, real ones, where we were so foolishly headed.

She ended her lecture with sudden silence. I nodded in appreciation for her concern. I reached out and took her small hand in mine and shook it vigorously, sealing my promise to heed her advice. Then we walked hand in hand to the river. Word of our departure had spread through the village. Nearly every Ifakaran had crowded into the clearing above our boats to witness the last curtain call of the amusing performance we'd staged in their village. Or maybe they craved one last glimpse of the foolish white people before they floated away, never to be seen again?

Our guides pulled their neatly coiled bowlines taut as our passengers climbed into the three paddle rafts and took seats, balancing on top of the raft's side tubes. Fighting back tears, I gave my *rafiki* one last heartfelt hug, then scrambled down the bank to climb aboard Richard's oar boat.

One by one, each guide gave a mighty shove and hurled themselves over the widening gap of water to board their raft. Cheers of excitement arose from the crowd on shore. I took their spirited farewell waves as a sign of good luck as our boats drifted downstream. Our expedition members waved back a final thanks for the generous hospitality the village had shown us.

My little friend, standing in the very front of the crowd of villagers, never stopped waving. I watched her until I could no longer make out her face. I raised my Mao cap in one last wave to her, then turned my attention forward, staring into the silty water.

I'd expected the few days I'd been in Tanzania to be wild and frightening. Yet I'd already made two friends: first Solomon and now this little girl. Their simple gifts of friendship had meant so much to me. So much for the spear-toting cannibals. These two had kind gentle souls.

A profound sadness filled my heart, knowing I would never see either again. A single teardrop leaked from my eye and rolled unseen down my cheek.

13

INTO THE UNKNOWN

OUR PAYING GUESTS OCCUPIED our three paddleboats. Two or three paddlers sat on each side and used a lightweight but nearly indestructible polypropylene paddle to propel the raft. A Sobek guide sat at the stern barking out paddling instructions: forward, back paddle, right turn, left turn. Leaning out over his seat on the rear tube, the guide's paddle blade served as both rudder and propulsion to steer the raft away from obstacles in the river.

Richard captained an oar-powered boat that carried much of our expedition gear. Lashed securely to its tubes was a rowing frame with two eight-inch-high pins on each side, around which two 10-foot-long wooden oars rotated. Suspended about eight inches above the raft's floor, snug between the cross tubes (known as thwarts) and braced between the side tubes, a varnished plywood deck supported our heavier coolers. Tie-down straps secured one cooler tightly in place behind the front thwart, the second against the front of the rear thwart. The latter served double duty as the rower's seat.

We had jammed bags of canned goods, nonperishable supplies, and tents into the bottom of the rear compartment of our raft. Our personal waterproof dry bags formed the second layer, while our kitchen gear, piled high, topped the mound. Several ropes crisscrossed over the mound and cinched to D-rings mounted on the interior side of the tubes to hold everything securely in place. Nothing would come loose should the raft, God forbid, flip or sink.

Richard's oars provided all the power and steering for our raft, so the game ranger assigned to our raft, Masengela, and I enjoyed the luxury of being curious spectators on this first day. Clutching his rifle, he climbed onto the high vantage point atop the pile of gear behind Richard. He settled in there like a cat claiming a sunny spot on a favorite piece of furniture.

The heavier weight of our raft floated it lower in the water, slowing its speed, so we occupied the sweep position in our flotilla. Conrad captained the lead boat with Magigi and his rifle on board. As we journeyed down the coffee-colored Rumemo, we kept our eyes peeled for signs of danger. Strangler fig trees and dense brush crowded the river banks, obstructing any view of the savannah lands and flood plains immediately beyond.

We passed a group of local men resting on the river bank in the shade under a leafy miombo tree. They stared as our boats floated past. One of them grunted some Swahili words that might have been serious advice, but more likely comments on the lunacy of us sharing the river's real estate with the creatures who dwelt in and around it. One man offered a limp wave.

We wouldn't cross the boundary into the Selous Game Reserve until a few miles after the Rumemo flowed into the Kilombero a short distance downstream. These guys were the last local inhabitants, the last vestige of civilization we'd see for 12 days. With a heavy sigh, I waved back at the man, then turned to face forward as we headed into the unknown.

A hundred shades of green colored the dense jungle that flanked both sides of our stream. Overhead, pure white cotton-ball clouds drifted with the breeze. Now and then they obscured the tropical sun, providing a temporary reprieve from its subtle toasting of our exposed body parts. Up ahead of our caravan, a flock of herons concealed in the riverside reeds took flight, noisily beating their wings to lift their bodies skyward.

Further downstream invasive grasses and papyrus edged outward from the shoreline, eventually clogging our channel. We no longer saw any route through this reed-filled swamp.

"What's up with this?" Slade gestured ahead at the thick tall grasses blocking our way. "This looks like a dead end to me." Sarcasm tinged his remark, directed at Conrad.

Conrad pointed to the current beneath our rafts still inching through the tangled growth. "Look. The water is still flowing. We just have to push our way through it. The Kilombero is straight ahead. I'm positive. Maybe with the small rains lighter than usual, the vegetation took hold and overspread the riverbed."

"We can't paddle through this. What's your plan for how we get past this?" Slade challenged.

A simple answer to Slade's question didn't exist. Less than an hour into our expedition, this roadblock had brought us to an unexpected standstill. With no obvious detour, I quickly realized the extent of our vulnerability. We had *no* way to summon help. The unbroken expanse of tall grasses and reeds obscured the boundary between the shoreline and the channel of water. Even if we could find the shore, it looked impenetrable. Considering our inability to paddle back upstream against the Rumemo's brisk current, Ifakara, the only village anywhere close to this obstacle, now seemed as distant as Dar es Salaam.

If we got in real trouble, Conrad and Michael both had signal mirrors. But these would only be functional with direct sunlight *and* if, by some miracle, a plane flew by overhead. *And* if the pilot saw the mirror flash. *And* if the pilot realized the flash was an SOS. Talk about a real long shot.

I wondered if the ploy of stranded sailors who used the smoky plumes of an SOS bonfire to attract rescuers might work. But in our predicament, we had neither an open sandy beach nor firewood, let alone anyone to spot the smoke. Out here, we'd have to solve our dilemmas on our own.

After a short, animated discussion with Magigi, Conrad announced the plan. The game rangers and some of the crew would slip into the water to pull the boats forward. Everyone else would paddle as best they could and watch for approaching crocodiles. Even sitting securely inside the raft, floating on top of the murky water, I feared what lurked in the depths below. How did they dare to submerge themselves into this reed-clogged mire?

Reluctantly sinking into the swampy ooze, a few crew and game rangers bravely clawed their way through the tangled mess. Masengela slid over the front tube of our boat into the water to do the dirty work for us. He tugged on our bowline, staying within an arm's length of the boat ahead. Richard and I stood on the sides of our raft, each using an oar as a pole to prod our way along the gap carved by the three boats ahead of us. Cuss words filled the air from those struggling to drag the fully loaded rafts through the tangle of submerged papyrus stalks and reeds in the murky water.

Despite using all my strength to push the blade of the oar against the resistant underwater vegetation, the raft moved barely an inch at a time. We pushed forward at a snail's pace until the swamp's vegetation completely engulfed us. It became impossible to tell which direction we were headed or how much more of this swamp lay ahead. Tracking the barely perceptible sluggish current proved our only clue that we remained on a course that would eventually reach the river.

After what felt like forever in the windless, stifling tropical heat, we emerged into open water, our first obstacle conquered. Thrilled to escape the swamp's claustrophobic grip, we celebrated with cheers and paddle slaps. Richard and I hauled an elated Masengela back on board. Then Richard steered our raft into the deepest, fastest-moving channel of the meandering Kilombero River.

But now, an upstream headwind hindered our progress, dashing Richard's hope for a break from the exhausting slog through the swamp. He leaned hard on the oars, coaxing our raft downstream, avoiding the shallows where underwater creatures might lie in wait. The opaqueness of the Kilombero's water created an impervious veil to what lay beneath. The repetitive rhythmic dip, pull, and lift motion of the oars felt almost hypnotic. I made sure we didn't lose sight of the three paddleboats, whose lighter loads with six paddlers each had propelled them far ahead.

Suddenly, a shiny gray head burst through the surface of the water 30 yards ahead, shuddering and spraying water from its nostrils.

"*Kiboko!*" Masengela pointed at the giant head that turned to stare at the strange floating object headed straight toward it. It inhaled and dove back under the surface.

"Oh my God, Richard, I've just seen my first hippo!" I fumbled, hurrying to pull my camera from its case, hoping the hippo would resurface.

My rookie excitement amused him. "Don't worry. By the time we're through with this trip, you'll be sick of seeing them. There's plenty more where that one came from."

"*Kiboko. Kiboko.* What a melodic name for such an enormous beast." My vocabulary of Swahili words had just doubled.

According to a book in Sobek's library about Tanzania's Selous wilderness, I'd learned that the only land mammal larger than a hippopotamus is the elephant, though hippos, populous and dangerous, ruled the waterways. It surprised me that hippos had been wandering our planet for eight million years. Only 30,000 years ago, they still roamed widely throughout Europe and North Africa. The Greeks gave them a name that meant "river horse."

Hippos knew exactly how to deal with the daytime's energy-draining heat. They rested submerged in the relative coolness of the water. During the cooler night hours, they'd traipse onto land to forage for food, an alarming thought since we'd be camping on the very riverbanks they'd pass through to reach grazing lands.

I had anxious second thoughts as we approached the spot where the hippo had submerged.

"Are you going to row right over it?" I asked.

"It looks pretty deep here, and it's probably moved out of the way."

Richard sounded confident there'd be no collision with the hippo. I had to trust him. He'd survived the last seven years of rafting hippo-filled rivers and was still here to entertain us with his tales. Sure enough, we floated effortlessly right over the spot.

I recalled hippos could remain submerged with their nostril flaps closed, holding their breath for up to five minutes. I suspected soon this hippo would need a resupply of air. Right on schedule, two minutes after we passed its original location, I heard a boisterous snort behind us. I whipped around just in time to glimpse its exposed head.

"There it is again." I pointed excitedly. The hippo had surfaced too far behind us for a picture, and within seconds, it disappeared back underwater.

Richard laughed again. "Maybe or maybe not. Where there's one hippo, there's a herd, and they all have to breathe."

"How many are in a herd?"

"Each herd has one bull. He keeps a harem of 10 to 30 females and their young. Lucky guy. His territory covers a stretch of river about the length of three football fields. Young males must behave submissively for the bull to allow them in his territory. If you see two hippos fighting, it's usually males defending their territory."

Over the next hour, hippo heads popped up all along our route. Hippos in shallow water dove for deeper water the moment they spotted us. Others resting farther away from our raft held their positions, eyeing us cautiously as we floated by. Their strangeness probably fascinated me as much as our presence in their river fascinated them. For the first time,

I saw them as wondrous creatures rather than dangerous predators as they snorted, grunted, and lumbered along in the water. I swore their adorable, small, round ears were waving a greeting to me.

Since the hippos had shown no aggression toward us, my fear of floating in the water with them began to wane. When I decided they looked rather cute, it was time for a reality check. Of course, they *looked* cute. But they could be killers as well. I changed my mind about asking Richard to row us a little closer so I could get better pictures.

14

MAKING CAMP

"LOOKS LIKE CAMP," Richard said as he cozied our boat up next to the other rafts. Their grounded bows already rested on the sloping bank of an open clearing. The crew secured bowlines around large boulders off to one side of the beach. Meanwhile, Conrad and Jim Slade explored the site to judge if it might be adequate for our first overnight stop.

Slade was a boatman's boatman, an overall top-notch athlete with the looks and muscular build of a male model. He'd been accepted by four Ivy League colleges but turned them down to attend the smaller prestigious Williams College, where he majored in economics. Slade innately understood the intricacies and motion of fluid hydraulics simply by watching how water plunged through a rapid. Selected as a Sobek crew member for their earliest expeditions, he had earned Richard's respect. As his right-hand man, they continued to plan and execute many first descents of rivers around the world.

Slade wouldn't merely take the safest route through a challenging rapid. He'd choose the most exciting run that would deliver his raft right-side up with everyone intact at the end. A serious-minded leader, he performed his job with class. He treated words like gold coins, spending them only when absolutely necessary. Richard liked to quote Slade as saying, "There are old boatmen, and there are bold boatmen, but there aren't any old bold boatmen."

By the time we arrived, the two had made their decision. Conrad whistled to get everyone's attention.

"Listen up, everybody. This is camp for tonight. We're lucky because it's a nice flat open clearing, but there's a hippo trail that cuts right across the middle of it here." He pointed at a trampled path leading away from the river into the savannah land. "We're going to set up the kitchen here on this side. You need to set up your tents on the other side of the trail. The hippos need a clear path to get to their food."

This tidbit of information alarmed me and several of our clients. Shocked looks belied the angst we felt at learning a highly traveled expressway to the best food in the neighborhood led right through our camp. I had no desire to get between a hippo and its favorite grazing spot.

Helping to unload gear, I passed kitchen equipment from the back of our raft to crew members setting up the camp kitchen. The legs of our kitchen table unscrewed to make it easily portable. Once reassembled, it provided a stable surface for our camp stove. We stashed thick plastic buckets packed with assorted pots, pans, kettles, unbreakable aluminum cups, and cooking utensils under or nearby the table. When emptied, they could serve as bailing buckets in the unlikely event a raft took in water or as containers for hauling bathing water a safe distance from the river. Overturned, they also served as comfortable seats.

People reclaimed their dry bags while I handed out tents. Everyone staked a spot for their sleeping quarters, well away from the hippos' trail. I quickly assembled my green nylon tent, wrapping each of the four corner loops around a hooked metal stake. My body weight as I stepped on each stake drove it deep into the mix of sand and silt. I could imagine how easily flood waters would inundate this beach during the river's seasonal heavy rains. When the mud-saturated waters receded, a fresh layer of silt would cover the riverbanks, a reminder of the river's temporary dominance over the land.

After stowing my sleeping bag, waterproof gear bag, ammo can, and camera case inside my tent, I unzipped the nylon flaps covering the mesh windows so the tropical breezes could flow through. Finally, to discourage critters from exploring my sleeping space, I zipped the doorway flap closed. Then I joined the guides in the kitchen to help with dinner preparation.

Purifying our drinking water took top priority. The treatment took a minimum of a half hour to kill the microscopic organisms that could make us sick. Earlier that afternoon, the paddle raft crews had filled four heavy-duty plastic five-gallon jugs with water from the middle of the river's current, where supposedly the water was clearer. Conrad kept in his possession the most valuable commodity we carried, our water purification kit.

Like a chemist in his lab, he opened the purification agent, one of several dark glass bottles filled with rust-colored tincture of iodine. Not only a potent killer of bacteria, viruses, protozoa, and fungus, iodine could also be toxic to humans if ingested in excess quantity. The prescribed formula was two drops per quart for clear water and up to 10 drops for seriously cloudy water.

Conrad judged that the Kilombero's water fell midway on the spectrum between clear and cloudy, so he settled on five drops per quart or 100 drops per five gallons. He filled the bottle top dropper and counted the number of drops it held as they fell into the first jug. Carefully, he squirted the medicinal-smelling dark orangey liquid by full droppers into all the water jugs. After capping the jugs, he shook them thoroughly to mix the iodine with the water. He took responsibility for purifying enough water each day so we had an adequate supply of treated water ready for use at all times.

A human can survive, at most, for three days without potable water. Conrad safeguarded our iodine supply as if it was gold, because if we lost or accidentally broke the bottles, we might face a dreadful choice: dying from thirst or dying from an untreated bacterial infection turned septic.

Our passengers, accompanied by the rangers, scattered nearby to search for dry fuel for a campfire. Meanwhile, Slade and Conrad dug a firepit and rimmed it with large stones. A hardware store in Dar had fashioned a steel grate to Conrad's specifications to serve as a grill for cooking over fire and boiling steel buckets of river water for after-dinner cleanup.

Michael and Bart shared dinner duty that first night. The rest of us took orders from them. My task took me back to hot summers in my early childhood, when I'd wash down a peanut butter and jelly sandwich with an ice-cold cup or two of sweet, fruity Kool-Aid. On this trip, our primary source of hydration would come from the iodine-laced river water made more palatable by flavoring it with Kool-Aid.

It fell on my shoulders to make up pitchers of Kool-Aid to serve with every meal. I always made extra so our group could keep their canteens filled day and night. Nothing was as crucial as continually replenishing the moisture robbed from our bodies by the relentless tropical sun. I selected a handful of lemon-flavored Kool-Aid packets from our plentiful stash. Conrad checked his watch and nodded. The required 30 minutes for purification had passed. As I filled three gallon-sized pitchers with the iodine-purified water, he offered a bit of advice.

"You should use two packets per pitcher. It takes more than you'd think to cut the taste of that iodine."

I tore open the packets and added the powder. After giving the mixture a good stir with a long-handled slotted spoon, I poured a mouthful of the lemony liquid into a cup and took a sip. Instantly, I spit it out. It tasted like citrus poison. It would be impossible to drink this foul liquid for the next two weeks.

"Honestly, Conrad, this tastes disgusting!" First the vile taste of kerosene mangos, now this toxic-tasting concoction; I'd never take a pure glass of water for granted again.

Amused by my shocked reaction, he pointed out my rookie mistake. "I think you forgot to add the sugar."

I felt like an idiot as I searched for the burlap sack of raw cane sugar. The Kool-Aid packet instructed to add one and a half cups of sugar per packet. Three cups of sugar in one gallon seemed like an enormous amount.

"Am I really supposed to put three cups of sugar in each pitcher?" I asked Conrad.

"Aw, you can probably get away with two."

I added the sugar and stirred it until the granules dissolved. Again, I poured a sample in my cup for a second taste. The sugar made the flavor tolerable. Only a trace of the yucky iodine flavor persisted. To totally mask the iodine taste, I figured it would take five cups of sugar per gallon, and at that point we'd be drinking syrup. We'd just have to get used to the faint taste of iodine in all of our beverages.

Our first dinner featured Michael's and Bart's culinary talents. Because they had used items from our limited supply of fresh food, it was one of our best meals. Their menu featured grilled beefsteaks, boiled white rice with spicy herbs, and a fresh fruit salad made with pineapple, papayas, bananas, and mangos. I went for everything but the mangos. There were no leftovers, only full bellies when we finished dinner.

Dining on the beach at dusk gave us front-row seats to watch the hippo antics playing out right in front of us. A few lumbered out of the river onto the opposite shore. When fully exposed on land, their massive size astounded us. Their giant barrel-shaped bodies rested on four short stocky legs. Four hippos of varying sizes casually wandered along the bank across the river.

"How much do they weigh?" Kate, the single slender, fair-skinned woman, asked Conrad.

"Those are average size. They probably weigh somewhere between 3,000 and 4,000 pounds. But there are bigger ones in this river that can weigh up to 8,000 pounds." That bit of information drew murmurs of astonishment.

"Wow, that means even a small one would be the equivalent of 25 of me," Kate said, surprised by the result of her calculation.

I quickly did the math for one of the bigger ones. It would take 50 of us to win a tug of war with one male brute. It finally hit me. This wasn't the Jungle Ride at Disney World. I was *not* one of the thrill-seeking amusement riders waiting for hours in line, having paid an exorbitant fee for a mechanical lifelike hippo to lunge from the lagoon, soaking me with water and coercing squeals of feigned fear. No, I had plopped myself down on a remote riverbank, 50 yards away from the real deal with no barriers to prevent the gray monsters from walking right out of the river into our campsite.

Conrad made a pot of iodine-flavored instant coffee for those who craved their fix of evening caffeine. We carried two pails filled with river water that we'd boiled over the fire to the kitchen to fill the wash buckets. One bubbled with dish soap; the other contained clear rinse water. A bowl for small food scraps that we'd incinerate in the campfire sat on the ground next to the soapy water. Each person had to scrape and wash their own dishes. Afterward, I helped to wash the cooking dishes.

Before calling it a night, Conrad summoned us to gather around the fire, where he delivered a safety talk.

"I told you earlier a hippo trail crosses through our campsite. Hippos come out of the water to forage for grass at night. That means there *will* be hippos wandering around out here later on. They usually come on land to feed for five or six hours. During that time, they can cover as much as six miles along these paths out into the surrounding grasslands. Depending on their size, they can consume anywhere from 80 to 150 pounds of grass every night."

Worried by this shocking disclosure, Martha asked, "What should we do if we hear one getting close to our tent? Should we yell at it?"

Before Conrad could answer, one of the younger men piped up, "Do we have to worry about one of these hippos stepping on our tents?"

Conrad held up his hand to pause the questions. "If you hear a hippo passing by your tent, the last thing you should do is scream. Just relax. Screaming will only wake up our entire group and might spook the hippos. They'll avoid any object that has a slope as steep as the side of a tent. They look for gradual slopes when climbing."

The other guides with experience around hippos confirmed they knew of no squashed tent incidents caused by a curious or spooked hippo.

Conrad continued, "Here's what you need to know to be safe. First, everyone needs to do their bathroom business before settling into your tents. On that large boulder, over behind your tents, there's a roll of toilet paper in a sealable plastic bag and a trowel. The rocks are tall enough to offer some privacy. If you don't see the trowel and toilet paper bag, someone's using them and doesn't want visitors. Very important—if you want to remain in the good graces of your fellow rafters—return the trowel and TP to the top of the boulder when you're done."

The group chuckled at Conrad's helpful hint: none of us wanted to be waiting with our legs crossed for someone to finish their business, when in fact the last bathroom user hadn't replaced the items to signal the coast was clear.

"Use the TP sparingly. If our supply runs out, you'll have to find yourself some smooth leaves. Let's hope that doesn't happen.

"Also, you need to put the roll of TP back in the bag and seal it so the night dew doesn't dampen the tissue. I can tell you it's no fun using soggy TP that disintegrates in your hand. Be considerate of your camp mates. Dig yourself a foot-deep hole in the sand, do your thing, and bury it along with any TP you used.

"Hopefully, if you use the toilet facilities now, you'll be able to make it through until morning. Fill your water bottles with Kool-Aid, if you think you might get thirsty. Once you're in your tent, fully zip it up and stay hunkered down inside until morning."

Martha raised her hand reluctantly. "What if I just *have* to go in the middle of the night and can't wait until morning?"

"If you really have to go, I'd recommend opening the door of your tent, hanging just your butt outside to do what you have to. You can clean it up in the morning. It's imperative that no one goes outside their tent when hippos are here in camp. If anyone in this group is foolish enough to ignore this warning and leaves their tent, let me tell you another fact about these animals. Despite their massive size and short legs, if provoked, on land, a hippo can easily outrun a human. Hippos have been clocked running at 19 miles per hour. Think about that for a minute. Can any of you run that fast?"

I knew the fastest marathon runners in the world could race 26 miles in about two hours, an average speed of 13 miles per hour. Since several recent marathon champs hailed from East Africa, I wondered if trying to outrun a hippo played any part in their training regimens. Absent any world-champion marathoners on this trip, Conrad's point had hit home like a dart piercing a target's bullseye.

By the time I crawled into my tent, I had the energy level of a dead battery. Sure, I worried about the hippos loitering in the river waiting for the opportunity to come ashore, but I found it impossible to keep my eyelids from closing. Still doubtful about the wisdom of pitching our tents on Hippo Beach, I drifted off to sleep.

An hour later, when the ground under my sleeping bag trembled, I woke, not daring to breathe. It happened again, and again, and again with greater force each time. Stunned, I realized why the ground was shaking, just as I heard a snort and a soft growl. The hippos had emerged from the river and were roaming through camp.

Where the heck was Michael when I needed him? His compact bivy sack didn't have sloped sides, but he'd wedged himself in between the tents of two other crew members. My heart pounded as I listened to the tromping hippos march closer to my tent. I prayed they just stayed on their path without veering off on a shortcut. I lay still, not moving a muscle or making a sound.

After two minutes, that seemed to last for two hours, of intensifying ground-quaking, I figured the hippos were only a few feet from my tent. Dying for a glimpse, but terrified to know how dangerously close they were, I sat up and flipped around into a kneeling position. I swallowed my fear and peeked through the mesh window at the far end of my tent.

Shimmery moonlight reflected off their wet backs, faintly illuminating their shapes. Hunger must have trumped their apprehension of the invaders on their riverbank. They plodded along their track like gymnasts on a balance beam, as if a misstep might spell disaster. I felt exhilarated, secretly spying on the fearsome hippo parade a mere 18 feet away. Once the hungry beasts passed into the forest on their way to the grasslands, it took a while for my adrenaline rush to dispel, but when it did, I fell into such a deep sleep I never heard the hippos return.

15

DODGING HIPPOS

FRESH EGGS SIZZLED in the cast-iron skillet while our passengers worked on repacking their tents and dry bags. The irresistible aroma of crispy fried bacon made my mouth water. I stirred up more Kool-Aid to fill our canteens to quench our thirst while on the river. At Conrad's call to come get breakfast, our clients lined up to fill their plates not only with bacon and eggs, but with homemade bread and jams, cereal, and fresh fruit, all washed down with kick-ass river coffee.

"So, how'd everyone sleep last night?" Conrad asked as we ate.

One by one, our clients shared their impressions of the hippo procession through camp.

"I thought for sure one was going to walk right into my tent," said one of the younger men, a first-time Sobek client.

Kate giggled as she confessed, "I live in California, so at first, I figured it had to be an earthquake."

"You just don't realize how frigging enormous the damn things are until you see them that close up." Frank, an older repeat client, winked at Conrad. "They amaze me every time."

"One word—unnerving! It took me forever to fall asleep after that," said Martha.

"I can vouch for that. She had a stranglehold on me and wouldn't let go. I don't know which scared me more, the hippo or her," Jay teased.

Even as we sat eating breakfast in the middle of their territory, the hippos seemed to tolerate our presence as they soaked in the river's cooling water with only their eyes and nostrils exposed. I took photos of some as they surfaced, grabbed a breath of air, and dove back under the water. Though we'd staked our claim on their shoreline, they showed no aggression toward us. Perhaps because we hadn't bothered them or interfered with their nighttime foraging, they remained docile. Whatever the reason, it didn't bother me. We'd be leaving soon, before we wore out our welcome. I hoped the herds we'd float by or over in the coming days would be equally tolerant.

Already, the air felt steamy. The heat would only intensify as the sun rose higher. Everyone helped by bringing personal gear and pieces of camp gear down to the rafts for loading. Everything went back into the boats in the same order we'd initially packed them.

"Last call for the john. Use it now or hold it 'til lunch." Conrad sounded like a mom getting the kids ready for a long road trip. After a last sweep of the area for anything missed in repacking, we climbed aboard the rafts and shoved off. The lazy current gently tugged us downstream.

After studying the topographic map, Slade and Michael determined this section of the river dropped barely three feet per mile, so there wouldn't be any rapids to worry about this day. The current meandered at a leisurely pace. Conrad planned we would paddle for three to four hours before stopping for lunch. As we drifted downstream, a variety of birds perched in trees along the river or waded in the shallow reeds along the river's edge. Some smaller ones even dared to perch on the backs of resting hippos.

At one point we passed the most hideous-looking pair of birds I'd ever seen. When the Creator designed Marabou storks, he must have been having a bad day. Their bodies, plump as large turkeys, perched on white, three-foot-long spindly legs. A foot-long, dagger-shaped, tan beak extended from their bald heads covered with ugly pink scaly patches. The pair of scavengers waded in shallow water, waiting for a small fish or frog to swim into their striking zone. As we approached, the Marabous unfolded and used their enormous 10-foot-wide wingspans to rise above the river like a fully loaded military transport plane, gradually climbing to safety high overhead.

I observed an ever-increasing number of hippos as the day proceeded. A few rested with just their heads above water, warning us with a wary stare not to come too close. I wondered if with their chorus of grunts and snorts they might be alerting their buddies downstream about the invaders headed their way. I'd been thinking about Conrad's talk on hippos from the previous night. Something he said bothered me, so I asked Richard about my concern.

"Yesterday, Conrad said hippos can eat up to 150 pounds of grass a day, and, well, I suppose that has to come out the other end. Given the number of hippos we've seen already, that seems like a massive amount of dung going into the river every day."

Richard cracked up. "You're worried about hippo crap? What you need to be worried about is hippo bites!"

"I am. But we're drinking that water. Are you sure those few drops of iodine are really effective at killing all the bacteria and parasites?"

"Well, explorers have been using iodine for the last century. Heck, I've been drinking iodine-treated water from these African rivers for years myself. I'm still alive."

I conceded he had a point. Who was I to question the logic of this man who likely drank more water from remote rivers than purified tap water every year?

At midmorning, a strengthening headwind led to a more strenuous workout for the passengers in the paddle rafts than they'd expected. Their guides kept urging them to paddle forward, despite their complaints about burning biceps and aching lower backs. They toiled in slow unison. Close to noon, the rafts ahead of us pulled up to a wide, sandy stretch of beach with easy access.

Conrad, Slade, and Magigi quickly inspected the area for signs of lurking wildlife. Satisfied the area presented no dangers, they motioned for us to follow. The paddlers gratefully deserted the rafts to enjoy a break on land. Some sprawled on the sunbaked sand for a quick nap. Others, tired of sitting sidesaddle on the raft tubes, stretched their legs with a walk down the beach.

Kate helped me cover a sizeable flat-topped rock on one side of the beach with a blue-checkered vinyl tablecloth. Our lunches were simple affairs, not requiring a complete kitchen to be set up. The crew laid out slices of fresh bread along with a spread of hard cheeses, jams, peanut butter, and Conrad's favorite canned sardines. A platter of cut-up fresh fruit served as the crown jewel of our midday buffet. As luscious as the fruit looked, I suspected it wouldn't take long for every piece to disappear.

Already, I'd earned the title of Kool-Aid Kid. For lunch, I mixed up a berry-flavored batch, while still cringing at what I suspected the iodine-flavored water contained. With all the lunch items neatly spread across the table, Conrad whistled, "Lunch is on. Come get it!"

Our passengers gathered around nature's impromptu table just as two of the men returned from a stroll to the far end of the beach. One of them, speaking with the spirit of a naturalist, broke the news of their discovery.

"Hey, guys, we found the remains of a hippo jaw way down there. You won't believe how gigantic the teeth are! If you want to see a hippo's self-defense mechanism up close, you need to go see the skeletal remains for yourselves."

Right away, the others inundated the men with questions about the location and what the jaw looked like. Basking in their newfound notoriety, they promised to lead the group to see the hippo jaw after lunch.

"Eat hearty and drink plenty of Kool-Aid," Conrad urged, gulping a swig from his own canteen. "We have at least four more hours on the river this afternoon, and the upstream wind may pick up." His prediction triggered groans. "We need everyone to have lots of energy."

It didn't take long for all the food items on our table to disappear. To top off canteens, I made a second batch of Kool-Aid we'd need to quench the inevitable thirst prompted by strenuous afternoon paddling. We stashed the sponged-off tablecloth and the bag of lunch trash back in the raft.

Then, with cameras slung over our shoulders, we trooped off to view the unexpected find. A huge lower jawbone lay partially submerged in the sand. Vultures or other carnivorous predators had picked the bone clean. Several months in the sun had bleached it white. At first glance, I clearly understood what made hippos so dangerous. The width of the U-shaped jaw measured 16 inches.

Two 10-inch-long incisors protruded straight forward from the front of the jaw like twin pointed dueling swords. Two still razor-sharp 12-inch-long canine tusks curved upward from the front corners of the jawbone. The incisors and canines, used for warding off attacks by aggressors, functioned as potentially lethal weapons. Behind them, six stubby four-inch-long molars lined the back sides of the jaw. They used the molars to grind the staple grasses of their diet.

Michael reminded us the missing top jawbone had equally sharp incisors and canines that were only slightly shorter than the bottom ones. He judged that this jaw might have been from a juvenile since adult hippo canines can be as long as 20 inches. He explained that the hinge connecting the jawbones can open as wide as 150 degrees, giving the impression of a whopping yawn, then slam shut with a force of 1,800 pounds per square inch.

A blow of that magnitude could easily crush a human, crocodile, or small predator in half. Now, I explicitly appreciated why Richard warned I should be worried about hippo *bites*. One well-placed chomp could leave four puncture wounds in a raft's inflated air chamber. A deflated tube would cause the raft to list, be more difficult to steer, and slow its forward progress.

Eager to cover several more miles before finding a place to camp, Conrad wrangled all of us back to the rafts. The paddlers dipped and stroked their paddles with renewed vigor as we continued down the Kilombero. In our boat, Richard pushed forward on the handles so the sunken blade paired with the river's force to propel the raft forward. The repetitive motion of rowing created a reassuring cadence.

I enjoyed the guilty pleasure of kicking back and watching the scenery, hoping for a glimpse of wildlife hidden behind the greenbelt of vegetation on both sides of the river. We'd been back on the river for a short time when Conrad ordered his crew to stop paddling. In the rafts gathered ahead of us, everyone appeared to be staring through a break in the dense foliage at an open area of grassland dotted with a few acacia trees.

From his perch atop the gear pile, Masengela spotted what had grabbed everyone's attention. He shouted, "*Tembo!*" Pointing to the distant beasts, counting, he repeated, "*Tembo, tembo, tembo.*"

A sizable herd of adult and juvenile African gray elephants trudged through the tall grass. If we thought the hippos were huge, these wild elephants were ginormous in size. We needed no further proof for why they held the title of the planet's largest land animal. Even their ears, habitually flapping to cool their bodies, were the size of twin bedsheets.

Prior to this Tanzanian trip, the only species of the Selous's wildlife I'd personally encountered was an elephant.

"I had a pet elephant once." I impishly peeked at Richard to gauge his reaction.

He gave me a cynical look. "Let me guess. His name was Babar." His assumption that *my* elephant must have been a toy version of the loveable elephant, featured in a beloved series of children's stories, irked me. The stories of Babar my mother read at bedtime fired my imagination. I loved joining Babar on his many exciting adventures before becoming king of the jungle. But *he* was not my pet.

"No… mine was real, and she was just a baby."

Richard still looked skeptical. He must have been thinking the searing sun and oppressive humidity were causing me to hallucinate. As the wild herd strolled along grazing on grasses, I explained to Richard how I had befriended a real elephant.

My encounter wasn't at a zoo but at a place called Temple Heights on the mid-coast of Maine in 1956. My mother, younger brother, and I had spent hours on the beach below my great-aunt's cottage that July afternoon searching for bleached white sand dollars and colorful seashells. With the incoming tide, our beach disappeared, so we gathered up our treasures and scrambled up the steep crushed-shell path to the dirt road leading to the cottage.

In the distance, we glimpsed an odd-looking gray creature roughly the size of a small cow shuffling toward us. I noticed it had a strange snout that curled and swayed. Coiled around its neck was a thick rope used to lead the animal by a young, dark-skinned stranger. We waited by the side of the road until they reached us, eager for a closer inspection of the creature.

The man leading her greeted us. "Meet Kitubinissa. She's friendly and very gentle. You don't need to be afraid. She's just a baby."

"Wow! That's the biggest baby I ever saw," I said.

Her towering four-foot height gave me the impression she'd be a giant someday. They'd come from half a mile down the road at the seaside summer estate of Horace Hildreth, a former governor of Maine.

In 1953, the Eisenhower administration appointed Hildreth to be the US Ambassador to Pakistan. In appreciation of his service, the Pakistani government gifted Hildreth this baby elephant as a token of its esteem. Protocol in those days demanded acceptance of a gift of this nature. So in 1956, Kitubinissa journeyed halfway around the globe to her new home in Maine in the company of her native handler, Cyril, whose job it was to feed and exercise her.

Gobsmacked by this creature, I summoned up the courage to touch the elephant's leathery flank, only to find the end of its trunk sweeping around to pause directly in front of my face. My brother screeched and fled for his life. But I froze, torn between awe and terror. With my mother's permission, Cyril hoisted me up to straddle her bareback for a ride through the community.

Nearly every day after our first introduction, I would see Kitubinissa and Cyril walking up the road to Temple Heights. As soon as they came into view, I would charge down the hill from the cottage to offer "my" elephant a handful of fresh green grass that I snatched from the lawn on my way to greet them. The tip of her trunk felt like warm velvet as she swept the gift from my hand. Curling her trunk inward, she stuffed the fresh greens in her mouth.

Her strangeness faded, and I adopted her as I would a pet, albeit a rather unique one. When our vacation ended, we left for home enriched by our time spent with Kitubinissa, not realizing she would be gone when we returned a year later.

By the end of summer 1956, she was no longer a baby. The Hildreths decided a zoo would be the best place for her to live, so they gifted her to the Stone Zoo in Stoneham, Massachusetts. My family lived only ten miles from the zoo, but I never knew that my dear little elephant lived so close by. One day in the mid-1970s, a few friends and I went to check out the small zoo that, despite its proximity, none of us had ever visited.

The zoo sheltered only one elephant. The information on the exhibit signage shocked me. It said the Hildreth family had donated this elephant to the zoo. My heart burst with joy. Kitubinissa *had* grown into a magnificent giant!

Elated by the surprise reunion, I told my friends, "I rode on that elephant." They laughed at my preposterous statement. None of them believed me until I explained she was just a baby when I had met her. For a while I watched her wandering around in circles in her small enclosure, lonely and bored. I never would have wished that fate for baby Kitubinissa. She deserved better than to be a captive object of curiosity gawked at by thousands of people.

Now, a decade later, I was witnessing a majestic herd of elephants living the life that Kitubinissa should have had—roaming wild and free. My heart broke for the solitary caged existence Kitubinissa endured. How proud the Selous herd looked ruling over their homeland. These full-grown healthy elephants were mighty enough to fend off all predators except one. Human beings. One of the densest populations of elephants on the continent surrounded us as we floated through the Selous, roughly 100,000 of them, all endangered by seriously out-of-control poaching.

We watched the herd until Conrad ordered us back to paddling. The wind always grew in ferocity in the afternoon, so he pressed us to make as much headway as possible before paddling became an onerous chore.

While Richard rowed on, I lost myself in a daydream. I became Katharine Hepburn's Rose while Richard assumed the role of Humphrey Bogart's Charlie aboard the *African Queen* navigating our way down a jungle river in German East Africa during the First World War. In fact, our route on the Kilombero and Rufiji rivers cut through the middle of what had been German East Africa, so it wasn't hard to imagine we were reliving a part of their route.

Hepburn and Bogart struggled fiercely to survive a constant stream of hardships, as their 1951 movie-going fans rooted for them to wreak revenge on the German bad guys. Shaking off my daydream, I was grateful that German soldiers were not on our expedition's list of hazards to avoid.

Over the years, hippos had often chased rafts captained by Richard, but none had ever taken a bite of one of his rafts. Other Sobek rafts weren't so lucky. Some bore permanent scars in the form of patches, proof of other guides' clashes with angry hippos on other African river trips. Richard's apprehension level had risen because our trip coincided with the peak season for hippo births. After a gestation period of eight months, a female gives birth to just one calf in late winter.

"Believe me, there's nothing as fierce as a mama protecting her calf," he explained. "If we get too close and piss off a female with a newborn baby, that mama will chase us for a mile, maybe more. And here we are in winter floating on a river with a much lower water level than we expected. We've got way less clearance over hippos resting below us on the river bed."

With tweaks to the right or left as needed, Richard kept our raft centered in the deepest channel of the river, tracking the route of the paddle rafts ahead. We noticed some passengers in the paddle rafts were needing to take frequent breaks from the repetitive forward strokes.

"This wind is brutal on the people paddling. It might be a good idea for you to trade places with one of them tomorrow," Richard said.

"Sure, no problem." I preferred being active, and I wanted to learn more about our paddleboat clients. I suspected my muscles were in better condition than theirs, having worked the past summer season as a paddleboat captain back in California. I didn't mind spelling a weary paddler for a few days so their muscles could recuperate while Richard chauffeured them downstream.

The river widened, splitting the current into multiple channels that poured around islands, some of which weren't more than rocky ledges or sand bars. Some channels reunited after a short distance to travel together as one, for a few more miles, before again splitting apart. Many of the shallower channels petered out in dead ends. Those we needed to avoid.

Slade and Michael kept the topographic map in their boat. They decided, when we had a choice, which channel ran the deepest. When they disagreed on our exact location, they'd search for notable topographic features. They also relied on Conrad's memories of his first trip down the river. Ultimately, they listened to their pure gut instincts. So far, their skills had kept us safe and on course.

Late in the afternoon, the blustery upstream winds took their toll on Richard. The load of gear piled high in the back of the raft acted like a sail, catching the wind's force. With this extra drag slowing us down, we fell far behind the other rafts. We had to stay within sight of them, or we risked entering a different channel that might bypass our campsite altogether. Richard, Masengela, and I would be fine because we carried the kitchen, tents, and our personal gear, but our clients would spend a miserable night without food and tents, though they had the iodine. We couldn't let that happen.

Richard made Masengela relinquish his spot atop the gear to sit in the front with me. That reduced the drag from the wind. Though he'd no longer have a downstream view, Richard spun our raft 180 degrees and adopted a more powerful stroke. Engaging all his back muscles, he pulled rather than pushed on the oars. Our speed picked up. With Richard rowing blind, Masengela and I became nervous navigators, directing him toward the strongest part of the current and away from hippos.

We'd nearly caught up with the other rafts when Conrad decided we'd paddled far enough for one day. We beached on a narrow but flat strip of land. Only half the size of our previous camp, there were no signs of hippo trails and plenty of driftwood scattered nearby for our fire.

Glad to be relieved of paddling, our clients quickly set up their tents, then set out to gather wood for our campfire. This would be Slade's night to take a turn as head chef. He drew Richard as his assistant. Slade easily spent more nights sleeping outside under the stars in the course of a year than sleeping indoors in a regular bed. He'd honed his culinary skills for prepping and cooking up hearty meals over a fire and a camp stove rather than using a conventional oven.

Richard, on the other hand, was an expert at using a well-oiled frying pan over a kitchen stove to whip up a batch of his favorite food—popcorn. When it came to cooking, Richard humbly took his cues from the experts. While the two men chopped onions and browned ground beef for a meat sauce, I did my Kool-Aid thing.

Meanwhile, Conrad mixed the ingredients for homemade bread, a pastime he found relaxing after a long physical day. Once he'd created soft pliable dough, he plopped it into a well-oiled Dutch oven to let it rise. He enlisted the help of our passengers to dig a baking pit in the sand, away from where people might walk. They lined the bottom with smooth river stones.

After the taxing effort of paddling for hours, Jay declared he was hungry enough to eat an entire hippo! Unable to resist this cue, Richard launched into an account about a trip he'd taken on a different African river. After several days on the river, as their expedition approached a small village, they saw the prone carcass of a slain hippo lying in a clearing next to the river.

"The villagers started stomping and waving wildly. For all we knew, those crazy hand gestures could have been warning us to stay away. Maybe they were worried we might try to steal their fresh meat. But they weren't heaving rocks or shooting arrows at us, so we pulled in for a friendly visit."

He recalled the entire village had turned out to celebrate the kill. The Sobek visitors watched the village men hack away at the creature's massive bulk with machetes. Once butchered, the village women threaded long strips of the fatty meat onto pointed wooden stakes. They sat in a circle around a glowing firepit, slowly rotating the meat-laden skewers until the outer surfaces charred to a crisp. As the fat melted, it dripped onto the red coals, sending up scented curls of smoke. The hospitable villagers offered their fascinated guests samples of the greasy medium-rare hippo steaks.

"So, what did it taste like?" Martha asked, captivated by his tale.

"You know, I think that was the best char-grilled steak I ever ate. Just wish I'd had some A.1. sauce to put on it." Richard smacked his lips, reliving that past moment. "Sorry you guys won't get to try that delicacy, since the Selous hippos are protected. But if you ever get the chance to try a hippo steak, I highly recommend it."

The idea of eating a hippo steak struck me as about appealing as it would be biting into a Rocky Mountain oyster, the polite term for a bull's testicle. But I had to admit that once I heard Richard's description of the flavor, maybe, just maybe, if I were starving to death, I might try it, drenched with A.1. sauce.

To ward off the rumblings of hungry stomachs, I put out crackers, cheese, and a tin of sardines packed in oil and tomato sauce for the group to snack on while waiting for the main meal to be served. We dined that night on spaghetti "al dente" topped with a flavorful, garlicky meat sauce and freshly grated cheese. Eating Slade's "Spaghetti Selous" by the glow of the campfire rivaled dining on "Spaghetti Bolognese" by candlelight in a fine Italian restaurant, both in quality and ambience. Absent a fruity Italian Chianti to pair with the pasta, the warmish grape Kool-Aid proved delightfully thirst-quenching. Almost.

After dinner, when our campfire no longer roared, Conrad transferred a shovelful of the radiant red embers into the bottom of the baking pit. He punched the bread dough down one last time, then covered the Dutch oven and lowered it into the pit. Carefully, he spread the oven's flat-lipped lid with another generous layer of glowing coals. Then he refilled the pit with sand, to bury the top of the oven 10 inches underground. Finally, to mark the exact spot of the baking bread, Conrad drove a stick into the sand.

The sounds of small creatures, maybe mice or frogs, rustled around the fringes of our camp as I drifted off to sleep. Hippos didn't disturb our camp that night, or if they did, I was too deep in sleep to hear or care.

16
THE STRUGGLE TO STAY HEALTHY

THE SUN CREEPING above the horizon served as nature's alarm clock. Nocturnal creatures concluded their recital of primal croaks, squawks, snorts, and peeps as the eastern skies lightened and the coolness of the night vanished. My dream-filled brain reacted to the sudden extra warmth tugging me into wakefulness.

Along with the rise in air temperature, I noticed another strange sensation. My bottom lip had puffed up and rolled outward. It felt ridiculously swollen, like a Hollywood starlet's lip enlargement procedure gone terribly wrong. Instead of deliciously kissable lips, the transformation yielded a frightfully repelling distortion.

I found my small travel mirror, but wasn't sure I wanted to look at my face, afraid of what I'd see. Curiosity finally won out. UGLY! My bottom lip looked hideous. Could an insect have bitten me while I was sleeping? Could I be having an allergic reaction to something I ate? As I puzzled over the cause of my distended lip, it occurred to me I had used plenty of sunscreen lotion on exposed parts of my skin, but none on my lips.

My own thoughtless neglect had left my lips vulnerable to a double dose of damage from the sun's ultraviolet rays: directly from above and reflected up from the surface of the river. My disfigured lips proved how much havoc the tropical sun could wreak. Fortunately, I had worn my sunglasses all day so my eyes hadn't swollen shut. Glancing once more in the mirror, I wasn't sure whether to laugh or cry. I could fit in perfectly with the freaks in one of those third-rate traveling circuses.

I hesitated to leave my tent, not wanting to face the others looking like I'd just come from Dr. Frankenstein's lab, but I needed to help with the morning chores. Conrad had started a pot of coffee when I joined him in the kitchen. One look at me and he nailed my diagnosis. "Got a little sunburn on your lips, hey?"

Mortified that it looked that obvious, I asked, "Looks pretty gross, huh?"

"You need to use some zinc oxide on your lips. Have you got some?"

"No, just regular sunscreen lotion. I didn't think I'd need it." I'd never used the thick, white, metallic-smelling cream, despite spending days out on the rivers at home. This was a first for me.

"You can use some of mine," Conrad offered. "The sun here near the equator is much harsher than you realize. It'll burn you to a crisp if you aren't careful. And don't worry. Your lip will be back to normal in a couple of hours."

I vowed never to forget this lesson. However, when two other people stumbled out of their tents that morning with swollen lower lips, I felt less self-conscious. We shared a good laugh about our altered appearances and vowed never to leave homes again without our American Express card and a tube of ZnO.

That morning, the early risers got to observe Conrad digging the Dutch oven out of the ground. He lifted the cover to reveal a perfectly baked 15-inch round loaf of golden brown crusted bread. As Conrad split the still-warm loaf in half, the fresh yeasty aroma made my mouth water. He placed one half of the loaf on our breakfast table, where it commanded center stage.

Enthusiastically, we devoured every crumb of the tasty homemade treat from Conrad's "Wild Flour Bakery." I smeared my slice with a spoonful of guava jelly and took a bite. My taste buds danced with delight! The crust was flaky-crunchy, while the interior had a soft, moist texture. Not even a Parisian bakery could serve a boule as tasty as Conrad's loaf. I looked forward to enjoying another slice from the second half for lunch.

Once we'd loaded the rafts, I took my place in Conrad's paddle raft, along with the married couple, a 40-year-old repeat client in great physical shape, and the head game ranger, Magigi. All day long, the river current teased us, first picking up speed in narrow sections, then slowing when the banks opened wider. Clutching a paddle and using my energy to help propel our raft forward provided a sense of satisfaction I lacked as a spectator in the oar boat. Though, in the slow sections, the constant paddling became monotonous. Every half hour, we'd switch from one side of the raft to the other, so we worked muscles in both arms equally.

The hippos I found so intriguing on the first day of our trip now seemed ordinary, if not boring. Those endearing wiggly ears and their variety of snorts still amused me, but they were everywhere I looked. Some lumbered along the shoreline. Some rested, half immersed in shallow pools. While we couldn't see them, we knew there were plenty of them completely submerged in the depths below us. Michael had an ambitious plan to count every hippo we passed for an informal wildlife survey he hoped to publish. Each evening, he'd announce the addition of a couple hundred more to his tally.

I felt safer in deeper currents based on Richard's rationale of staying where the greatest headroom over the submerged ones existed. I dreaded heading into the wider shallower sections. I found it nerve-wracking not knowing how many hippos were hiding in those murky depths, each having the advantage of sensing us before we spotted them.

It reminded me of childhood birthday parties. After donning a blindfold, an adult would spin us around and tell us to go pin a tail on the picture of a donkey taped to a wall in front of us. Whoever came the closest to placing their tail in the right spot won. Here on the river with my eyes wide open, the silty water acted like a blindfold. I had no way of knowing if the waiting hippos were straight ahead or a safe distance away. The opposite strategy won the game in the Selous: being farthest away from the hidden target. So far, we were winning.

Conrad gave us rest breaks when he spotted different species of wildlife. My first glimpse of an African fish eagle convinced me I was seeing an American bald eagle. With dark brown body plumage and a snowy-white head, breast, and tail, the two species closely resemble each other. However, the bald eagle makes its home only in the Americas, while the fish eagle's territory is strictly in Africa.

Fish eagles roosted on branches of the taller trees along the river. One dazzled us with its hunting skill by swooping from its high scouting perch, gliding just above the water's surface, then plunging the sharp talons of its feet into the water to clutch the body of an unsuspecting fish. I marveled at how adeptly the fish eagle, flapping its eight-foot wingspan, rose from the river, tightly clenching its wriggling, shiny-scaled prey. Back on its perch, it finished the kill using its yellow, razor-sharp, hook-shaped beak. I suspected the fish eagle enjoyed its prized catch that morning as much as I had enjoyed Conrad's freshly baked bread.

In this wilderness, animals lived and died by Charles Darwin's theory of "survival of the fittest." Out here, one was eating or being eaten depending on skill and luck. The fish eagles earned my admiration for their success at fishing. Nearly every time one dove, it scored a meal for itself.

Conrad pointed out a couple of Nile crocodiles sunning themselves on the left bank of the river. Since they perceived our bulky rafts as ominous, unknown predators, they hoisted their bodies up on short leathery legs and hustled to the edge of the river. Helped a bit by gravity, they slid down the sloped bank and disappeared. Their entrances into the water rivaled the perfection of an Olympic high diver, dissolving through the surface with barely a wrinkle to mark their entry.

These antisocial, heavily armored crocodiles presented a constant danger. They rarely made their presence known unless they felt threatened. Since they cleverly remained concealed underwater when other predators passed nearby, knowing exactly where they lurked proved impossible.

When hunting for food, a crocodile launches a stealth attack on its victim, clamping onto a body part with the 68 conical pointed teeth lining the sides of their powerful jaws. After rendering prey helpless, the crocodile drags it under water and patiently waits for it to drown. Ironically, being drowned may be a kinder death than being eaten alive, but I had no desire to experience either.

In a ploy to bring positive mojo to his risky debut African expedition, Richard decided to name his fledgling company Sobek, after the ancient Egyptian god of crocodiles. The Egyptians believed Sobek possessed magical powers to grant safety and deflect evil influences from the watery domain of the crocodile, especially in the Nile River.

Conrad insisted no one go more than ankle deep in the river to wash up. Even then, he required a bather to have a spotter close by to watch for stalking crocs. Most of us opted to dip buckets into the river by leaning out over the back of an anchored raft. After clambering back along the length of the raft, spilling as little water as possible, we carried the buckets onto dry land, where we could wash up out of harm's way.

The irony of getting "cleaned up" meant using water saturated with brown sediment (and diluted dung!) to deposit a fresh layer of river silt on our bodies while rinsing away the grime and sweat-caked coating. Constantly wearing this thin coat of muck bothered our two women clients, but since we were all in the same boat (pun intended) it was simply a matter of accepting mottled, dirty skin and hair as our fashion style for the trip's duration.

AS WE STASHED what remained from our lunch break back in the rafts, we heard a loud commotion of stampeding hooves and bushes being trampled. The tumultuous crescendo racing upstream toward us startled everyone.

Conrad shouted, "Get over here and get in the rafts. Now!"

We tumbled into the rafts as if our lives depended on escaping whatever was coming for us. He suspected it was a herd of elephants, but as we pushed away from the shore heading for the middle of the river, the racket-makers came into view.

"Cape buffalo," Conrad said as the herd of a dozen muscular beasts crashed through the brush along the overgrown riverbank. "It's good we got out of their way when we did. They are the meanest, nastiest animals out here. They're also the stupidest damn animals on the planet!"

Conrad's description proved apt as we watched their straight-line method of getting from one point to another, bulldozing anything in their way. Those buffalo weighed between one and two tons. Across the top of their heads, they sported a rack of sharp curved horns with a fused base forming a continuous bone shield in the shape of a waxed mustache.

Watching the carnage wreaked by the charging herd, Conrad said, "The damn things are temperamental, unpredictable and aggressive. They use their horns to gore and rip apart their adversaries. You want to know what can kill a lion? You're looking at them. They're called Black Death by big-game hunters because they're so dangerous. If you ever catch sight of one of these buggers, run the other way. If you can look them in the eye, you're in big trouble."

Floating in the river away from the shore protected us from a stare-down with these killers. They stampeded through the spot where we'd had lunch and kept charging straight ahead like marathon runners going for gold.

AFTER THREE HOURS of struggling against wicked afternoon headwinds, we camped that night on a wide-open sandy beach. The river widened in this spot, forming a spacious still pool, home to at least a couple dozen hippos. Conrad pointed out two different hippo trails leading across the beach and advised it would be unwise to stake a tent on top of either. I set my tent close to a small boulder on one side of the beach. I figured the solid rock might offer some protection. A little wishful thinking couldn't hurt.

Not far away, Martha and Jay sat together on the sandy bank. I couldn't tell if she was angry, hurt, or unhappy. Jay wrapped one arm around her shoulders to comfort her. When she began to cry, I warily approached them.

"Is something wrong? What's happening?" I asked gently.

Martha turned away and buried her face in Jay's chest to hide her tears.

"My wife's just having a hard time being out here in the wild," said Jay.

"What can I do to make things easier?" I asked.

Between sobs, she whined, "It's just so hard. I didn't think it would be this hard. I didn't think we would have to paddle nonstop in this vicious wind all day. It didn't say that in the brochure. This trip should be fun. It isn't fun *at all*. My arms hurt. My back hurts." Her tears continued to flow. Her bravado at the start of the trip had vanished.

Even though I'd strengthened my muscles with a summer's worth of paddling, I'd found the afternoon paddling strenuous and tiring. I recalled she had taken more frequent rest breaks than the rest of us. I could imagine how much her body must hurt. "Would you like to ride in the oar boat where you won't have to paddle for a day or two?" I offered.

She turned to face me, tears still streaming down her cheeks. "Jay wants to be in the paddleboat, and I don't want to be apart from him.

Your company should have motors for these rafts, so we don't have to work like slaves all day." I didn't expect such a feisty response.

"That's not how we operate. We don't ever use motors." I worried that if she was miserable now, what would she think when we reached the two waterfalls and had to portage all of our gear through the jungle to bypass the 250-foot drop in the river?

"I don't think your company planned this trip very well. Other safaris we've been on have brought portable showers where at least we could get clean every day. And I think your brochure is full of false advertising." Her ranting only fueled her discontent.

"Let me mention your concerns to Conrad and Richard and see what we can do." She must have realized she wasn't going to get any of the things she wanted, but I didn't want her to get any ideas about suing us for breach of contract either.

I strolled back toward the kitchen and discreetly asked to speak with Richard and Conrad, away from the rest of the group. They followed me to the far edge of the beach, where we sat on some smooth boulders. I told them about my disturbing conversation with the couple. How she seemed so upset, I worried they might sue us.

Richard chuckled dismissively. "They aren't going to sue us. Don't worry about them."

"But she's sitting over there crying, Richard, and we've still got days left on this trip. Maybe you should talk with her."

Again he laughed, and I noticed Conrad agreed with him. They apparently intended to ignore the situation.

"Brenda, there's something you've got to understand about these types of trips. Yes, some people find them tougher than they expected, but no one has ever sued us because they were miserable in the middle of a trip. This magical thing happens at the end of a trip as they look back on all the challenges they survived. Once they get back home and start telling tales, they'll likely embellish about how they almost got eaten by a hippo and how a pack of wild lions were roaring just outside their tents. Their friends will be in awe of them. At that point, they will realize what a truly grand adventure they've had. It's like childbirth. All the unpleasant memories fade away." Richard's explanation sounded dubious to me.

Aware of this adventure travel phenomenon, Conrad said, "He's right, Brenda. This happens all the time. I've gotten letters from clients who, while on their trips, claimed they hated every minute, but afterward they recalled only glowing memories, insisting it had been the best trip they ever took."

Reluctantly, I accepted what they said, figuring they'd been getting people through the hardships of these trips for many years. But somehow, I couldn't imagine Martha ever wowing her friends with fantastic tales of her adventures in the Selous. Richard reassured me, "Really, don't get yourself all worried about them. They'll be fine."

THE HIPPOS VISITED our beach that night on their way to fill their hungry bellies. Shortly after dinner, my full belly started feeling queasy. Resting on my sleeping bag in my tent, the gurgling waves of discomfort in my stomach and abdomen grew stronger. I took inventory of what I'd eaten over the past couple of days.

Was the iodine-treated water irritating my stomach? Could it be the zinc oxide on my lips that I'd done a good job of licking off during the day? Was something I ate undercooked? The only other plausible cause that came to mind was that worrisome hippo dung in the water. But no one else had gotten ill from drinking the Kool-Aid.

Like a Slinky toy, waves of cramps slithered through my innards. I predicted it wouldn't be long before I'd suffer an inevitable explosion of sorts. What I needed more than anything was a clean white porcelain throne on which I could park myself. That wasn't going to happen.

I certainly couldn't relieve myself outside my tent for fear of finding myself face-to-face with a hippo. I remembered Conrad's instruction in an emergency, "Just stick your butt out through the tent door."

I sat up and looked through my gear to see if I had any tool to help me dig a hole. I had one essential piece of equipment that every river guide carried—my buck knife. That would help to loosen the ground, but the trowel and the TP I really needed rested at least 10 yards away.

As the urge to go intensified, I unzipped the mesh door of my tent and hurriedly dug into the ground just outside with the blade of my knife, every few second stopping to scoop out the loosened soil with my hands. I prayed I would finish the damn hole in time. When it reached a depth of nearly eight inches, I knew I'd run out of time. I dropped my shorts and prayed my butt hovered directly over the hole.

The relief was immense. I retrieved a small pack of travel tissues from an outer pocket of my backpack, cleaned myself, and carefully covered the mess I'd made with the loose sand I'd scooped from the hole. I hoped no one else heard the sounds of my distress.

I lay back down on my sleeping bag, only to have a fresh wave of cramps take hold in my gut. Horrified, I realized I would need to repeat the process a second time. The pressure building inside me once more was agonizing. As much as I just wanted to lie there and cry, I began digging a second hole, off to the side of the first one. I had no time to spare. In my misery, I empathized with Martha. This was no fun at all.

After I had filled and covered the second hole, I realized I had Imodium tablets with me, just in case the Tanzanian version of Montezuma wreaked his revenge on me. Immediately, I washed down a dose with some of the Kool-Aid in my canteen. I flopped back down and begged God to let me fall asleep. I felt wasted. Mercifully, after a while, sleep came.

A few hours later, my rumbling guts woke me. I got one more chance to practice the fine art of toileting, designed to avoid hippopotami in the dark. If practice makes perfect, I was well on the way to becoming an expert. Once I'd dug a third hole, I'd used all the real estate immediately outside my tent door, leaving me to fret about what I would do if the need arose again before daybreak. From now on, I vowed to bring a supply of TP and a full bottle of Kool-Aid into my tent each night before bedding down. The Imodium finally kicked in, and I slept peacefully until morning.

17

HIPPO HELL

BY THE FOURTH DAY, our clients knew exactly what to expect. We'd paddle for hours dodging hippos, keeping watch for ripples near the river bank where crocodiles silently slid beneath the water. The relentless sun's blistering rays reflecting off the water prompted us to slather a fresh coat of sunscreen onto our skin every few hours.

Our clients stopped complaining about the inevitable upstream winds that intensified every midafternoon because Mother Nature obviously wasn't listening. When our arm muscles wearied from tedious stretches of paddling, we dug deeper for energy reserves within ourselves, finding strength we never knew existed. When monotony tempted us toward complacency, we resisted. Inside the boundaries of the world's greatest zoo, brimming with wildlife that could easily end any of our lives in a heartbeat, we remained vigilant.

Instead of the stealthy carnivores I presumed would hide behind every rock, bush, or underwater ready to pounce, the wildlife had tolerated our presence as uninvited strangers, floating through their world. Initially, my fear of being so close to these wild animals nearly paralyzed me. Bit by bit, it had subsided to a justifiable wariness. I realized they weren't likely to make a meal of me when a short hunt would provide their favorite foods.

During the morning, the river widened and split into several channels, while the banks on both sides of the river steepened. The tangled green fringe bordering the river's edge thickened.

At midmorning, our rafts entered a calm stretch where the river pooled in a sizeable basin behind a rocky ledge that stretched across the width of the river. Several narrow channels breached the ledge where their currents cascaded into a much smaller pool, which drained back into a single downstream flow. Both the surrounding land and the water in the upper pool teemed with hippos, more than we'd seen in one location before.

Hippos don't tolerate other hippos invading their territories. They get very cranky and aggressive. Off to our left side, in shoulder-deep water, one large male sparred with another, each with their jaws strained wide open. They jousted, thrusting their razor-sharp incisor tusks forward with the finesse of practiced swordsmen in a duel while bellowing thunderous warnings of imminent attack. The hippos' aggression resembled a *paso doble*, the passion-filled, intricate dance to death between a taunting matador and his enraged bull.

We hadn't witnessed hostility this violent between hippos until now. But dozens of them crammed together this densely created the perfect conditions for territorial sparring. Four more giant objects entering their already overcrowded space did nothing to reduce their irritation. Heads with nostrils flaring randomly popped through the surface of the pool, grabbed a breath of air, then dove below the surface again. My internal fear meter rose higher.

Each raft carefully threaded its way around the hippos in the upper pool. Bart's raft took the lead. He stood to survey which of the chutes ahead looked passable. None was terribly wide. He needed to find a channel wide enough and deep enough for the rafts to pass through without grounding on the ledge or a submerged snag. Our safety depended on this critical feature. If a raft became trapped in one of these channels, we'd be offering ourselves up to the hippos on a silver platter.

From the front of our raft, Magigi, the head game ranger, surveyed downstream searching for the best route to prevent us from embarking on a collision course with any of the hippos. He shouted, wagging his arm, directing Bart to take the channel furthest to the left. That chute dropped over the ledge into the lower pool a safe distance away from several waiting inhabitants, cautiously eyeing our approaching caravan.

Lagging 50 feet behind in Conrad's raft, we watched Bart's boat line up with the far left chute. On his command, "Forward paddle," his crew dug their paddles into the water and powered straight ahead. Just as they entered the chute, the raft bottomed out, snagging on a shallow underwater outcropping of the ledge. I watched in horror as a moment later, a few yards below them at the bottom of the channel, a hippo surfaced and stood glaring up at the stationary raft.

Bart screamed, "Back paddle! HARD back paddle!" His paddlers strained with every ounce of energy in their adrenaline-flooded bodies. I knew their lives depended on getting ungrounded and back into the upper pool. Somehow, they had to get away from that hippo. With exaggerated rocking and jiggling, the raft finally dislodged from the snag. Bart, hyped up from the horribly close call with disaster, kept screaming, "Back paddle! Keep it up! Hard back paddle!" I could see the raft fighting against the rushing current as it began to creep backward.

Slade, captaining the closest raft to Bart's, mobilized for a rescue before the menacing hippo could climb onto the ledge for a taste of the peculiar prey just above it. Astonishingly, Bart's crew, determined to save themselves, employed muscle strength normally reserved for superheroes. After several minutes of frantic paddling, the back end of their raft finally leveled out in the calm water of the upper pool. Bart pivoted the raft away from the swift current siphoning into the chute, safe for the moment.

Watching this terrifying drama unfold, I had the surreal feeling of being on the set of a Spielberg action-adventure movie, standing right next to him as he directed the cinematic shot. In search of a safer route, Bart ordered his passengers to paddle further to the right in the upper pool to scout other chutes. The hippo they'd just aggravated stood its ground, guarding the left-hand side of the small lower pool. But 100 feet away, on the far-right side of the lower pool, stood another group of three gray monsters in shallow water, watching our movements.

There wasn't any point in hanging out in the upper pool, hoping the beasts below would graciously move aside to offer us a safe route through the lower pool. With so many hippos, we might have to wait there for days. We needed to move beyond this risky area, even if it meant each passing raft might invite the wrath of this herd of hippos. The decision to push forward became our only realistic yet perilous option.

Slade picked a middle channel leading into the center of the lower pool to try. The chute appeared wide enough and deep enough for his raft.

"Let's go," he commanded. Seconds later, his raft dropped into the chute and effortlessly slid into the lower pool.

Joyful shouts of celebration erupted, but abruptly stopped when a massive shape rose out of the water on the left side of the raft. Before its head even broke the surface of the water, Slade screamed, "Hard right! Now!" The raft pivoted to the right while Slade continued to shout commands to his crew to "thread the needle" between the surfacing hippo on one side and the group of three on the other.

Slade's bellowing, combined with the turbulent water churned up by his passengers' frantic strokes, obviously startled the hippo. It quickly dove back into the depths of the pool. Slade shouted for his crew to keep paddling to the far side of the lower pool.

Bart's rafters, still pumped up with adrenaline, paddled over to follow the route Slade had just taken. I couldn't believe that Bart would dare the hippo, submerged under water at the end of the chute, to resurface. Did he assume the racket made by Slade's paddlers had sufficiently spooked the hippo to retreat to safety elsewhere in the lower pool?

"Let's do it. Hard forward," Bart commanded. His bold decision captivated my attention. His paddlers dipped and mightily pulled their paddles through the water. The raft sped down the chute. Not daring to breathe, I watched the raft coast smoothly into the lower pool. Not a single gray head popped up as Bart's raft raced across to the far side of the pool where Slade's boat waited.

In just seconds, it would be our turn to run the gauntlet. After witnessing two harrowing near disasters, my internal fear meter's indicator needle spun past the maximum mark, snapped itself off the pivot pin, and fled in sheer terror into oblivion. I gripped my paddle with white knuckles. My heart pounded, and a knot the size of a cantaloupe squeezed my stomach. "Third time's a charm" popped into my mind. What if Slade's and Bart's unwelcomed intrusion had so aggravated the lower pool's now-invisible hippo that *our* raft would be the victim of a blind-sided ambush?

Out of the blue, a deafening explosion somewhere behind us shattered the river's peaceful surrounds. The thunderous reverberation left us stunned. It didn't sound like a gunshot, more like a blast. Whatever had caused the explosion must have been closer to Richard's boat, following 50 yards behind us.

I spun around and saw Richard's arms pumping as if someone had pressed his fast-forward button. His flying oars churned up a wake behind the raft. It reminded me of the animated cartoon character, the Road Runner, churning up a cloud of dust in a mad dash to escape Wile E. Coyote.

Until that moment, I honestly believed Richard feared nothing. On expedition after expedition, he emerged unscathed and victorious over every hazardous challenge he faced. Now he looked terrified as he fought for survival. His visibly shaken passengers peered over the right-hand side of the raft into the river. Masengela, from his perch atop the gear in the raft's rear compartment, held his rifle with both hands, the butt end snug against his shoulder. He aimed at the surface of the water behind them.

Richard shouted for the clients to pull a quick-release knot to free a spare 10-foot oar lashed to the side of the raft.

"If it gets close, whack it on its head," Richard yelled to the white-knuckled man, hoisting the spare oar.

Only then could I see that one section of the raft's side tube had collapsed. Just then, an outraged, bellowing hippo surfaced 10 feet behind the deflated section of the boat. In hot pursuit, with its jaws stretched wide open, it threatened another chomp.

Richard's heroic rowing continued to widen the gap between the beast and the raft until the hippo lagged far enough out of range to prevent it from delivering a second disabling blow. With five of its six outer tubes still intact, the raft only listed a few degrees to the damaged side. However, if the hippo caught up and chomped a second time, the raft would be in serious danger of sinking.

Richard had been far enough behind the other rafts that he hadn't seen the near disasters Bart and Slade had faced navigating the gaps in the ledge. Nor had he been close enough to see the hippos waiting in the pool below. He rowed with all his might to stay ahead of the foe chasing him. Conrad knew that for all of us to get through this stretch unscathed, our raft had to slide into the lower pool before Richard's raft reached the top of the channel.

"We've got to go now!" Conrad ordered. "Follow Bart. Hard forward!" We jammed our feet under the side tubes for more leverage and dipped our paddles deep in the water and stroked. The raft lurched forward as Conrad steered straight toward the chute. "Watch out for hippos in the lower pool!"

I'd never been this scared or felt this helpless. The cascading water tugged our raft through the gap in the ledge and flung us down the chute into the lower pool. On both sides, the hippos glared at our intrusion but stood motionless.

A wave of dread drenched me as I turned around to watch Richard's wider and heavier raft attempt to follow us down the same chute. Our smaller paddle raft had barely squeezed through the chute. Alarmed, I asked, "Conrad, is Richard's raft going to fit through that channel?"

"It's got to. At least he's got a bunch of speed coming into it."

I couldn't bear the thought of his raft getting stuck in the chute with that livid hippo in close pursuit. Silently, I invoked the great deity Sobek to use his magical powers to save the passengers and gear of the raft bearing his name. I trusted Sobek had Richard's back. Unable to pause for even a second to scout the route through the ledge, on blind trust alone, Richard followed our raft's route.

He nosed his raft into the chute, shoving the oar handles to the floor. This raised the oar blades high enough not to catch on the exposed rocky outcrops along the sides of the narrow channel. It wasn't pretty. Both sides of the ledge grabbed at the raft, but never got a firm enough grip to stop it as the raft pinballed down the chute. My heart didn't stop pounding until Richard's raft careened safely into the lower pool.

We hurried to regroup with Bart's and Slade's rafts already beyond the pool, slowly drifting in the Kilombero's main current. Looking back upstream, one lone, defiant hippo had climbed up onto the ledge. With its jaws opened aggressively, it let out an ominous, throaty groan clearly meant to warn us, "Get the hell out of my space and don't come back." Triumphantly, it watched our rafts disappear downstream.

"What happened back there?" Slade asked as he eyed the flaccid tube on Richard's raft.

"A massive hippo surfaced right next to us and tried to take a chunk out of our raft. When its tusks pierced the tube, it exploded."

Richard flung his arms wide open, reenacting the rupture. "I'm sure the last thing the damned hippo expected was a blast of pressurized air gushing down its throat."

He figured the surprise explosion momentarily stunned the beast into releasing its grip on the raft. That tiny jump-start allowed him to outrun the hippo. Though they can be speedy on land, a hippo's forward progress in the water drops to only five miles per hour.

"Hey, I bet you scored a new personal best back there. It looked like you were cranking those oars at a speed of at least 10 miles per hour." Richard accepted Bart's joke as a compliment.

Slade steered his raft alongside Richard's boat to get a closer view of the gash in the limp tube.

"It's too risky for you to row much further with a tube that damaged. Let's look for a safe spot to pull over where we can have lunch and get a patch on that puncture."

At the next beachy-looking spot on the riverbank, we took a break. We partially unloaded Richard's raft and dragged it up onto solid ground for the repair. A survey of the damage revealed how close Richard had come to a disaster. The hippo had only wrapped half its jaw around the tube, resulting in a single puncture four inches away from one of the internal baffles that partitioned two tubes into separate compartments.

Had the beast landed a full-on attack, it would have also punctured the adjacent tube, resulting in the collapse of one whole side of the raft. With one-third of the raft's buoyancy lost, Richard would not have been able to outrun his pursuer. The furious hippo would have had a field day.

Guides prefer to fully deflate rafts when repairing damages. Resting the damaged tube flat on a hard surface maximizes the contact area of the patch with the surface of the tube when applying the patch. When done at an overnight campsite, the adhesive also has ample time to cure. But this lunch spot wasn't large enough for an overnight camp, and Conrad wanted to cover more miles to stay on schedule.

Bart, who prided himself on boat repairs of all sorts, found the repair kit and set about crafting a patch to repair the four-inch-long rip. With a piece of sandpaper, he vigorously roughed up both the outer surface of the aging tube around the tear and the inner surface of the rectangular patch he cut to size to cover the gaping hole. After sanding, he cleaned both surfaces with isopropyl alcohol to remove any debris that might interfere with achieving a solid seal, then smeared an adhesive superglue on the two prepared surfaces.

With a perfectionist's precision, he centered the new patch over the tear and pressed the two layers together. All the seal needed was time for the bonded pieces to cure. Ideally, a major patch requires eight hours in direct sunlight before reinflating the tube full of air. We had just four hours of blazing tropical sun left before sunset. In the worst-case scenario, Bart felt confident the repair would hold enough air until he could properly patch the leak.

Lunch felt like a victory celebration. As we ate, Richard and his passengers entertained us by recounting every gripping detail of the attack they had just survived.

Kate, who'd hoped for close encounters with the wildlife, had gotten her wish. "Its eyes flashed with fiery indignation. I thought those tusks were going to slice right through the side of the boat. I'm still jittery from the adrenaline rush," she told us.

"Yeah, but that enormous head came up only three feet from where I was sitting. That crunch and explosion scared the shit out of me," the quick-witted client who'd manned the whacking oar explained. "I thought we were goners, but I wasn't giving up without a fight!"

With everyone safe for the moment, Richard's demeanor turned gleeful. He had challenged the enemy and defeated it in its own territory. He'd added one more accomplishment to his long list of "firsts."

After lunch, Bart reinflated the repaired tube using the high-volume hand pump. The patch seemed to hold despite the short curing time.

Michael offered to row the oar boat so Richard could kick back and relax after the morning's excitement. I rejoined them in the oar boat to help monitor the vulnerable patch. That afternoon passed pleasantly. The lack of any confrontation with wildlife and the unexpected absence of upstream winds offered us a much-appreciated respite. I enjoyed the lazy float as the current carried us undisturbed for miles. This effortless style of rafting must have been what Huckleberry Finn meant when he said he felt "free and easy and comfortable" living on a raft.

Stories of the hippo attack continued around the campfire that evening. Several of our group discussed the steady increase in the number of hippos we saw each day. Frank, who'd seen plenty of hippos on his previous trips, couldn't shake the image of the angry hippo standing on the ledge glaring as we retreated. "That hippo was damn upset. I'm sure glad we don't have to go back through that section again. I don't think we'd make it. He'd be waiting for us and probably shred us and our raft to pieces."

Richard agreed. "That groan was a warning for sure. The bulls will fight each other if one thinks another bull is encroaching on his territory. I gotta tell you, watching hippos fight is pretty ugly. They're like warriors engaged in a fight to the death. Each wants to drive its tusks through the other's face. All that thunderous bellowing and the vicious whacks of tusk striking tusk are deafening. It's like being at a hard metal rock concert."

As I waited for sleep to come, dreadful images replayed in my mind: the jagged ledge, two pools, too many hippos, and a partially deflated raft. Today, I'd gotten only a tiny taste of the real dangers this wilderness posed. Well aware that at least 50 miles and days of walking separated us from medical help left me both sobered and humbled in this place of extreme isolation.

Reliving the scenes of threatening hippos caused enough adrenaline to seep into my body that the sleep I needed so badly dangled just beyond reach. My thoughts wandered to my friend from the train, skeptical Solomon. What would he say about our near-disastrous morning? I could imagine his brown eyes, wide with astonishment, and his head bobbing at our misplaced bravado scolding us, "I warned you."

18

A DAY WITHOUT DRAMA

A FEELING OF UNEASINESS weighed on me as we readied our rafts for our fifth day. Barely a third of the way into our journey, we'd already survived one hippo attack. Michael's informal wildlife survey count totaled over 1,000 hippos. Deeper within the interior of the Selous, I wondered if the river might become so clogged with these beasts it would become impassable. We didn't have a choice. We had to keep going, staying in deep water and weaving our way through the ever-present clusters of hippos.

Each morning before we shoved off, the guides pumped more air into the tubes of our rafts to "top off" the pressure inside. That morning, as Michael pumped additional air into the tubes of Richard's raft, the hippo-bitten tube hissed quietly, a telltale sign the patch had failed.

"Hey, Richard, your patch is leaking air." Michael's discovery gave rise to an impromptu discussion between the guides on what to do about the leak.

To replace the patch that morning would stall our departure by at least a couple of hours while the adhesive glue cured. Conrad, eager to log more miles that morning before the afternoon heat riled up headwinds, argued to delay the repair until we reached our next camp as long as the leak didn't worsen. Richard came up with a makeshift temporary solution.

"Brenda, you've gotta ride in my boat today. I need you to keep pressure on this patch so we don't sink," he said.

There was no way his boat would sink because of the miniscule whiff of air escaping from the side of the patch. But over time, with no intervention, the amount of air leaking from under the patch might grow. As the peon of the Sobek crew, it wasn't my place to debate the plan. Naturally, the tedious task of applying pressure on the patch got delegated to me. In a rookie attempt at creativity, I asked Richard, "Can't we just cover the patch with some duct tape?" That alternative might free me from being anchored to that one spot all day.

"Go for it if you want, but I doubt it'll work."

I found a roll of gray duct tape and carefully layered several strips around the four edges of the patch. I felt confident this solution would work. Once we launched, I kept an eye on my ad hoc improvement, smug in the belief the duct tape had stopped the leak. Less than an hour later, the subtle hiss returned.

"Damn, it's leaking again!" How foolhardy had I been to think duct tape could cure a leak that superglue failed to stop?

To his credit, Richard merely shrugged his shoulders. Mercifully, he didn't say aloud, 'I told you so.' Now I was stuck applying pressure with the heel of my hand to the edge of the patch where the air was leaking. It didn't take long before my hand developed a cramp.

"Tell me if you see any hippos under the water up ahead," said Richard. After the excitement of the previous day, I could understand why he wanted to give a wider berth to any visible river dwellers.

"Don't worry. I will," I answered. I too wished to avoid a reenactment of the previous day's encounter. "But it's pretty hard to see much through the silt in this water."

"Watch for bubbles breaking through the surface."

We both knew at any moment our raft might pass over hippos lazing on the river's bottom, once again provoking a dramatic turn of events. Richard had scrounged a couple of spare paddles from the other rafts, so oar boat riders could lend some extra paddle power to keep pace with the faster-moving paddleboats. To give my cramped hand a rest, I gripped a paddle and found a rhythm: dipping it in the water, pulling back, lifting it up, bending forward, dipping, and pulling back. Repeating the rote motion allowed my mind to wander.

The oppressively steamy weather and the consistency of the river's landscape day after day contributed to a feeling of déjà vu. Dense foliage lined the riverbanks; behind it, savannah grasslands stretched forever. An endless number of hippos and crocodiles remained our closest companions. In the distance, elephants trumpeted.

Gigantic white puffy clouds filled the otherwise azure blue sky. Their shapes morphed from the silhouette of a raven to the shape of a turtle to an outline of a sailboat. Mesmerized by their cotton-candy appearance, I imagined what fun it would be to stretch out on top of one of those fluffy cotton-ball clouds. Drifting along with a bird's eye view, I could look down on the elephants, waterbuck, and gazelle grazing in the grasslands, hidden out of sight beyond the forested banks of the meandering ribbon of water below.

My initial oversized fear of being in proximity with these "killer" animals shrunk a bit each day as I realized their sole purpose in life was not to pounce on humans and bite off their heads. I welcomed the relief this new insight offered. Each day, the odds grew better that I'd return home with all my body parts intact. Still, it would be foolish to let my guard down the farther we intruded, uninvited, through their homeland.

With fear no longer consuming my every waking second, my focus turned to the intricate details of these dazzling surroundings. In the complete absence of manmade background noise, I joyfully listened to the astonishing clarity of nature's enchanting sounds: the gentle fluttering of warm breezes tickling leaves on tree branches overhead; melodic riffs of unfamiliar birds, each chirping unique refrains; the tapping of tiny wavelets drumming against the raft's tubes during a gentle drop in river elevation. How ironic it was to feel such serenity and a deepening connection to a place I had so feared and wanted nothing to do with only days before.

With nothing to hinder our passage and no unexpected surprises, my uneasiness from the morning proved warrantless as we pulled into camp that afternoon. Even the leaking tube hadn't slowed our progress, though it had become softer as the internal pressure dropped. The day had been anticlimactic in a stress-free way for the first time on our trip.

Michael and Bart hauling the oar boat out of the water to repair the stubborn leak ranked as *the* memorable event of the day. Together, they worked to peel off the failed makeshift patch, then prepped the torn tubing for a proper repair. This time, they attached a new industrial-strength patch.

"You really didn't expect that first patch to hold, did you?" Michael kidded Bart.

"Ya know, this boat's so old it's hardly worth the effort to repair it. Look at how many other patches are on her," Bart replied. "She's been a workhorse for sure. Been on all of our Africa exploratory runs. Fended off more than her share of croc and hippo attacks."

"Well, this new patch ought to do the trick. In fact," said Michael, "I'd be willing to bet the original tubes will disintegrate before this new patch even shows any sign of distress."

"No argument there, bro."

19

RAPIDS AND RAIN

AT BREAKFAST THE NEXT MORNING, Conrad told us we'd reach the start of the first of our two portages sometime that afternoon. "I'm not sure how long it'll take us to reach the upper falls, so everyone needs to be listening for anything that sounds like water cascading over a waterfall. Slade and Michael are tracking where we are on the topo maps, and it looks like we still have a way to go. But the last thing we want is to have a raft drop into that crazy section of the river."

He detailed his plan for avoiding this perilous obstacle.

"Once we get close to the top of the falls, we'll be leaving the river. We'll pack up all the boats and equipment and carry them overland by foot until we reach calm water below. Everyone's going to help by carrying your own personal gear."

I'd guided on a few rivers where a small dam or natural obstacles halted our forward progress. These required a short portage of usually less than 100 yards over well-worn paths created to ease the job of transporting gear. Even then, it was time-consuming grunt work. The exponentially greater difficulty of the portages on this trip would test both our physical and mental strength.

Late that afternoon, we finally heard the distinctive gurgling and splashing of the river pouring over the upper cascades of Shiguri Falls, a series of nonstop short but perilous drops that stretched on for half a mile. Quickly, attention turned to locating a safe place to pull off the river. Failure wasn't an option. The idea of a raft dropping into the foaming froth, flipping and dumping its occupants to swim through these jagged rocky rapids, made my heart race. The simple odds were heavily against a swimmer surviving. Anything that made it through the upper cascades would be battered and broken. Swept along by the strong current for two-and-a-half more miles, it would then plummet nearly 200 feet over Shiguri's main waterfall.

Almost as soon as we began searching the shoreline, a landing place appeared just ahead on the left. As our raft touched land, I leaped ashore with the bowline, grateful that the upper falls no longer posed a death risk. With only a few hours of daylight remaining, the guides decided to portage the gear we needed to establish a campsite below the upper falls before dark. I helped unload the rafts and gather the essential items we'd use that night. Those items would go with us on the first trip.

"Look, it won't be possible to portage everything tonight," Conrad said to the group. "We'll secure the rafts up here and come back for them in the morning. I need everyone to carry their own personal gear and a piece or two of kitchen gear we've laid out over here."

He showed how to string four or five dry bags along the length of an oar through their buckled straps, with enough space left at each end to hang a couple of baskets or buckets. A loaded oar weighed between 100 and 120 pounds. By hoisting each end onto the shoulders of two oar carriers, the weight became evenly shared. They could carry much more gear suspended safely above the ground than either could lug by themselves. This freed up other passengers to carry tents, sacks of food, cooking pots and pans, and lanterns. The crew hefted the heavier items crucial for setting up the campsite.

"Drink some water now before we start," said Conrad. "The game rangers will be up front using their machetes to clear some of these vines and other crap out of the way. It'll be a crude pathway, at best. Keep your eyes on the ground where you're walking so you don't trip over any roots. There will be plenty of them. Take your time. Holler if you have any problem or need a break."

He whistled to signal the start of the trek. Bearing our loads, we headed single file into the dense growth for the half-mile hike following the freshly carved path. Sweat streaked down the rangers' faces as they worked nonstop slashing and hacking at the undergrowth that blocked our way. We progressed at the speed of a leisurely walk, based on the time it took for the three rangers to clear the path.

The shoreline vegetation choked off the breezes that drifted upstream over the open river, leaving us to struggle with stifling heat and humidity. An earthy dampness drenched my clothes. Every so often we glimpsed frothing white water pouring over the cascades, plunging into eddy pools below, collecting for a moment, then surging over the next drop in its path. If anyone questioned the necessity of this tedious grind overland as opposed to rafting this section of river, this visual evidence of the river's chaotic behavior supplied an undeniable answer. The downhill hike took a grueling hour to complete.

At last, we arrived at a pleasant patch of sandy beach. Our relieved passengers began setting up tents, while the crew members huddled and came up with a plan. I would stay in the camp with the passengers and one of the game rangers while the rest of the crew and the other two rangers used the remaining daylight to make a quick trip back to the head of the falls. They'd retrieve the rest of the supplies and food we'd need that evening and double-check the ropes securing the rafts to sturdy trees and each other.

Slade led the crew back up the path at a brisk trot. While they carried out their retrieval mission, I assembled the parts of the kitchen we carried with us on the first leg of the portage. Our passengers divided up, some scavenging to collect firewood while two dug a firepit and rimmed it with stones. Having a cleared trail to use cut the hiking time of the second trip nearly in half. When the crew returned a little more than an hour later, a fire roared in the pit, and I'd assembled a rustic-style charcuterie platter of cheese, nuts, dried fruits, and cracker snacks for all to enjoy along with our Kool-Aid cocktails.

After we finished with the cleanup from dinner, Conrad previewed what still lay ahead.

"If any of you found today's 'short' portage difficult, let me warn you—this was just a dress rehearsal for our next much longer and daunting portage around the major falls. That one will tax your physical and mental limits. You've got to be prepared for this challenge. Get as much rest as you can and stay hydrated."

It didn't take long for the group to agree on the merits of an early bedtime, both to restore our ebbing energy and to heal our sore muscles. Just before succumbing to slumber, I realized Conrad had been wise to set some expectations with the group about how difficult the coming days would be. I also wondered how many of our passengers would have opted to do this trip if they had truly known the strength and endurance it required. I even wondered if *I* had the physical and mental fortitude to complete the long portage. I'd soon find out.

RAIN ARRIVED in the middle of the night. The downpour drumming on the sides of my tent vaguely annoyed me. Even in a groggy state of semi-consciousness, I knew if the weather didn't clear, we'd have to finish our first portage and pack the rafts in the morning with all of our gear still sopping wet. Fortunately, sleep drew me back to dreamland before I could fathom whether it was really raining or just raining in my dream.

It was still dark when the grumbling of a couple of guides, waiting for the coffee water to boil, woke me. Raindrops continued to pound on the walls of my tent. Ugh. I always dreaded waking up to a gloomy rainy day, even from the comfort of a soft dry bed. But this was a disaster. During the night, rainwater seeping under my tent had soaked through the floor.

Down by my feet, where my sleeping bag extended beyond the end of my foam sleeping pad, my toes squished into the wetness that saturated the end of my bag. Fuck! Not just my sleeping bag, but everything touching the floor or side of the tent was soaked. This was *not* an auspicious start to the day. The thought of wringing out sopping wet items, scrunching them into a plastic garbage bag, and then into my dry bag, made me want to weep. But I didn't have a choice. Richard's advice that the most-hated moments of a trip disappear once the trip was over had better be true because right then I was suffering through my worst moment.

I crawled out of my waterlogged shelter to join the others under the steel-gray, moisture-laden sky as they discussed the plan for the day. I hated the thought of plodding up the trail in the pouring rain to help retrieve the boats. Not being as physically strong as the male crew members, they left me behind to prepare a simple hot breakfast for the group. What a relief! After caffeinating themselves with a few cups of steaming coffee, the men of the crew and two rangers set off to complete the portage of the four rafts left upstream overnight.

The nighttime deluge had turned the soil into slippery mud, far from the ideal surface over which to be hauling 100-pound rafts. The boats needed to be deflated, flattened, folded in thirds, and each rolled up tightly around the shaft of a ten-foot wooden oar. Ropes cinched tightly around each rolled-up boat anchored it to the oar, guaranteeing it would stay in place. A crew member at each end of an oar would then hoist it onto their shoulders for the trek.

A few of the passengers helped me rustle up a large pot of steaming oatmeal mixed with chopped dried fruit and nuts, sugar, cinnamon, and a few tropical raindrops for a truly exotic flavor. When the boat-carrying caravan returned, thick brown mud smeared their bodies, clothing, and the surfaces of everything they brought with them. As the muddied crew members gobbled their bowls of hot oatmeal, nature's shower of warm rain trickled down over them, rinsing away most of the muddy grime.

Meanwhile, we started the nasty task of breaking camp in the rain. Taking down tents soggy with mud and debris-coated floor bottoms and stuffing them into their tent bags created the optimum environment for foul, white mildew to coat everything. Slogging through the morning chores in the rain and mud sucked. Big time, with a capital S, *Sucked*!

I tried awfully hard not to be grumpy, but if I'd had a time travel machine, I would have hit the return button and transported myself back home in a heartbeat. Truth be told, I suspected many of my fellow rafters would have opted to join me. By the time we had the rafts loaded and ready to shove off, my waterlogged hands looked shriveled the way an Olympic swimmer's skin wrinkles after they've practiced for long hours in the water.

20

THE SHOT

AS WE FLOATED DOWN RIVER that morning, my gloomy mood matched that of the moisture-laden gray clouds drenching us with their tears. At least the pelting rain obscured the sun, so we didn't have to worry about burning to a crisp. A persistent clamminess filled the gaps between raindrops. Nothing would dry until the rain stopped and the sun's rays reappeared.

I took a place in Conrad's boat to allow a couple of soaked-to-the bone passengers, dispirited about paddling in the rain, to hunker down in Richard's boat. I didn't mind paddling in the rain. It reminded me of getting drenched during spirited water fights between rival rafts on Stanislaus River trips back home. While paddle blades skimming the river's surface could throw up a frothy swath of spray, the best weapon in our armory was a bailing bucket half-filled with river water flung at point-blank range over our adversaries. On a sunny summer's day, the drenching felt refreshing, but this was not a sunny summer day.

Over the next half hour, the sky to the east brightened. Finally, the rain petered out. Perhaps as a concession for the 12-hour soaking we'd endured, Sobek, our guardian god, had commanded the sun to reward our grit by gracing us with its brilliant rays. Given the uncertainty of what lay ahead, we all needed this sign of hope. Just maybe, we wouldn't be doomed to wear wet clothes for the rest of the trip. My spirits soared. If we made camp early today, it would give our waterlogged possessions a chance to dry out before the sun set.

A short whistle from Bart's lead boat interrupted our celebration of the sun's return. In hushed silence, his paddlers gazed intently toward the six-foot-high steeply inclined river bank on our right.

"*Simba!*" Magigi whispered.

Sprawled across a wide-open stretch of sand, not even 10 feet from the edge of the river, lay a stunning, tawny lioness. We stopped paddling and drifted in complete silence, hoping not to spook her. She raised her head and struck the same pose my female house cats assumed when lounging contentedly on my sofa cushions. She displayed no sign of fear, while my body involuntarily prepared for an impending fight-or-flight.

I couldn't decide whether it petrified me to approach this frigging close to a real lioness or if it thrilled me to have the chance to photograph the encounter. Awestruck, my trembling fingers fumbled to get my camera out of its case. She remained motionless as each one of our rafts floated by. With a regal aloofness, she observed her audience, accepting the homage we bestowed upon her. With the majesty of a benevolent queen, it seemed the lioness demanded the devotion of all creatures in the jungle kingdom she ruled. Or was she so dumbstruck by our curious flotilla passing by that she dared not to move?

It blew my mind to be so close to a wild lion that I could look directly into her gleaming amber eyes. For a moment, I fantasized about stroking this lioness's sleek, fur-covered, 250-pound muscular body. At least 25 times larger than my cat back in California, I couldn't help but think this lioness, too, might enjoy a tickle behind her ears. The unfamiliar clicking of cameras frantically capturing shot after shot of her majesty must have puzzled those sensitive ears.

Once we moved beyond where our voices might scare her off, Conrad and Magigi filled us in on some lion facts we ought to know, given the undeniable proof we'd invaded their backyard.

"We probably woke her up from a nap. Lions love to sleep whenever they aren't hunting, even up to 22 hours a day. They hunt mostly at night with lionesses doing 90 percent of the hunting," Conrad explained.

I couldn't resist pointing out, "So that common adage about 'doing a lion's share of the work' is complete bunk. It should be 'doing the lioness's share of the work.'"

Conrad winced at the correction. "You best not be wandering in the grasslands at night because their eyesight is six times better than a human's vision at night, and they can hear prey up to a mile away. They have this funny little organ on the roof of their mouth that allows them to taste the air for nearby food. Once they've caught a scent, they like to sneak up on prey before the chase starts. With a short distance sprinting speed of 50 miles per hour, they can easily take down any animal that gets separated from its herd."

"She didn't even growl at us. That surprised me," said Jay.

"It's still early in the day. She might have been sleeping off a full stomach from the meal she ate last night. Lions don't kill just for the sport of killing. They hunt when they're hungry or if they feel threatened, but there aren't many animals that pose a threat to a lion out here."

Not quite convinced, Martha pressed further. "You don't think she saw us as a threat, even though we passed so close to her? And it's a little late to be asking this now, but can lions swim?"

Conrad laughed. "Let me put it this way. If Mark Spitz and a lion went head-to-head in a swimming race, I'd bet on the lion, particularly given that a lion can leap as far as 36 feet."

"That's a hell of an advantage off the starting platform," said Jay. "Christ, that means she could have leaped right into our raft if she'd wanted to. Thank God she wasn't hungry."

I agreed with his unnerving assessment. "Thank God she held her serene rock star pose while our 17 cameras madly clicked away."

BY LATE MORNING, the current was sweeping us downstream to a rendezvous with the biggest obstacle we'd encounter on this trip. The massive Shiguri waterfall would thwart our passage just a few miles ahead. Gauging the speed of the river's current, we calculated we could reach the waterfall in an hour. But as we proceeded, the river banks strayed apart, forcing the water to spread across a broader channel, slackening its speed.

Conrad explained, "We've got to do the portage along the left-hand shore of the Kilombero because another river, the Luwegu, dumps into this river from the right just a short distance above the falls. At that point, the volume of water instantly doubles and speeds up before cascading over a ledge a quarter of a mile wide."

I imagined the swift current dragging us toward the deep abyss where our little gray raft would teeter before being sucked over the edge, like a strand of spaghetti into the mouth of a ravenous Italian. I tried to block that thought, waiting to hear how Conrad expected us to survive.

"So any attempt to portage along the right-hand side of the falls would be ludicrous. There's not much distance left between us and the lip of the falls. One thing I can guarantee is that no one would survive that drop. It would be instant death."

The image in my mind grew more vivid by the second. I pictured our raft plunging airborne over the falls for the final few seconds of our lives. We'd crash onto submerged shattered rock slabs fallen from the face of the ledge long ago. The crocodiles trolling the swirling pools of water below would devour our remains in a frenzy.

To avoid being sucked into the more powerful currents sweeping straight toward the falls, Conrad's strategy relied on navigating through shallower water along the left shore.

"Isn't that risky if we get too close to a submerged hippo?" I asked.

"Maybe. But we've got a more immediate problem to deal with. A couple miles above Shiguri, the river gets clogged by a labyrinth of braided channels surrounding a hundred densely foliated islands. Like randomly scattered puzzle pieces, they dot the river right up to the drop. Every choice we make in this maze is going to be critical. We've got to find a safe landing space on the left-hand shore above the falls."

Not long after Conrad's "pep talk," the river began splitting into competing channels. Conrad signaled the other guides to gather their rafts in a huddle, where he stressed the rafts had to stay close together while we searched for an opening along the shoreline. We proceeded cautiously as the guides deliberately chose the channels they calculated would keep us close to the left bank.

Further complicating matters, thick bushes and tangled vines blocked any view of what lay beyond the rocky islets. The river had split and braided so many times in the past half hour we'd lost track of how close we were to the left bank. Each time we arrived at a fork in the current, where the left channel proved too narrow or shallow, our anxieties ratcheted up a notch as the river forced us to divert to the right, away from our left bank escape route.

"Stay to the left!" Magigi ordered, each time an opportunity to shift to the left presented itself. We hardly needed his guidance, since the consequence of not finding the solid left shore in time was obvious. The approaching falls seemed to frighten Magigi a hundred times more than facing a pride of starving lions, not that either scenario would end well.

Everyone listened for the faint sound of cascading water, a telltale sign we might be nearing the falls. Only a few feet separated the bows from the sterns of our rafts. Bunching together in this manner guaranteed that no raft would accidentally stray into a different channel.

We floated along one stretch where the deeper channels of three forks in a row veered to the left. Shifting that far to the left over such a short distance improved my outlook, but still I worried how many more forks to the left we'd need before we found the true left shore. I gained some comfort from silently repeating Magigi's mantra "stay to the left" in my mind.

Time dragged on, while tensely we paddled in a current that had no forks splitting the flow. Conrad cautiously speculated we were running parallel with the river's true left bank. Even the vegetation on our left wasn't as tangled and dense as the vegetation on the rocky island to our right.

Given how much time we'd been on the water since leaving our camp that morning, the falls had to be close now. The guides agreed to pull over at the next spot on the shore large enough to serve as a potential campsite. From there, a scouting party would go on foot along the river to determine how close we'd landed to Shiguri Falls.

Ahead, our channel split again. Our obvious choice would be to stay to the left if possible, clinging to the solid shore line. The narrower left fork, only 20 feet wide, looked deep enough to accommodate our seven-foot-wide rafts. From upstream, we could see this channel swerved slightly to the left, funneling into a chute that gently dropped two feet. A clump of thick brush overhung the river at the top of the chute, obscuring the shoreline just beyond it.

Slade's boat, in the lead, slid into the chute as his crew stroked in unison. Our three other rafts hung back while he checked the route for any snags or shallow spots that weren't obvious from upstream. We watched their perfect run straight down the chute. It was clear of obstacles.

Abruptly, Slade stood and pivoted to face us. Staying calm and in control, he yelled back to us, "Hey, guys, hold up a minute. There's a huge hippo standing on the left bank just behind those bushes." Passing so close to the river titan left Slade's paddlers wide-eyed with astonishment. One of them later swore he'd felt a warm draft of its exhaled breath on his face.

Quickly, I did the math. A huge hippo was likely 10 feet long. If it charged directly into the water, the gap between the hippo's mouth and the near side of a raft would be just three feet. Only if the hippo suddenly halted its charge, while we paddled like hell, would we stand a chance of sneaking by. But we'd already witnessed that a charging hippo doesn't stop.

From a safe distance downstream, Slade observed the hippo, while upstream the guides strategized our next move. What a paradox! Our greatest fear had shifted in a heartbeat from the possibility of our rafts plunging over Shiguri Falls to the real prospect of this hippo plunging into the water to block our passage, or worse. Slade reported the hippo didn't seem agitated, but its puzzled stare remained laser focused on his lingering raft.

Conrad, Richard, and Bart hashed over whether Slade's boat might be enough of a distraction to allow a second raft to quickly dash past the beast. They debated whether paddlers could generate enough speed to elude the hippo's attention until the raft was past its striking zone. Even if this high-risk maneuver worked, it would still leave two rafts stranded upstream.

The guides brainstormed alternative options. Where we waited, at the top of the rocky island that formed one side of the channel, we still had the chance to divert our three rafts to the channel running along the right side of the island. Whether we could get back to the left shore before the falls became the critical question. No one favored taking that risk. Or we could engage the hippo in a standoff. Waiting for it to abandon its position on the riverbank would require Slade's boat to stay within sight of the hippo, possibly provoking the animal to attack the boat loitering in its territory.

Both game rangers with rifles, in the rafts upstream, had no clear shot to take the animal down. But unless an animal directly attacked us, none of us wanted an animal to be killed. We might need every one of those precious bullets in the days ahead. Slade yelled up to us again, "The thing hasn't budged, but neither is it showing any signs of aggression."

Since the hippo still appeared to be in a docile mood, Michael, our wildlife expert, proposed sending a second paddle raft down the channel. Richard's and Conrad's boats with the armed game rangers would go last in case further close encounters with our rafts sparked hostility by the hippo.

Michael pumped up the paddlers in Bart's raft with practical advice. "You guys need to hug the far-right side of the channel. Keep the maximum space between you and the hippo. Paddle like hell, but stay calm and don't splash. You've gotta be really quiet 'cause you've got to get as far down that chute as you can before it realizes you're in striking distance."

Bart's paddlers readied for their evasive run, and when Slade waved for them to proceed, all paddles dipped in unison. Down the chute they went at full speed. Bart riveted his attention on the hippo, which, upon seeing another object hurtling past, shifted its focus to the closer raft. When they were directly in front of the puzzled hippo, Bart, struggling to keep his voice calm, urged, "Paddle like your life depends on it." The enormous head swiveled suspiciously, following the movement of Bart's boat. When they reached the spot where Slade's boat waited, Bart's paddlers rejoiced with silent high fives.

Now that two rafts had successfully dodged the befuddled hippo, Conrad argued with contrived confidence that his boat should proceed next. Though he never voiced his rationale for wanting to go third, I believed he figured our chances of surviving were better if we ran the gauntlet before the ongoing invasion incensed the hippo to attack. My mind fixated on the theoretical three-foot gap, silently protesting the obvious, "This is frigging insane." Magigi loaded his rifle with one of the precious bullets stashed in his pocket.

Conrad reiterated how critical it was that we paddle in unison and paddle hard. I heard him say, "Let's go!" Somehow, the paralysis in my brain magically failed to affect my arms as I leaned forward and dug deeply into the water with my paddle. Our crew paddled with a single-minded goal, as if our raft was a scull leading by half a boat length, on its way to winning the ultra-prestigious Head of the Charles River Regatta Race. As we entered the hippo's field of vision, its gaze refocused directly on us. We weren't supposed to be in its territory. Never had it laid eyes on the oddity of these fully loaded inflatable rafts.

Docile? No way. The fire in this beast's eyes reminded me of an enraged bull, provoked by the swirling red cape of a matador moments before charging to its death. The hippo's muscles flexed with tension; its ears twitched angrily as just 15 feet separated us from the resentful monster. Each millisecond lasted forever as we watched for any sign of defiance. Each moment of the beast's inaction carried our raft farther from its reach. It snorted, groaned, and pawed at the ground but did not advance. It let us escape. I wanted to shriek for joy. Instead, I stifled my exuberance lest it trigger the hippo to attack.

Only Richard's boat remained upstream. On board with him were Michael, two passengers, and Masengela with his loaded rifle sitting high atop his perch in the rear. Slade relayed to Richard that the hippo had become more agitated but still held its ground on shore. We all prayed it would just stay put. Wasting no time, Richard pushed forward on the oars before the creature decided to put an end to this intrusion. The appearance of a fourth raft even larger than the others must have tipped the scales for the hippo. Its patience tested beyond a breaking point, seconds after the oar boat passed, it flung itself into the water with a tremendous splash. Richard rowed furiously, gaining a slight lead.

The hippo surfaced in shoulder-deep water. It bounded twice in chase. As its jaw started to open, the ear-splitting crack of a rifle firing echoed through the air. Stunned, the hippo stopped, still focused on its prey. Maroon liquid spurted from a circular splotch on its head between its eyes. Shocked silence ensued as everyone in our group struggled to comprehend what had happened. The dazed hippo stood motionless, watching our rafts float away. After a minute, it turned upstream and dove under the water.

21

THE UNANSWERED QUESTION

WE HAD NO WAY OF KNOWING whether Masengela's bullet had killed or just wounded the charging hippo. Either way, the guides wanted to put some distance between the place where the hippo disappeared into the water and our next campsite. No one wanted to be the target of a wounded hippo's revenge. But we also knew that soon we'd run out of shoreline.

The anticipation of the immense watery abyss not far ahead, eager to swallow our rafts, left everyone edgy. Trepidation caused my body to tremble and electrified my nerves. Knowing our lives depended on getting off the river upstream of it, we floated on in silence, each one of us hypervigilant for the tiniest sight or sound of the treacherous hazard.

Even Conrad, the only one who had previously navigated around these perilous falls, seemed nervous. I watched him anxiously searching the left shoreline for a place the rafts could land. The tangle of jungle greenery claimed every inch of land right to the water's edge. Our only option was to keep going.

If we couldn't find a clearing on the shore, we'd have no choice but to choose a spot to hack away at the bushes and vines to make our own clearing. With only four machetes, the backbreaking work to clear a space large enough for our rafts, our tents, and a kitchen could take hours and still leave us with ground so rough it would be impossible to stake a tent on it.

Another five minutes of sweating bullets ticked off. Then, miraculously, a small opening along the shore appeared. Conrad whistled and beckoned for the rafts to pull over at that spot, cramped for a campsite, but the space would suffice. I finally exhaled a huge breath I felt I'd been holding for the last hour. We were safe for now. But the adrenaline that had pumped through my body combined with the emotional toll of the day left my body feeling wasted.

Slade and Michael pored over the topo map, trying to determine the actual distance that remained to reach Shiguri Falls. Their triangulations showed the 200-foot drop to be within a half mile downstream. To validate their calculation, Conrad and Richard set off on foot in search of the falls, accompanied by two of the game rangers who slashed away at liana vines and dense brush to clear a narrow pathway.

Less than an hour passed before the scouting party returned. They reported they didn't actually see the falls, but they could hear the deafening roar of the cascading water. Rather than taking the chance of finding a camp closer to the precipice, we would stay put and set up camp right where we'd come ashore.

From the brink of the falls, the topo map showed the river's edge turned to steep canyon walls for a distance of two or two and a half miles until the height of the canyon gradually tapered back down to river level. A grueling downhill trudge would be required to reach a spot where we could lower the boats back into the water. The plunge of the Luwegu and Kilombero currents over Shiguri Falls marked the end of those two rivers and the start of the Rufiji River.

Conrad figured this second portage would take at least two days, starting with the onerous task of clearing an additional two miles of track ahead of those hefting the heavy loads. He estimated we would need a minimum of three trips to haul all of our equipment. It would take two trips just to transport the four rafts. For this longer, more strenuous portage, we'd move the boats and heavier equipment to the lower camp first.

I had the strength to do a two-person heave of our deflated and rolled-up 100-pound rafts into the back of the equipment truck at the beginning and end of every California river trip. I could also man-handle our largest food-filled coolers into proper position in the raft. But carrying this bulky equipment over two miles of rugged terrain required a different sort of strength and endurance, the kind I feared I might not have.

My outlook rallied when I learned I wouldn't be part of the crew assigned to portage the first load that afternoon. Bart and I stayed behind with our clients, got our campsite set up, and prepared dinner, while the rest of our physically fit, heavily muscled Sobek colleagues tackled the heavy lifting. I felt no guilt dodging the arduous first leg of the portage. The heavy lifters were all getting paid for their work while I was working for free.

Magigi leveraged his seniority, dispatching the other two rangers to assist with the initial portage while he remained with us. Several passengers weary of watching Masengela's repeated exuberant reenactments of shooting the hippo expressed their displeasure with his antics. Magigi understood that removing his adrenaline-wired subordinate from camp for a while might be prudent. A trek through the jungle might be the perfect solution to temper his euphoria.

The crew and rangers loaded up for their first trek. Rather than heft the weighty rafts on backs and shoulders for such a long distance, Conrad had brainstormed a more efficient method for transporting the rafts. He commissioned a hardware shop in Dar to construct two inexpensive carriers using spare parts. Each contraption consisted of a large metal basket mounted on top of a bicycle wheel's frame.

The vital question was whether the raft carts would be a help or a hindrance. First, each raft had to be wrapped securely around an oar before being placed into the baskets. The guides would use both protruding ends of the oar to steer the cart by pulling from in front or pushing from behind. The last step entailed lashing each raft tightly into the carrier's basket so the guides could use the ends of the oars to lift the cart over roots, fallen trees, or large stones along the route.

It looked awkward, but in theory the crew would exert less energy rolling the rafts with the help of the wheels. They also lugged as much of the heavier equipment as they could manage. Importantly, they carried an extra jug of water to replenish their canteens along the way.

The two rangers who accompanied the crew led the way with slashing machetes mercilessly hacking at vegetation blocking the way. Since no one knew how long it would take them to reach a viable campsite downriver, the crew carried a couple of tents and a small supply of canned food in the event darkness delayed their return trip. As they disappeared behind the green and earth-colored woven drapery of vegetation, I silently wished them a safe trip and prayed we'd all be back together soon.

My job as Bart's helper and hostess to this eclectic group of travelers offered me the chance to show off my cooking skills while mingling with our clients. Each had coughed up nearly $4,000 for the privilege of roughing it with hippos and bushwhacking miles through the jungle lugging their personal gear. The shared extraordinary experiences of the last few days had bonded us as a group. The obstacles we faced forced each of us to dig deep for strength and the willpower to keep going. I hadn't heard Martha complaining once in the last couple of days.

Before this trip, I never imagined such a world existed. Each day away from "civilization," the strangeness of my surroundings and each spontaneous moment of discovery steadily chiseled a new image of the world for me. This remote piece of the planet still existed exactly as it had for hundreds of thousands of years, untainted by human exploitation. Passing through this remnant of nature's original creation astounded me.

I delighted in the honor of sharing this space with exotic wild animals I'd only seen in pictures. To watch them roam wild and free filled me with a commingled sense of joy and fear. In the truest sense, for the first time in *my* life, I, myself, experienced the total freedom in which these animals lived. Here, I'd escaped from all the duties, responsibilities, obligations, commitments and promises that overwhelm a life in the civilized world.

In this place, I found moments where I could simply stand still and let my senses grasp the novelty of this habitat: the earthy smell of the jungle, the strange expressive sounds of its wildlife, the enticing beauty of its untouched natural essence. With survival being my only challenge, the Selous became my teacher. It taught me about nature's delicate balance and the fragility of every living thing.

While I was no longer *terrified* of the African wilderness, I remained apprehensive of all the dangers that surrounded us. But I appreciated the rare privilege of each day I spent in this game reserve. To be one of a miniscule number of humans to ever pass through the heart of this pristine paradise was a precious gift. I finally understood why Richard had been so insistent that I come on this trip. Gradually, gratefulness replaced the dread that had consumed me at the start of this uncharted exotic journey.

Despite the unsettling incident with the charging hippo, we'd avoided plunging over the falls, and I took comfort knowing we'd soon be past the toughest part of the trip. Since the crew had not returned by dinnertime, Bart confidently assured us they must have camped downstream for the night rather than risk trekking back in darkness when the alpha predators would be hunting for their nightly meal.

Conversation during dinner was noticeably subdued, as we missed the jovial bantering of our absent crew members. As darkness enveloped us, we gathered close around the dancing flames of the fire. Heat from the ruby coals on the floor of the firepit radiated outward, raising the temperature of the still-sultry evening air. The glowing light they cast formed a mantle of safety against the invisible creatures that watched us from the black depths beyond our fireside circle.

Our discussion meandered around to the one troublesome question on everyone's mind that night. The animal-loving Kate posed it aloud to the group.

"Do you think Masengela killed that hippo or just injured it? I just don't understand. It stared at us for so long, then turned around and dove under the water. Could it still be alive?"

One of the older male passengers echoed her dilemma. "I'm wondering the same thing. Maybe it didn't die right then, but ultimately, the shot might have been fatal."

Heads nodded in shared concern as another passenger steered us to the elemental question that would haunt us forever. "I wonder if we could have escaped from the hippo without the ranger needing to shoot it. I wish he'd waited just a few seconds longer to see if we could have gotten away."

Either Magigi knew enough words of English or he could sense from our conversation we were reliving the unforgettable event we witnessed that day. He shook his head slowly from side to side, though it wasn't clear if he knew whether the shot had killed or just injured the animal. His wrinkled brow and the despondent look on his face made me wonder if he regretted his subordinate taking the shot.

Bart voiced a more depressing thought. "I just hope it wasn't a mother protecting a baby hidden in that bush behind her."

Being so absorbed by the drama of what the huge hippo would do next, none of us would have spotted a baby tucked out of sight behind its massive body, but one explanation for its reluctance to charge sooner might have been leaving its offspring unprotected on the riverbank.

"Oh, dear, I so hope we didn't orphan a baby hippo by killing its mother," Kate cried, shocked by Bart's hindsight possibility.

An undeniable emotional bond linked us forever by the tragedy we witnessed that day. Each of us bore an invisible smudge of shame for what Masengela had done, not that any of us could have prevented his impulsive act. To move on without certainty about the hippo's fate left us uneasy. With heavy hearts, we rose from our seats by the fire to refill canteens before heading off to our tents.

From upstream, a prolonged wail of horrendous agony shattered the night air. Earsplitting in intensity, the final exhaled vocalization exploded with despair, suffering, and ultimate surrender before death snatched the injured creature from this world. We stood in shocked silence, no longer needing to wonder about the fate of the unlucky hippo whose path we'd crossed.

CONTRADICTORY THOUGHTS DUELING inside my brain foiled the sleep I struggled to find. If they had their way, hunters and poachers would annihilate every living creature in this reserve, robbing them of tusks and horns, meat, and skins. Sobek had obtained a permit to travel in this vast reserve based on the premise we would not intentionally harm any of its wildlife. Yet today our expedition killed one of these animals in its home territory.

The hippo had charged into the river behind the last boat, but only after being provoked by the intrusion of our four rafts on its turf. Did it really need to be shot? Had any of us been in mortal danger? The hippo that days ago bit Richard's boat had done more damage than the hippo today, but Masengela had not fired a shot at that one.

If it had been up to Richard or Conrad or Slade, would they have given the kill order? Actually, I didn't remember anyone yelling to shoot the hippo. Masengela just leveled his rifle and took the shot. Were we any better than the poachers? My inability to answer any of these questions left me frustrated and angry. I was furious with Masengela's elation at his kill. How could he consider himself a protector of the Selous wildlife if killing animals gave him such joy? That night, I prayed for God to protect these magnificent animals if human beings wouldn't.

22

THE LONG PORTAGE

BART THUMPED THE SIDE of my tent with his open palm as the first rays of light brightened the morning sky.

"Hey, it's time to get to work." Bart's wakeup call was a rude intrusion into the pleasant dream world I'd entered to escape the traumatic events of the previous day. I wasn't ready to leave this new world yet.

"I'll be there in a minute." I stretched out my arms. A minute later, I pulled myself into a sitting position.

I chose the word "efficiency" as my mantra for the day. For the portage to run smoothly, it would take all hands to break down this camp, divide up gear to carry, trek for hours, and reassemble camp below the falls. We faced a full day of tedious, energy-sapping, willpower-testing work. Bart and I hustled to get breakfast prepared and served early so we could pack the kitchen to be ready for transport as soon as the rest of our crew returned.

Just as we began serving pancakes with syrup and jam, leaves rustled and branches snapped in the dense foliage just beyond our camp. Everyone instantly froze as the sounds grew louder. I held my breath, muscles tensed, frantically wondering what I'd do if a wild herd invaded our camp. A minute later, the noisemakers tromped into our kitchen.

Richard led the troop, in a sweat-drenched Sobek T-shirt and a navy blue bandana folded across his forehead to hold back salty sunscreen from dripping into his eyes. The rest of the crew trailed in single file behind him. All they carried were the two empty basket-bike-wheel contraptions, so they had jogged a good part of the way back.

"I told these guys that if we hustled, we might make it back in time for breakfast." Richard grinned, never one to miss out on a homemade cooked meal.

As our group devoured pancakes and steaming cups of jungle joe, Conrad told us what to expect. "The first half mile or so is really rough. We figured it might be easier going if we could break out of the damn tangle of foliage by going inland to open savannah. We had to climb uphill over some seriously rutted terrain, but it was worth it."

Antsy to know more, Bart questioned our crewmates about logistics. "So, will it take one or two more days to complete the portage? I assume you found a place to camp below? How long did the first leg take?"

Slade, with his signature dry wit, responded, "All depends."

Bart's impatience flared. "What's that supposed to mean?"

"All depends on if we run into elephants again."

"Really?" Bart's tone sobered. "How many elephants? How'd you get around them?"

Richard described the encounter. "I think the noise we made spooked the herd. They charged in our direction, while we were still fighting dense brush, so we never saw how many there were. We just dropped what we were carrying and raced back downhill. We hunkered behind some big rocks and waited 'til we heard them stop chasing us. We stayed put for a good half hour to let them back off and wander away."

I didn't think I could handle a run-in with elephants on this most strenuous day of the journey. If our macho crew had been so panicked by the herd they'd encountered, there'd be no way my shorts would stay clean if I encountered a stampeding herd. I couldn't even imagine trying to outrun a fully grown elephant. I wondered if the god Sobek's special powers extended to warding off elephant attacks.

"The good news is we found a delightful campsite under a gigantic baobab tree with a fantastic view out over the river. You guys will love it." Richard knew we could use a little motivation. Now at least we had a reward to look forward to at day's end.

After breakfast, Conrad handed out assignments. Bart would replace Conrad on the crew, who were leaving immediately to wheel the remaining two rafts and other rafting equipment lugged or strapped to backs. Two of the more athletic male passengers volunteered to work with the crew. Their extra manpower all but guaranteed we would complete the portage by late afternoon. After dumping their first loads of the day at the new camp, they'd hustle back for a second trip.

While we were still together, Michael gave a safety talk about the risks of traveling overland. He warned our lives might depend on knowing which animals we could scare off by making noise and throwing rocks and those with whom we should be silent and motionless.

We had floated for a week on a river dense with creatures that could easily kill a human being, yet despite two close calls, all of our body parts were intact. This bolstered my confidence that if we did nothing to provoke them, they would be gracious enough to share their domain with us.

Traversing overland was a whole different ball game. On terra firma, the list of wildlife not to perturb was much longer, the creatures much larger and potentially more lethal. Of a hundred ways to die in the Selous, my imagination conjured in exacting detail an elephant tusk goring through my chest, the clenching of a lion's canines deep into my neck, and the flesh being ripped from my body by a frenzied pack of wild dogs.

Slade offered his own warning about the arduous hike ahead of us. He painted a realistic scene of the path's challenging terrain winding through jungle and savannah. And he set truthful expectations of the physical demands that would tax us to our limits for the next several hours. We had four bullets left for protection as we set out to trek through this sprawling reserve. I felt smothered by the same fears of the unknown that had paralyzed me before we launched on the river. Again, with no exit available, I just had to face my fears and walk straight into them.

Conrad stayed behind with me to finish packing the kitchen. When that was done, we helped our passengers figure out how they could best carry or strap their personal gear around their bodies. Those with extra space in their packs filled them with pieces of camp gear. With everyone loaded to the max, we set off on the portage. Conrad and Magigi led the way while the rest of us labored to keep up. We resembled the famous Boston Public Garden ducklings lined up single file behind them.

We trudged through the jungle, mostly following a narrow, rutted game path. Still, vines and branches frequently impeded our way. In my left hand, I gripped the handle of my personal ammo can. My right hand clenched the handle of our first aid ammo can. Strapped to my back were my stuffed waterproof bag and tent. I hadn't been hiking long when the exertion of carrying 50 pounds of weight left me winded.

I questioned whether I had the physical strength and stamina to finish this trek. I worried about our older passengers, who, despite being quite fit, had to be finding this jungle march exhausting. Conrad noticed the pace of our group slowing as the distance between each trekker widened. He stopped to let us catch up and catch our breaths. My lungs gulped in air, my heart pounded, and sweat rolled down my flushed cheeks.

As the group closed ranks, Conrad reminded everyone of the importance of staying hydrated. Our bodies were expelling large quantities of sweat as our lungs strained to breathe in the hot, heavy air. He suggested we all drink some of our favorite lemon-iodine-flavored

Kool-Aid. He topped off the canteens that needed refilling from the water jug he was carrying.

I gulped a few generous swigs of Kool-Aid, grateful for this brief rest. While I intended to take photos of the sights along this portage, with both hands full, I couldn't retrieve my camera from my ammo can without stopping. I took advantage of our break to unpack my camera and shoot a few random frames before we continued on.

"Hey, guys, we're already halfway through the hardest part of this hike. In another 20 minutes, we'll be out of this dense stuff and on a level open plain. The going will be much easier, mostly downhill from there." To further encourage us back into motion, Conrad promised us a sit-down break once we reached the savannah. It worked.

With cameras re-stashed and pack straps adjusted, Conrad signaled it was time to move out. I couldn't help thinking about the theme song of the 1960s TV show *Rawhide* that launched Clint Eastwood's career. Frankie Laine crooned while valiant cowboys rounded up stray cattle and steered them back into the herd they were driving. We were human wanderers who had strayed from our civilized herd, reluctantly moving into parts unknown. I just prayed that Conrad would shepherd us to safety rather than the slaughter that awaited *Rawhide's* unsuspecting cattle. We fell back into position and pressed onward, still uphill but over less steep terrain. Were we lost, I wondered, when Conrad's promised 20 minutes felt like an hour had passed? Still, we plodded forward.

Stepping out of the canopied jungle into the open savannah grasses felt like entering another world. There were still plenty of trees, but they only dotted the landscape. Here, the sun beat down on us from a cloudless sky. A warm breeze wafted across the plains, cooling my wet T-shirt and helping to evaporate the beads of sweat on my forehead.

We could no longer hear the roar of Shiguri Falls, nor discern a break in the horizon where the river carved its route toward the Indian Ocean. It was out there to our right, but the natural landscape provided the perfect camouflage to conceal its whereabouts. Conrad found a comfortable rest spot under the shade of a 40-foot tall, umbrella-shaped acacia tree. He opened a can of nuts and passed them around to give our waning energy a boost.

We rested on a high point of land where the view stretched out for miles. In the foreground, distinct flora in many hues of green gradually darkened to less distinct layers of green-blue landscapes. In the far distance, the haze of the midday heat blurred a band of lighter blue. This idyllic setting away from the river provided a well-needed reprieve from constantly scanning the surface of the water for any trace of hippos and crocodiles.

Magigi held his rifle at the ready and watched the low brush and tall grasses that surrounded us for any signs of stealthy motion. The last thing we wanted was to become a tasty meal. Out in the open, our scent scattered by the wind notified the alpha predators of our presence in the vicinity, but during the scorching daytime heat, most of the wildlife nestled in nooks of cool shade to nap.

Except for the gentle rustling of tall grasses swaying with the breeze and the occasional bird song, complete solitude surrounded us. Even our voices sounded harshly intrusive, where otherwise only the sounds of nature interrupted the stillness.

Every minute of every day back in the United States, noise bombarded me: sounds of motors running, TVs blaring, jet engines above, telephones ringing, alarm clocks buzzing, and the piercing sirens of police cars and fire engines. It took the absence of the civilized world's daily ruckus for me to be amazed by the serenity of the natural sound of our planet. Just these few days of living in the wilderness forced me to question whether what we humans considered "progress" was, in fact, making our world better. Or was our "progress" paradoxically destroying a habitat meant to ensure the survival of both humans and all the other creatures and plants with whom we share the planet Earth?

I wanted to sit under this shady tree for the rest of the afternoon to contemplate how little I knew of this world and how much I still had to learn. Even now, it seemed surreal to be sitting in the heart of Africa. The terrifying impressions of Africa my imagination had conjured were so opposite of the Africa that surrounded me here in Tanzania.

While still a dangerous place, I was quickly learning how to avoid the obstacles that could ruin my day. At first, it bothered me not to know what was happening back in America, but after a few days, I accepted that whatever happened in the civilized world was completely irrelevant to my survival and happiness as the days passed. I had never experienced this type of independence. I loved the simplicity of just living in the moment.

An entire afternoon of work still needed to be tackled when Conrad announced that our peaceful interlude was over. I snapped a few more pictures before repacking my camera. Everyone topped off their canteens. Then we were ducklings once more, trailing Conrad through the African bush. The trek through the open savannah meandered gently downhill around clumps of scrub brush and away from thorn trees.

About an hour after resuming our hike, we crossed paths with the first group headed back to retrieve the rest of the communal gear left at the upper camp. Despite the chain gang work of transporting heavy gear, the returning crew seemed to be in good spirits.

"Hey, how's it going?" Conrad greeted them.

"Great. It's a hell of a lot easier hiking when you're not loaded down with 80 pounds of gear," Bart admitted.

"It would be a real killer if we were doing this in reverse," said Richard. His headband was so soaked with sweat that an escaping trickle rolled down the side of his cheek. "Luckily, we have gravity on our side for the weight-bearing hike downhill. At least most of the way."

Conrad asked what we all wanted to know. "So, how long did it take you to get to the lower camp?"

Slade gave a wry response. "I think what you're really asking is how much farther is the camp?"

"Exactly," said Martha.

"After we rested up for a bit, rehydrated, and made sure we'd secured everything…" Slade glanced up at the position of the sun. "I'd say we've been back on the trail for 40 minutes."

"Did you catch sight of any game this morning?" Conrad asked.

"Nothing except birds so far." Michael's voice registered disappointment. "At least it's a break from counting damn hippos."

"You guys got enough water?" Conrad offered refills from his jug.

With everyone in good shape and no reports of wild animal encounters, our groups continued on our separate ways. Based on Slade's information, we'd already covered over two-thirds of this tortuous detour. That knowledge lifted everyone's spirits.

A short while later, Conrad and Magigi led us back into the denser vegetation, heading roughly in the river's direction. I wondered how Conrad knew where to reenter the jungle so we wouldn't get lost. Then I noticed him watching for freshly hacked V-shaped notches in the trunks of trees. The crew had marked their previous day's route with their machetes. The notches, along with Conrad's memory of other natural landmarks, provided a visual road map that would lead us to the place where they had stashed our equipment and rafts. Slade's calculation proved accurate. We emerged from the jungle, tired but jubilant.

We stood at the top of the riverbank, gazing for the first time over the Rufiji River. Below the falls, a considerably wider river with a noticeably swifter current raced past us. In the distance far upstream, I could see only a portion of the spectacular cascading waterfall, larger than Niagara Falls. It took only a heartbeat for me to understand why Conrad was so insistent that we not stray anywhere near the rim. Had we not found the safety of the left-hand shore, we would have plummeted to a certain death.

Instead, we were safe. Until that moment, I'd only been able to imagine the immensity of the elusive Shiguri Falls. Now standing below it, I was stunned by the actual sight. Judging from the superlatives flowing from the mouths of my companions, so were they. Millions of gallons of water tumbled in several distinct cascades with a thunderous rumble, disappearing into the swirling mist below, to crash on boulders that had once belonged to the cliff face over which the water spilled.

Such power and tropical beauty took my breath. This moment alone made every pain and agony of reaching this point on our journey worth it.

THE CAMPSITE OUR CREW had deemed suitable covered a flat landing on top of the riverbank, towered over by a massive baobab tree. It wasn't as tall as the giant redwoods back in California, but its trunk certainly equaled or was larger than many redwoods. Just as with the redwoods, I felt insignificant standing next to it. Instinctively, I wanted to fill my arms with part of its trunk, but I held back lest my raft mates would change my nickname to Brenda the baobab hugger.

An article I'd read before coming to Tanzania contained facts that surprised me about these giants of the African grasslands. The most remarkable feature of the baobab is its longevity. Most live well over 1,500 years. Nicknamed the 'Tree of Life,' every part of the tree is useful in sustaining life in this harsh habitat. Both humans and animals seek the refuge of the baobab for shelter, food, and water.

The rotund trunk resembles a tower, wrapped with a smooth, shiny bark, which can grow up to 82 feet tall. The top of the trunk sprouts Medusa-like tangled branches. The swollen trunks, which have measured as much as 46 feet in circumference, can store as much as 1,000 gallons of water during the rainy season to tide it over through seasonal droughts.

The fibrous bark of the baobab is used to weave cloth and make rope. Packed in its edible leaves are many of the same nutrients as spinach. Its fruit, tucked inside hard-shelled, six- to eight-inch-long pods, has a citrusy flavor and contains six times the amount of Vitamin C as oranges. Locals swear by the medicinal properties of its ground seed powder to cure whatever ails them.

The nearest market was between 50-100 miles away from where I stood, but directly in front of me was nature's own Quick Mart. If I ever got lost in this wilderness, I would find myself a baobab and live under its branches. The need to help set up our camp finally drew me out of my Robinson Crusoe daydream under the majestic guardian baobab.

Those of us who only made the portage one way knew we owed an enormous debt of gratitude to the guides who persevered, covering the portage route five times in the last 24 hours. We hustled to get everything done in camp so when they arrived, they could kick off their hiking shoes and relax for the rest of the day. They deserved to be served an ice-cold beer for their efforts, but that amounted to wishful thinking.

As we waited for our colleagues to complete the final leg of the portage, Conrad whipped up a batch of fresh bread, while I chopped potatoes for dinner. Suddenly, an idea popped into my mind. I knew how I could get to hug the baobab without everyone assuming I'd lost my mind.

"Hey, guys, who wants to help me measure how wide that baobab trunk is? If a few of us stand against the tree with our arms outstretched, we could see how many people it would take to circle the base completely."

Conrad rolled his eyes, poking fun at my proposal. But one by one, the passengers stepped up.

"I'll help."

"Me too."

In less than a minute, several people had volunteered to take part in my scientific experiment. My wingspan stretched close to five feet with my arms extended and my chest splayed against the smooth bark. Another person lined up next to me, his fingertips barely touching mine. Each additional person who joined the chain stood against the tree with their fingertips connecting with the adjacent persons' fingertips. The trunk required five huggers with a two-foot unfilled gap. With a circumference of at least 27 feet, our tree still needed more time to grow before it set any record.

Our group effort amused Martha and Jay, who had settled comfortably together on a log next to the firepit. Finally, she stood up and headed our way. "Let's finish putting a bracelet on this darling," she said before closing the gap to complete our embrace.

I hoped the baobab could feel our respect and admiration for how huge it had already grown. As I clutched at its mass, I swore energy from the tree flowed into my body. As someone who had never seen a baobab before, it felt strange, in a good way, to feel such a powerful affinity with this tree.

WITH ONLY A FEW HOURS of daylight remaining, the returning crew members trudged into camp and promptly dumped their loads. Relief flooded through their exhausted bodies with the day's work done. Inflating and reloading the rafts could wait until morning after everyone had rested and replenished their energy. The sweaty, grimy, heavy-lifters group took soap and towels and strolled along the shore, until they were just out of sight, to clean up before dinner.

The sun hung low in the sky, casting rich colors onto the clouds above while we ate our evening meal. Even nature seemed to celebrate our successful portage around Shiguri Falls with a magnificent golden sunset that deepened to oranges, blues, and purples before the last vestige of brilliance took a final bow and slipped below the horizon.

We'd endured eight days of no hot showers, eight days of not sitting on a toilet, eight days of wearing a coating of light brown silt on our hair and skin, and eight days of back-breaking paddling and hauling gear over rough terrain. Amazingly, we weren't just surviving; as a team, we were thriving in the Tanzanian wilderness.

Less than two weeks ago, my distorted views of Africa had me convinced I would be in mortal danger of being captured by cannibals or pounced on by bloodthirsty wildlife. I deemed my chances of survival here were 50-50 at best. But in the past few days, my brain, like a sponge working overtime, had been soaking up the true essence of Africa. I'd found people much like myself except for the color of our skin and discovered wildlife that tolerated our presence on their turf. Each day, new experiences continued to blow my mind. Even who I was, my persona was changing in an affirmative, empowering way.

23

THE LION ROARS

THE CLANKING OF COFFEE POTS woke me from a deep slumber, devoid of dreams. Crawling out of my tent, I noticed some of the crew were beginning to reassemble the rafts, so I hurried over to help. They had spread the flattened boats open on the ground and were using two high-volume pumps to inflate each raft's six segmented tubes by hand.

Inflating the rafts wasn't difficult, but took time. Pumping in five-minute shifts, like workers on an assembly line, we furiously pushed and pulled on the pump handles until our biceps burned with the strain. Just before one pumper's strength failed, another guide stepped up to take over. For this task, a large crew was an advantage; many hands would get our work done quickly.

As each tube swelled with air, the pump handle fought the greater resistance on its way down until it would go no further. The test of a properly inflated tube was no give at all, from the weight of a person sitting on it. We inflated as much air into the tubes as our strength allowed in the cool shade under the baobab tree. Once we launched on the river, the energy of the sun's rays beating on the neoprene surface would heat the air inside the tubes, further increasing the internal pressure.

We'd fully inflated three of the rafts when Conrad hollered, "Breakfast is ready."

The morning's feast featured slices of Conrad's yeasty, crispy-crusted bread with fruit jams, plate-sized flapjacks drenched with honey, and crispy slices of fried SPAM. While we ate, we viewed the powerful flow of the Rufiji's current racing past our campsite, raising everyone's hopes that finally, the need for back-breaking paddling was over. The number of crocodiles occupying the river below the falls had doubled from the amount we'd observed for days on the Kilombero. We watched them, watching us, as we ate. More intense vigilance would be required to avoid passing too close to where they hid lying in wait.

The smaller size of the crocs didn't mean they were any less intimidating than hippos. Their medieval appearance with thick spiked body armor and long snouts filled with razor-sharp teeth made us wary of trespassing through their territory. It didn't help that they were masters of stealth, making slick, noiseless entries from land into the river's depths at the first sign of an approaching raft.

The sun had climbed to nearly overhead before we shoved off from our cozy shelter under the baobab tree. Immediately, the powerful current grabbed and dragged the rafts downstream. Its ultimate rendezvous with the Indian Ocean was still 370 miles away.

Instead of letting the river do the work of transporting us while we kicked back and enjoyed drifting along, Conrad pushed us to cover as many miles as we could that afternoon to stay on schedule. Grudgingly, we hoisted our paddles and resumed our rhythmic stroking. Between the powerful draw of the current and the efforts of our paddling, we were flying. For once, paddling wasn't a chore.

However, Chaucer's famous quote "All good things must come to an end" proved true that afternoon, when an angry hot wind gusted upstream, slowing our progress to a crawl. Despite our strenuous efforts, we barely inched along. I gained a new respect for nature's brute strength:

the power of water currents to move us forward, and the power of the wind to nearly halt our progress.

While the sun still hovered well above the horizon, we spotted a campsite on a wide-open plateau three feet above river level. Unable to forgo such a luxury, we paddled up to the most spacious campsite we'd encountered yet. At some of our camps, merely finding enough level spots for all of our tents had required the skill of a jigsaw puzzle master, but at this site, each tent could claim triple the prime real estate of earlier sites. This decadent excess delighted us.

We needn't rush to cook dinner either, since daylight would hold off the fall of darkness for at least two more hours. Our passengers, who by now had perfected their fire-building skills, got right to foraging for dry wood. Some collected driftwood, scattered along the riverbank by the last flood, while others hauled fallen branches from the edge of the forest that rimmed our campsite. The mound of fuel for our fire quickly expanded in size.

"You oughta collect extra firewood. Since we're setting up camp earlier than usual, we'll need fire for a few extra hours tonight. I'd hate to send you off searching for firewood in the forest after dark." Conrad's suggestion sent the wood gatherers scurrying out to forage even more wood. Just the mention of entering the forest after dark should our fire dwindle too early powerfully motivated us to over-stock our stash of wood.

After stirring up three fresh pitchers of Kool-Aid, I joined the scavengers on the river bank to help haul a couple of armfuls of drift-wood. Within half an hour, the woodpile had more than doubled in size. Two of the men dug and rimmed a firepit, then coaxed small flames to devour the starter pieces of dried grass and twigs.

"Come on, baby, burn." They blew onto the sparks, encouraging a brilliant flash to erupt.

"There it goes. We've got a flame." We all cheered for our colleagues, who'd ignited the fire on their first attempt.

They added larger logs in the shape of a pyramid, crisscrossing over the expanding flames in the center of the pit. Before long, the fire roared with fury, greedily consuming all the fuel they fed it.

Conrad finished kneading fresh bread dough and put it aside to rise. Clutching Kool-Aid cocktails, along with plates of cheese and crackers, we gathered around the blazing campfire. The mingled aromas of bubbling macaroni and cheese and slices of pan-fried canned ham sizzling on the camp stoves made our mouths water.

Yellow and orange flames danced wickedly. Their snapping, crackling gyrations mesmerized us. Wisps of smoke, fragrant with the scent of fresh charred wood, rose only to be carried away with the gentle breeze. Colors of the fading sun reflected across the rippling wavelets on the surface of the river. Our surroundings discreetly melted into blackness as we finished dinner. Nature invited us to use her darkness as downtime to rest our bodies, but I hadn't been in bed by 7 p.m. since I was five years old. Nobody was ready to leave our congenial group relaxing around the firepit.

With full bellies and another successful river day notched on our belts, life was feeling pretty damn good. Enjoying the shimmering radiance of the glowing embers, a few of our group sipped on nightcaps from their private stashes of spirits. Others helped themselves to the freshly brewed pot of coffee Conrad placed on the flattest stone of the fire ring.

Utterances of nocturnal creatures arose from behind the jet-black curtain that obscured our view of their hiding places. We listened to the cacophony of anonymous chirps, trills, and peeps, the first act of the jungle's enchanting evening that would eventually lull me to sleep.

Without warning, a thunderous roar exploded in the night air. The game rangers snatched their rifles. We sat frozen, paralyzed with fear. The lion's unmistakable warning reverberated with the volume and clarity of surround sound, making its location impossible to determine.

"Holy shit! Where the hell is it?" shrieked one of the younger men, visibly unnerved by its proximity.

"What should we do?" Martha asked. Trembling, she clung to Jay, who appeared equally shaken by the invisible roar.

To have produced a roar that deafening, it seemed the lion should have been sitting among us in our campfire circle. The idea of a wild lion wandering this close to us stunned me. Magigi barked something in Swahili.

Conrad shouted, "Let's get some more wood on this fire."

At once, many hands piled additional logs on top of the glowing embers. The peacefulness of only moments earlier again pivoted to sheer terror as a second deep-throated roar reverberated through the darkness, stretching on for what seem like minutes.

As Magigi continued his animated conversation with Conrad, my mind drew a vivid image of a ravenous full-maned lion stalking just beyond the circle of light cast by the campfire, patiently scouting which one of our group to pick off. Ironically, the idea of our tents being so spread out now seemed like a huge mistake.

"Listen up, everyone." Conrad instantly had our attention. "Magigi thinks a pride of lions made a successful kill out behind us in the savannah. The males always eat first until they are sated. Magigi thinks those roars were the males expressing their pleasure at having devoured fresh meat. Once they're full, they won't hunt again until they're hungry."

"But how close to us are they? It sounds like they're just over there." Another client pointed toward the blackness behind our tents. "What if they come prowling around here tonight?"

"I know it sounded so close to us, but Magigi says you can hear a lion's roar up to five miles away."

There was no way that lion was five miles away, I thought, after hearing that wild roar. I hoped to God that Magigi knew what he was talking about.

Kate, tightly gripping her hands in angst, asked, "Don't some prides have 20 or 30 lions in them? What if some of them are still hungry? What if they *do* come into our camp tonight? What should we do?" By openly confessing how petrified she felt, she was venting the fear we all shared.

To calm us, Conrad used some mathematical logic. "So, if a pride of 20 lions brought down a waterbuck weighing 400 pounds, there would be 20 pounds of food for each lion, more than enough to satisfy their gluttony. They're not much different from humans at Thanksgiving. What do you guys do after polishing off a turkey dinner with all the fixings? Take a nap. Right? So they're probably going to curl up and sleep off their feast tonight."

While Conrad delivered his Lion Behavior 101 lecture to us, Magigi huddled off to one side with his two colleagues. Their lively Swahili conversation involved much head nodding.

What Conrad told us made sense, yet I wondered how I could ever fall asleep with the sound of those fearsome roars echoing in my brain. I suspected I wasn't the only one struggling to feel safe.

"Look, Magigi and his men are going to keep the campfire blazing all night. They'll take turns standing watch. Most wild animals perceive fire as a 'threat' they will flee from. Also, these lions are not familiar with humans. We're an unknown menace they'd rather avoid. We're all going to be fine."

I so wanted to believe Conrad's reassurances, but even more I wanted to know what I should do if Magigi was wrong and the lions paid us a midnight visit. Unlike the hippos that avoided contact with our tents, razor-sharp lion claws, the length of a human finger, could easily shred a tent's nylon walls. I worried about Conrad's nonchalance toward the possibility of a midnight lion attack. Yet, as I studied the unconcerned demeanors of the other guides, I realized they too had accepted Magigi's assessment that we would be safe.

Worries persisted and doubts continued to be voiced by several of our clients. Knowing that wild lions lurked in our neighborhood proved utterly nerve-wracking. Recognizing I had absolutely no control over the actions of these lions led to a profound sense of helplessness. I inhaled deeply, focusing on staying calm. I silently prayed for serenity in the face of danger.

Strangely, I felt more secure climbing into the illusory security of my tent than sitting outside around the blazing bonfire. At least its walls provided a flimsy barrier between any approaching hungry lion and the tasty treat it sensed close by. I lay awake, replaying the ferocious roars over and over in my mind. These roars, and the fatally wounded hippo's death howl, forever etched into my memory, would echo with startling clarity for as long as I lived.

24

TIME FOR REFLECTION

BEFORE SETTING OFF DOWNRIVER the next morning, we filled buckets with baseball-sized rocks, ammunition to lob at any crocs that stalked our rafts too closely. The "croc rock" weapons, combined with vigorous paddle slaps on the river's surface, served as our primary defenses against the more aggressive crocodiles. So many of them inhabited this stretch of river that Michael lost count of their numbers for his informal wildlife survey. The plop of rocks cratering into the water and whacking paddle blades that sent spray in all directions served as a frequent audible warning for the crocodiles to stay away.

Rapids with frothy white tops speckled the surface of this stretch of the Rufiji, like bumpers on a pinball machine's field of play. After days of floating on flat water, I welcomed the excitement of these two-to-three-foot-high stationary waves. Rather than dodging the swells of water, the guides steered straight into them. One by one, the rafts coasted up, crested, and then slid down the backside of each wave. The sprays of water they spit on the paddlers didn't dampen their gleeful squeals one bit.

Usually when I approached easily runnable rapids of this size on California rivers, I'd encourage my paddlers to slip over the side tubes into the water to body surf the gentle roller coaster rides of these playful waves. But on the Rufiji, with daily sightings of crocodiles on the increase, keeping everyone inside the boat remained our top priority.

The river's width fluctuated repeatedly over the next two days, first growing to half a mile wide, then shrinking in on itself. In narrow sections, the current sped up. The boost in velocity translated to fewer hours of paddling and extra time in camp for chores and kicking back. One afternoon in camp, I filled a bailing bucket with river water and soap. Though I scrubbed forcefully, each attempt to remove the accumulation of fine beige silt trapped in the cotton fibers of my dirty clothes failed. My light-colored T-shirts grew a dingy shade darker with each washing.

After spreading my theoretically "clean" clothes on nearby rocks to dry, I found a sandy spot on the river bank to sit and write in my journal. I thought about how the jumbo-sized sack of fears I had carried for most of my life had prevented me from seeking encounters with unknown people, places, and things. For the first time in my life, I saw how severely stunted my sense of wonder about our world had become by my need for safety. At least a third of my life had passed, sheltered in a comfort zone I dared not step outside. Only now could I see the dangers I imagined lurking in the outside world were trivial compared to the treasures waiting to be found.

Richard craved meeting new people and exploring unknown places. Energized by those discoveries, he celebrated by sharing his newly gained tidbits of knowledge with others. I tried to swallow my envy while all the other guides and our clients welcomed fresh adventures with a zeal that matched Richard's. Why had I grown up to be so fearful? It had never occurred to me that fear could be a personal choice. I'd always accepted fear as an essentially intrinsic part of my being.

With the soothing gurgle of river water providing an acoustic backdrop, I puzzled over this revelation. Why had I become afraid of so much? Were my insecurities acquired as a child? My mind drifted back in time, trying to pinpoint the earliest times I recalled feeling afraid.

As a young child, I dreaded dinnertime the most. The position of my seat at our kitchen table trapped me in a corner. My father, who sat next to me, blocked my only escape route. His daily consumption of alcohol before dinner led to a drastic transformation in his demeanor by the time we sat down to eat. One moment, he'd carefully smooth my blob of mashed potatoes into a far more delicious-shaped cat or bunny. Seconds later, the tiniest whine or bit of fussing by my younger brother or me launched him into a verbally abusive tirade. Nearly every dinner of my childhood ended with my father ranting.

"Just once. Just one time. I'd like to finish my dinner in peace and quiet. But you damn kids, you ruin it every… single… night!"

Cheeks flushed red with anger, barely halfway through his meal, he'd throw down his napkin like a referee throwing a penalty flag and bolt from the table to pass out on the living room couch. Sometimes his rage left Mom helpless. She sat motionless in place, with tears streaming down her cheeks. The memories of cowering in the corner, filled with guilt and fearful of my father's wrath, reminded me of how confused his unpredictable behavior left me. A novel awareness struck me. If I could be *that* afraid of the person I loved most, it suddenly made sense how I could be afraid of anything.

When not under the influence of alcohol, I knew Dad adored me. Before I was old enough to comb out my own hair after a bath, he'd gently unravel the snarls in my wet, tangled mop. The way he patiently combed out my hair so tenderly never hurt compared to the faster, often painful yanking technique my mother employed. Years later, Dad and I designed gold-medal science projects together. Best of all, he built us a magnificent treehouse complete with bunk beds, windows, a door, electric lights, and a retractable stairway.

I hated how alcohol altered his mood. Staying out of trouble meant always being on guard for the first sign he'd picked up the bottle. I learned early in life to disappear when multiple martinis turned his mood foul. At those times, I fled to the safety of my bedroom. I constantly worried about not being good enough, smart enough, or worthy enough in my dad's eyes.

I toed the straight and narrow line, attempting to gain his approval. As long as I stayed within his strict parameters of proper behavior for a girl, I could avoid his ire. Even when I presented him with a report card loaded with A grades, he'd lecture me about working harder and always doing my best, as if something better than straight As existed.

I wiggled my toes into the warm sand where I rested. Maybe I *had* learned from my parents that if I behaved within the boundaries of "normalcy" (not running off to the African wilderness) and following society's expectations (finding a job, getting married, and raising a family), I could avoid trouble. But here in Tanzania, I could explore and learn things about our world which I would never experience at home. I felt seeds of courage sprouting, but true bravery remained a distant dream. I still doubted I would ever have the "stuff" that it took to be a pioneer.

But damn it, here I sat in the African jungle, a place I had sworn on my life only weeks earlier I would never go. I was freaking actually here! Even if Richard practically had to drag me. How could I ever learn enough to be confident about exploring the world on my own? The treasured chance to hang out with Sobek guides would help. I vowed to soak up every lesson they shared from their own exhilarating expeditions. An impatient eagerness coursed through my body as I decided I was done with running away from the unknown. I'd never experienced such a startling epiphany.

THE FOLLOWING DAY, we watched for another landmark on our downstream journey: the Great Ruaha River joined the Rufiji River as one of its major tributaries. Richard initially considered staging our expedition on the meandering Ruaha to avoid Shiguri Falls, but its course primarily flowed through a monotonous ecosystem of wetlands. We expected a river of substantial width with an impressive water flow emptying into the Rufiji from the left. At the location marking the confluence of the two rivers on the map, we found nothing but a puny stream, about a tenth the size of the Rufiji.

"This can't be it," Slade said. "Let me see the map."

"Well, it doesn't show any other waterways entering within miles of where we are now," Michael muttered, passing the map over to Slade.

After studying it for a minute, Slade chuckled and yelled back upstream to Richard. "Hey, man, this is your Great Ruaha." He gestured toward the confluence of the two rivers. "Sure glad we didn't go with your first choice. It would have been a suicide mission."

Slade's comment caught everyone's attention, not just Richard's. As we drifted past the "Not So Great Ruaha," it shocked us to see how many hippos crowded into its narrow stream bed. After seeing the tributary, Richard winced, knowing that running the Great Ruaha would indeed have been a fool's errand in these current conditions.

"Might be an interesting run in June, after the long rains," he yelled back to Slade. "It should be really cranking then. It's all about the timing."

Slade shot an *"are you out of your mind"* look back at Richard but passed on challenging him further.

I took advantage of being back in Richard's boat to get his advice. "You've traveled all over the world. What are your favorite places? I'm thinking I should plan a few more trips."

Richard's stupid grin showed my question amused him. "Hah, so the wanderlust bug *has* taken a bite of you."

Blushing, I admitted, "I guess so."

"Here's the thing. I'm not sure I have a favorite. I've been amazed by the unique aspects of all the places I've explored. I guess I'm just not the type of person who's apt to go back to where I've already been. I hope to spend my whole life trying to see amazing places on every part of this planet. Even then, I might not get to all of them."

Responding that he loved the entire world wasn't particularly helpful, since I knew so little about the world. I needed specific names of places to check out. I tried again.

"What are a few places you think *I* might find interesting?"

"While you're working for Sobek, take advantage of our trips, for sure. Or you could use my technique. Stand a world atlas on its spine. Let it fall open. Whatever map plate it opens to—that's your new destination."

I'd inadvertently discovered his secret "Bangs technique" for selecting new travel destinations. It perfectly suited his fearless, adventurous spirit.

"But what happens if the atlas opens to a place with a name you can't even pronounce?"

"Ah, but that's the fun of it. That's when you go to the library to research all you can about the place: its history, places of interest, language, climate, food, and transportation. All that information helps to shape an itinerary."

God, he made it sound so easy. Especially when he often journeyed to unexplored and barely accessible places. It would be years before I'd even consider adopting the "Bangs technique," if ever. It made sense for me to stick with Sobek's catalogue of expeditions to plan my next travels.

LATER THAT AFTERNOON, before they needed me in the camp kitchen for dinner prep, I grabbed my camera and strolled along the riverbank. I noticed tiny, delicate, pink blossoms sprouting from the short spikey stalks of sturdy river grass, covering sections of the shore above the high water mark. Further on, a single flower growing close to the ground with bright red daisy petals caught my attention. The vibrant colors stood out amidst the dominant palette of earthy colors, greens, browns, grays, and blacks of the surrounding landscape.

I concentrated on capturing the blossom on film until an out-of-focus object fluttered through my camera lens's field of vision. My attention diverted to following its flight. It came to rest on one blade in a tall clump of dried grass. The butterfly's light tan wings with small white dots tucked between brown squiggles provided perfect camouflage. After a minute, it took flight and hovered around me. I offered my outstretched arm as a resting spot, but this skittish butterfly didn't seem inclined to test it as a perch, preferring to remain just outside my reach.

My attempts to entice the butterfly back into a resting pose I could photograph proved futile. But they drew a chuckle from Magigi on his return from a scouting trek further downstream. He pointed toward the butterfly and gave it a name.

"Kipepeo."

"Kipepeo." I repeated the melodic name.

"Kipepeo." It rolled off my tongue like satin. After rafiki, it became my second-favorite Swahili word.

Could this be a prophecy that one day I too would flutter as freely as this splendid butterfly if only I could break free of my own protective chrysalis?

25

STIEGLER'S GORGE

THE 12ᵀᴴ AND FINAL DAY of our river journey arrived, bringing with it a slew of bittersweet memories. We had survived in the wilderness. Hell, we had thrived in this pristine pocket of Planet Earth devoid of humans. Already I could imagine the joy of drinking liquids not infused with iodine, the thrill of showering in silt-free water, and an end to the constant vigil for hippos and crocs. Yet I knew I'd be saying farewell to a place whose solitude, beauty, and timelessness had bedazzled me.

A few hours of rafting brought us to the start of the Pangani Rapids. For the next five miles, Rufiji River whitewater swirled through Stiegler's Gorge, named in honor of Franz Stiegler, a German civil engineer. In the early 1900s, he worked as a surveyor for the government of German East Africa, then engaged in building the Tanganyika Railway to connect Dar es Salaam with Lake Tanganyika.

In 1907, Stiegler became the leader of the Rufiji Expedition, charged with taking measurements of water levels and flows to assess the navigability of the Rufiji. That July, his expedition camped at Pangani Rapids. Stiegler met his demise in February 1908, 60 miles upstream from the gorge. While hunting, he wounded an elephant. The enraged beast charged Stiegler and hurled him into the air, killing him.

The gorge formed a bottleneck as the water from the sprawling river upstream constricted through a narrow rock-lined passage only 150 feet wide. The increased velocity of the current resulted in a series of standing waves known as haystacks, breaking waves that crest and fall back on themselves. In places, piles of hefty rock slabs lined the river bank, while at other points, cliffs rising as high as 300 feet formed the boundary of the gorge.

WE STOPPED FOR an early lunch on the last sandy beach before entering the gorge. While we rustled up the fixings for lunch, Richard, Slade and Conrad hiked up to explore a rarely occupied safari lodge on a bluff overlooking the entrance to the gorge. Conrad claimed guests could only reach this remote lodge in a bush plane capable of landing on a grass runway or, in our unique case, by raft. Chances favored they would find the lodge vacant. The rest of us had already finished eating when the three men trotted back onto the beach. They quickly scarfed down some food, while the rest of us readied the rafts to launch into the gorge's turbulence.

We whizzed through the gorge like a fire engine headed to an inferno. The larger waves thrashed the paddleboats, teeter-tottering them from side to side. In unison, paddles dug into the water, stroking furiously to maintain forward momentum, while the guides hollered commands and steered from the rear to avoid capsizing. The crests of the haystack waves drenched the paddlers. Cheering and whooping erupted each time they escaped from the grip of a wave that threatened to stall their raft. To stay upright, the rafts had to plow into each successive wave perfectly head-on.

If a paddle raft slid sideways into a haystack, the wave could grab the downstream tube, lift it, and roll it over, spilling some tasty human morsels into the roiling brown water. Voracious crocodiles waited for these treats. Our heavily loaded oar raft floated lower in the water, plowing a steadier course through the whitewater. The leverage from Richard's long oars helped stabilize our raft as we cruised down the highway of high rollers, safe from the soaking spray.

Part of me wanted to be yahooing with the wet and wild paddlers. Riding the Rufiji's best whitewater rapids offered the same fun challenge of trying to stay upright on a bucking bronco. But the advantage of being in the oar boat was staying safe and dry. The rollercoaster haystacks didn't worry Richard or me. But Masengela lay trembling on top of the gear pile, clenching the tie-down ropes for dear life. The terror of being tossed overboard as we rode up and over wave after wave paralyzed his entire body. Had a hippo surfaced along the side of our raft, I doubted he'd have released his grip on the ropes to raise his rifle.

Little vegetation grew below the high-water mark left at the peak of the rainy monsoon season, in April and May. The height of the peak surge crested at 30 to 40 feet over our heads, making the waves we were surfing look like harmless ripples.

After an hour of energizing excitement, the gorge spilled into a vast floodplain resembling an immense swamp. The endless expanse of tall grass covering the land blocked our view of what lay inland from the river. The homogenous landscape offered few distinguishing features: a few dead trunks of palm trees drowned by past floods and the occasional crevice carved in the grasslands where small channels emptied into the Rufiji.

"Now we've got to figure out which of these channels leads into Lake Tagalala." An unusual sense of urgency in Richard's voice raised an alarm. Rarely had I heard Richard sound anything but full of confidence. "If we get lost out here… well, we just can't get lost out here."

"But we have the topo map. Doesn't it show the channel on it?" I asked. I didn't understand why Richard seemed worried.

Sheepishly, he admitted, "'Fraid not. That map isn't current. Several years ago, an extreme flood diverted the Rufiji's main current into and through Lake Tagalala. When the flood receded, it left behind a narrow waterway that continued to drain water from the lake back into the river bed."

"You're telling me we don't have an accurate map?" I glared at him in disbelief. "How do we know where to go? It all looks the same out here." What was he thinking?

"That's why we went up to that lodge back at lunch. Conrad knows the owner, Andrew Stevenson. He's a bush pilot who knows this area well. Luckily, we found him there, and he offered to take me and Slade up in his plane so we could study the river below the gorge. We figured we could scout out the lake and find some landmarks to help us find the right channel leading into it."

"Wait a minute. Are you saying we're depending on what you and Slade saw from the air to make sure we get the right exit ramp out of here?"

"That's about it. We looked for unique landmarks that would help lead us to the Tagalala access channel."

My pulse raced. From my experience scouting rapids from the shore, I knew how tricky it could be to rotate a 90 degree side view of rapids to the view I'd see going straight forward on the river. Imagine seeing only the side of a stranger's face, then having to recognize that person from a front view. Richard and Slade had been high overhead, looking down at the twists and turns of the river channels as they skirted sandbars and islands. The map they memorized of landmarks from high above would look much different from the view of those landmarks at river level.

"Could you see the channel when you were in the air?"

"Oh, yeah, it's there and open all the way into the lake."

"What landmarks are we looking for?"

"Pink earth banks and two large channels."

That sounded pretty vague to me. For half an hour, we passed several channels so narrow that any attempt at navigating even short distances into their reaches seemed unwise. At each, Richard and Slade agreed these outlets probably dead-ended. Downstream we ventured. Richard and Slade focused intently on every detail of the riverbanks we passed.

"What if you can't find the right channel? Or we end up going beyond it?" I fretted that our final destination might be the Indian Ocean.

Richard frowned. "Trust me. We'll find it. Slade and I both took notes. Don't worry. We'll find it." After a pause, he muttered under his breath, "We have to find it. It's our only way out of here."

Since we'd entered the floodplain, I'd yet to see any suitable area for a campsite. Clumps of tall grass and short bushes leaned precariously over the edge of deeply eroded mud banks. The decision to carry as little weight as possible to ease the effort of portaging prevailed over packing additional food for extra river days. Paddling back upstream against the powerful current would be impossible. That old, familiar wave of anxiety washed over me. Locating the elusive channel that led into Lake Tagalala felt as critical now as finding solid shoreline above Shiguri Falls had been days before. Once again, failure was not an option.

A short while later, we approached an island that had an odd pinkish hue to the soil on its bank.

"Hey, Richard, look over there. Does that look familiar?" Slade pointed at the pastel-colored banking.

Floating closer, Richard nodded his head. "That looks like it. I think that's the first landmark."

"Okay, so if we stay left, the third large channel should be it." Knowing they'd found the first crucial landmark, the worried expression on Slade's face softened to a smile.

"Wait a minute. My notes say it's the second channel after the pink banks." Slade's quizzical look at Richard showed he thought Richard was mistaken.

"No, man, it's the third channel on the left just after another patch of pink earth. Remember?"

Unconvinced, Richard glanced again at his scribbled notes. I listened with concern to the volley of conflicting memories between these two men, knowing our fate rested in their hands.

"Are you sure, Slade? If we pass the second channel and the third one isn't the right one....."

Richard's unfinished remark felt like a gut punch. The precariousness of our situation became clear. We were literally a couple of hours away from the end of our trip, but only if we could find our way into Lake Tagalala.

"I'm sure it's the third channel," Slade insisted.

Richard still looked uneasy.

"Really, it's the third channel." Slade refused to back down.

"Okay, let's stop when we get to the second channel and see how it looks. You may be right," Richard conceded reluctantly.

Having accomplished several first descents of wild rivers together, Richard knew Slade was conservative, observant, and nearly always right. For that reason, he trusted Slade's instincts, at times, over his own. Again, we continued downstream. Reaching a second major channel, we paused at its entrance.

"What d'ya think?" Slade asked Richard.

"I don't know. This one is narrower than the other one."

"Remember, we haven't seen the second patch of pink earth yet, just upstream of the third channel."

"I remember a second pink patch, but I only remember two large side channels. Maybe we missed one of the pink patches." At least Richard and Slade agreed on something.

"We've gotta find the other pink patch, I'm telling you," Slade insisted. After one more review of his notes and a moment of soul-searching, Richard nodded. "Okay, let's go on."

Getting out of this swamp depended entirely on Richard and Slade merging their memories to find the right Tagalala channel. I sure as hell had no desire to spend another night drifting down this river, when an indoor cot and a freshwater shower beckoned so close by.

We floated along in silence, all eyes glued to the sometimes gray, sometimes beige, silty river's edge, searching for any change in color. When solving a problem, Slade's silence and concentrated focus were expected character traits. Richard's silence, though, was especially worrisome. His cheerful optimism seemed as lost as our expedition.

We drifted past more clefts the size of drainage ditches, obviously too small for our rafts to enter. Suddenly, from the lead raft, Slade yelled, "Look over there! Does that bank look pinkish?" He directed the question to all of us, hoping for affirmation. Heads swiveled to where his arm pointed.

In the last raft, Richard and I couldn't yet view what Slade was seeing, but people in other rafts nodded. Slade waited for all the boats to catch up. When I finally saw the second patch Slade had identified as pink, I agreed the hue of the soil appeared lighter, but pink? Even pinkish? To me, the light cream color didn't at all resemble the more distinctly pink shade of the bank on the small island we passed earlier. I glanced at Richard to gauge his reaction.

"Could be." He managed a half-hearted response.

"Let's go then. The channel should be just a little farther up on the left." Slade remained adamantly positive we were on the right track.

It wouldn't be long until we'd know whether his recall had served us well. I tried to feel hopeful, if only to suppress the much worse alternative, but I could tell Richard was dwelling on what our next moves would be if Slade was wrong. The irony of our situation made me laugh. I'd handled every obstacle I'd been worried about since the start of the trip: the wildlife, the portages and health issues. Though I *had* been concerned we might never make it to the "end" of the trip, I'd never once considered that we might get stranded in a gigantic swamp—unable to find our takeout location.

Shortly later, a larger channel appeared on our left. Slade's passengers cheered at the sight. Clumps of bright green Nile cabbage rosettes that only thrive in still waters or slow-draining streams floated on its surface. This channel had to be the link between the river and the lake. Slade gave an ecstatic fist pump, then turned around and beckoned for the other rafts to follow. Confidently, he ordered his paddlers to make a left turn into the channel.

"If this *is* the channel, it'll only be another two miles until we reach the lake." Finally, I heard concession in Richard's voice. Now committed to this route, we just needed a lake to appear.

The paddle rafts had plenty of room as they navigated further up the channel. As soon as Richard steered our raft into the channel, we saw it would be a much tighter squeeze for us, given the long oars protruding from the sides of our raft. Masengela came down from his perch, and we both took up paddles to help Richard power the raft forward. The further up the channel we went, the narrower it became, not at all what we'd been hoping for. I didn't voice my concern that we could be heading into a dead end.

Off to our right, a commotion erupted as several snow-white cranes, spooked by our close approach, flapped their way aloft from hiding spots in the tall grass. Was it an omen of sorts? An encouraging sign we were on the right path? Or a warning that we should turn back while we could? The clumps of Nile cabbage had gotten so thick they nearly covered the entire width of the channel. Every time Richard raised an oar out of the water, the blade dripped with clinging plants. The strenuous work of pulling our paddles through this floating mass of vegetation made my biceps burn. Close to tears, I expected we'd soon discover how wasted this exhausting effort had been.

Conrad's raft, with rifle-toting Magigi, led our procession, a shrewd precaution should we find ourselves face to face with a predator in these close quarters. Hope faded as this channel closed in on our rafts. I feared it would soon disappear altogether, as we slogged on toward the inevitable dead end. The thought of backtracking down this channel out to the Rufiji knotted my gut. I felt so defeated.

26

BEHO BEHO

FINALLY, RICHARD'S OARS became unusable in the narrow channel. They rested on top of the side tubes, safe from snagging on the waterway's edges. Up front, Masengela and I paddled our hearts out to keep up with the other rafts. Ahead of us, Conrad's boat plowed through the narrowest section of the channel. Just when I felt sure all of our efforts had been futile, Conrad's boat coasted into a much wider body of water. I just hoped the promising scene in front of us wasn't a mirage. The hopeful news traveled quickly from raft to raft. I dug my paddle deep into the water and pulled with every bit of my strength as our boat inched through the last narrow section into wide open water.

Conrad retrieved his pair of binoculars and scanned the distant shoreline. After a minute, he lowered them and pointed to the opposite shore.

"Damn, we did it. This *is* Tagalala!"

"Told ya so." Slade couldn't resist claiming credit.

"Wahoo!" Conrad's fist pumped like a gold medal winner.

Adding to the chorus of elated cheers, I squealed, "We made it! Yahoo! I survived!" I'd beaten my own odds by reaching the end of our river journey, still alive. I *would* get to see my friends and family again.

Far in the distance, parked on the opposite shore, a group of 4x4 safari vehicles waited. They would transport us and our gear to the rustic Beho Beho camp, where we'd spend the next two nights. But only if we managed to cross the lake, giving a wide berth to its hefty population of hippos.

"Let's go for it!" Conrad's joyful urging wasn't needed.

Every paddle dug into the lake's still water with a newfound burst of energy, as we raced to the finish line where the vehicles waited. The drivers walked down to the edge of the lake as we approached. How odd it felt to see humans again. I waved to these complete strangers as though they were long-lost friends.

Conrad shouted out, "Salaam." Waving, they returned the greeting. "Get the bowlines ready to toss to these guys when we touch shore." Then in Swahili he directed the men on shore to catch the bowlines and give our rafts a tug-of-war heave ho onto land. One raft at a time, the drivers pulled us ashore. Like lemmings, we leaped from the tubes of our temporary floating quarters one last time. I felt sad saying goodbye to the river, but the expedition had been physically and emotionally exhausting. I was ready for a couple of easier days on safari.

"Let's get these rafts emptied and packed up. We'll clean them out later." Conrad instructed us to unload our gear, passing items to our passengers, who carried them to the vehicles. There, the drivers stashed every item securely inside or on top of the vehicles' roofs. Our combined effort proved the theory that many hands make for light work. It took an hour to cram everything, including us, into the safari 4x4s.

The unforgiving natural contours of this terrain cruelly dashed any hope of a dirt "road" to lead us out of this wilderness. Our drivers did their best to keep the vehicles' tires centered in the deep ruts carved into the ground without scraping their undersides on the hump of hard-packed earth piled high between the ruts. Before reaching impassable sections, the drivers steered off the track onto smoother land until they had safely bypassed the hazard.

We bounced along the rutted track, the jolt from every bump vibrating along our spines and into our joints. I spent the hour-long ride to Beho Beho camp dreaming of sleeping on a cot with a roof over my head, eating a sumptuous meal prepared by the camp's resident chef. Best of all, I could almost feel the clean water dissolving the caked silt from my body.

From a distance, Beho Beho, the local word for pioneer, reminded me of a classic overnight summer camp. This tiny enclave perched high on the side of a hill overlooking an expansive valley had been the first camp ever built in the Selous's northern sector. Originally constructed as a hunting camp in 1972, the builders chose a site in the cooler highlands of the Selous away from the Rufiji floodplain. It featured a breathtaking view of the migratory routes of many wildlife species.

At Beho Beho, a collection of rustic cabins surrounded a central lounge and dining room. Its remoteness and lack of electricity and any means of communication with the outside world made it a near-impossible sell for high-end tourism. So in 1977, the safari company that originally built the hunting camp sold the property to the Bailey family. Initially, they intended it to be their private home in the African bush. Over time, they selectively rented the camp to groups of tourists looking for an authentic, no-frills, up-close wildlife experience. It perfectly suited the needs of a band of crazy adventurers who had just spent 12 days roughing it in the wilds of the Selous.

When the first vehicle pulled up in front of the main building, Conrad hopped out to search for the manager. While he checked us in, we checked out the spacious open veranda that extended out from the dining room. Gratefully, I plopped down on one of the well-cushioned chairs, strategically placed for game viewing, while sipping a cool drink.

There wasn't enough time to indulge in this luxury before Conrad emerged, holding a list of room assignments.

"Okay, listen up. I'm going to tell you which hut you've been assigned to. You need to gather up all your personal stuff and take it to your hut. Don't worry about getting cleaned up because we're going to a fantastic hot spring where we can swim and relax before the sun sets. We'll spend about half an hour there, and when we get back, dinner will be ready. All you need is your bathing suit and a towel. Safari vehicles will leave here in 20 minutes."

Conrad's surprise revelation met with murmurs of delight. The thought of sliding my sweaty body into a tropical hot spring to soak off two weeks' worth of grime was irresistible. I hurried down to my hut, the last one at the end of a narrow dirt path to the right of the main building. The lower half of its walls, constructed of stone embedded in a clay concrete, formed a sturdy base for the upper walls made from stalks of stacked bamboo laid horizontally. This construction design allowed a constant flow of cooling breezes to penetrate the interior.

A thatched palm roof, already graying from exposure to the weather, overhung three of the hut's walls by a couple of feet. Over the front entrance, a five-foot overhang created a cozy porch shaded from the midday sun with just enough space to accommodate two chairs. A wooden front door opened to the interior, where two screened windows on each side of the door offered panoramic glimpses of the valley below.

I stepped inside, surveying the room's sparse contents. It contained the bare necessities, but not much else. I had a choice of two single cots. Dangling over each, a mosquito net wound into a loose knot hung from a beam above. A kerosene lamp with smudges of black soot on its mantle rested on a wooden table nestled between the beds. Neatly folded on each cot, I found clean but threadbare sheets, a gray wool blanket, and a towel whose pile had thinned from repeated washings. Unbleached cotton muslin curtains hung over the windows held to one side by a hook. Released for nighttime privacy, they would completely cover the window.

I pulled out my camera, water bottle, and towel before stashing my gear and ammo can on the floor under the cot nearest the door. As I followed the path back to the main lodge, I began to skip at the thought of feeling bright, squeaky clean for the first time since leaving home. Once our group reassembled on the lodge's verandah, Conrad explained where the latrines and outdoor communal showers were located. A few people scurried off for a closer inspection.

Inside the dining room, we discovered an earthenware container with a spigot near the bottom, which dispensed boiled filtered water. Conrad gave us thumbs up to ditch the iodine-flavored Kool-Aid left in our water bottles. We refilled them with cool, purified Beho Beho water. Fresh thirst-quenching water had never tasted so good.

Again, we piled into the safari vehicles headed back to the hot springs near Lake Tagalala. As we arrived at one of Nature's original spas, late afternoon shafts of sunlight streamed low on the horizon, warning the sun would set soon. The near side of the hot spring offered a gently sloping entry into the clear, dark pool of water. 10-foot-high boulders covered with shaggy moss and surrounded by leafy ferns rimmed the rest of the hot spring. A few delicate flowering orchids grew from cracks in the boulders. The greenery created the ambience of an exquisite lush oasis secreted in the middle of the vast dry savannah.

Conrad set some ground rules as we gathered around the spring.

"So, we've got about half an hour to enjoy a soak before it gets dark. I don't need to remind you that dusk is the optimum time for critters to be hunting for food and water, so we might have a few visitors show up. We all feel really grungy and want to get clean, but we can't use soap or shampoo in this spring because it *is* a watering hole for the local wildlife. If the drivers see any animals approaching, we'll need to get back into the vehicles quickly. Got it?"

Martha pointed at the pool. "How deep is it? And is there anything in there that could be dangerous to us?"

"You mean like a baby crocodile or a water snake?" teased Conrad. "Only one way to find out." He boldly waded chest-deep into the dark pool. Immediately, we followed, splashing our way into the 102-degree warm water. The hot spring accommodated all of us, but just barely. Our moods turned euphoric as our bodies slithered beneath the warm, velvety water. The action of clamoring into the water roiled up some of the bottom sediment, but it quickly settled back to the bottom.

As giddy as toddlers in a bathtub, we marveled as our skin color lightened by several shades. I submerged my head momentarily, scrubbing with my fingertips to loosen the crust of river silt from my scalp. Relaxing in the hot spring, I decided this had to be the most beautiful spot on the planet. In that glorious moment, every one of my senses rejoiced. I willed my brain to engrave this scene forever in my memory. I figured God had given us this reward for our perseverance in traversing some of the wildest terrain on the planet.

As the sky darkened, the lead driver motioned to Conrad that we needed to start the trek back to Beho Beho. Knowing we'd need an enticement to lure us out of the cleansing water that had sent our spirits soaring, Conrad sweetened the deal.

"If we want to eat our dinner while it's still warm, we've got to head back now. The good news is I think you'll all enjoy the mixed grill barbecue they're preparing for us."

The image of fresh grilled meat persuaded us to climb back into the vehicles. I wrapped my towel around my wet shorts and T-shirt. With any luck, the clean water had also dislodged some of the silt embedded in my clothes. After air-drying them overnight, at least I'd have a cleaner outfit for our safari day.

By the time we arrived back at the lodge, darkness had fallen. Staff carrying lanterns led us back to our huts, where inside, kerosene lamps already blazed. The meager light cast by the burning wick flickered on the walls, barely providing enough light to move around in the hut. I swapped my wet clothes for dry ones, dug out my headlamp, and followed the path back to the lodge. The intense blackness of this cloudless night set the stage for a spectacular display of twinkling stars. The Milky Way, clearer than I'd ever seen it, spanned the dome overhead as a dense band of trillions of stars. I paused, literally starstruck, to admire the breathtaking sight. It humbled me to think I was gazing at starlight that originated 2.5 million years ago. At that moment, with full clarity, I saw that my life was but an insignificant blip in the universe. This otherworldly panorama defied description with words. It had to be experienced.

"No one's going to leave here hungry tonight. If you do, it's your own fault," Conrad said as the servers placed platters of grilled meat and chicken along the length of our communal table. The red meat looked like steak from a cow but might have been water buffalo or zebra. Whatever meat it was, I couldn't wait to savor its taste. Other platters piled high with rice and vegetables and a variety of fresh tropical fruits followed. I took advantage of the potable water they offered, while many of my colleagues made a sizeable dent in the camp's inventory of local beer.

With no worries about tents to set up or a kitchen to clean after dinner, most of our group gathered on the veranda to reminisce about the good and trying times we'd been through on our journey. Cooler breezes than we'd encountered at our river camps drifted through Beho Beho's open-sided lounge. Conrad and Richard asked questions to solicit our group's impressions of the river trip.

Favorite memories of the trip easily spilled out as each of us recalled a moment or two we'd especially enjoyed. Kate spoke first, extending her arms outward. "Seeing all this fabulous wildlife, especially seeing and hearing the lions!" Several heads nodded in agreement.

I'd been thinking about how our up close and personal encounters with wild animals had drastically changed my perceptions of them. "Never in my wildest dreams did I imagine sharing wide open space with hippos, lions, elephants, or any wild animals. I'd only seen them in zoos, confined behind thick panes of safety glass or secured in steel-barred cages, living in a tiny fraction of their natural habitat. As an American taught to be fearful of them, those practices seemed necessary and acceptable. Now, after seeing them free in their own environment, I question the ethics of isolating and—let's call it what it is—jailing them. Honestly, I don't know if I'll ever be able to go to a zoo again."

Other heads nodded as the disturbing reality of what we all now knew hit home hard. A moment of silence passed before the next person spoke.

Martha quickly offered a compliment. "Conrad's fresh-baked bread was my favorite thing."

One of the younger men agreed. "You can't buy bread like his at any bakery around where I live."

Winking at Conrad, Martha openly flirted. "I guess I'll just have to pack him in my suitcase and bring him home with me." At that, Jay feigned a startled look, drawing chuckles from the group.

"Our camaraderie as a group was just great," one of the older men added. "Everyone pitched in to help in any way they could. I feel like I can build a bonfire from scratch anywhere on the planet. Guess that's a good life skill to have."

"Well, maybe not in the middle of the Sahara Desert," Bart teased, triggering another round of laughter.

"For me, it was running a river I've never been down, in a place I consider the real Africa." There was a wistfulness in Michael's words. "It really bummed me out when I couldn't join Conrad two years ago on the exploratory trip, so I *really* wanted to be on this trip."

"Yeah, it's always satisfying to add a new notch to our belts, isn't it?" Slade echoed Michael's sentiment.

A gradual crescendo of nocturnal animal racket filled the air around us as Richard tackled the tougher question. "So what did you guys not like about the trip?"

"That's easy. Feeling dirty the whole trip," said Martha. "I know it's part of the whole wilderness experience. It's just that I've never felt so unclean for so long." Her pet peeve seemed to be shared by several other people, including me. "But after that hot spring bath, I feel like a whole new woman!"

One of the older male clients spoke up. "Aw, I can deal with being dirty. What got me were the portages, especially that long one. That felt like doing hard labor, and I only had to make the trip once. I feel bad for you guides who had to make the trip several times."

Jay asked, "Have you ever thought about putting a small motor on each raft to help us battle those brutal upstream afternoon winds?" I figured his wife had put him up to asking that question.

A little perturbed, Conrad laid out all the reasons that would be a bad idea. "Do you have any idea how much extra gas we'd have to carry, *and* portage, for each raft? The noise would scare off wildlife we might otherwise have the chance to see. God only knows what the hippos' reaction to the motors might be. If we built less time into the schedule for using the motors and one or more of them failed, we'd be up shit's creek. Nah, we're not going to use motors."

Slade agreed. "Yeah, it was more difficult than I thought it was going to be." The other guides chimed in to share their perspectives on the afternoon winds and the portages. In hindsight, they agreed the scaled-back exploratory trip, using a crew of only four muscular men in two oar-powered rafts, transporting far less equipment and food, made the expedition seem more feasible to pursue.

Conrad pointed out, "The fact of the matter is that the normal guide-to-passenger ratio is three to four clients per guide. To make this trip work, including the game rangers, the ratio is one to one."

Richard looked crestfallen. "This is an amazing trip everyone should experience, but I realize the long portage is a big ask for clients who are not expecting the hardship it involves. And if the only way to run it commercially results in us losing money, then we need to take a hard look at whether we should continue to offer it."

The unique journey on this river system had required more resources than any other commercial trip run by Sobek, and on a scale of logistical difficulty, it placed near the top. It was a dilemma for me personally, being responsible for Sobek's finances. If we couldn't make money running the trip, I'd be in the heartbreaking position of having to advise against operating any future river trips through the Selous.

"Well, we'll do a complete analysis of the trip back in California." Richard shrugged his shoulders, looking around the group. "It may just turn out that all of you have been on the one and only river trip through the Selous that we'll ever run. Consider yourselves lucky."

27

SAFARI DAY

THE SOUND OF A GIANT SLEDGEHAMMER repeatedly pounding the ground outside the wall of my hut forced me awake and out of my cot to investigate. Daylight brightened the room, but I didn't hear any voices or activity of humans outside. Every couple of seconds, the earth shuddered. Something brushed against the bamboo wall. Whatever was out there had to be huge. Then I heard the swish of grass being ripped from the ground and the crunch of it being ground to a pulp.

The creature rounded the corner of the hut, then paused. I drew one side of a curtain open just enough to peek out, staring directly into a rusty-brown eyeball sunken in wrinkled folds belonging to a giant gray elephant. Standing in front of my hut, its six-foot-long trunk uncurled, reaching toward the window, sniffing for scents of the creature that belonged to the eyeballs hiding inside the hut. I'd never had such an unusual wakeup call.

Sensing no imminent danger, the elephant casually sauntered away from my hut, stopping every few steps so its trunk could snatch another helping of field greens, hungrily shoving them into its mouth. I pulled the curtains wide open to watch my unexpected visitor amble along on its morning foraging stroll. Its long tail, sporting a clump of thick hairs at its tip, resembled the bristles of a hairbrush. It swished at a cloud of pesky flies surrounding its girth as its hindquarters lumbered in the exaggerated sway of a runway model.

This elephant showed no fear as it wandered the grounds of Beho Beho. Pleased that I'd resisted the urge to scream, I stood transfixed in awe that a wild elephant had passed within five feet of me. For the second time in my life, I'd had a close, intimate encounter with an elephant.

I dressed in a hurry before heading to the dining room, still wary about what other animals might cross my path on the way there. Far in the distance in the valley below, I saw a herd of grazing gazelles. But my walk to the lodge lacked any surprise animal encounters. A simple continental breakfast of fresh fruits, breads, and jams provided us nourishment before we departed on our early-morning game drive. The odds of seeing wildlife significantly increase when driving through the bush in the early morning before they hunker down to wait out the heat of midday.

Armed with visored hats, sunscreen, full canteens, cameras, and binoculars, we eagerly divided up and climbed into three, open-sided, canopy-topped safari vehicles. The driver of the vehicle I'd boarded briefed us on what we'd be looking for and lectured us on safety protocols while in the bush.

"You must stay in the vehicle at all times. No wiggling hands or feet over the sides of the vehicle. If we're able to approach wildlife, stay quiet. Do not taunt or provoke them as they might decide to attack. My job is to deliver you back to Beho Beho camp in one piece."

His stern warning reeked of irony. He was lecturing us about proper behavior in the bush as if we were first graders, while I felt I'd just earned a college degree in wildlife survival from living on my own in the African wilderness for the past 12 days. But honestly, I had to admit our safari vehicle felt much safer to me than our rafts.

We drove out across the valley following dirt tracks, searching for species of game we hadn't yet seen. Our first sighting was a lone warthog.

These African porkers are fierce-looking, with two pairs of short tusks protruding from the sides of their snouts and bristly hairs forming a crest that runs along the top of their backs. While they are in the same family as domestic pigs, they are smaller and nimbler. Both love to use their long snouts and hooves to dig holes in the ground. As our driver left the track to get closer, the warthog raced off into a wooded area.

A few minutes later, we emerged into more open grasslands. "Have a look over there." The driver pointed ahead to our right while he brought the vehicle to a stop. A herd of wildebeests galloping for their lives stampeded away from a pride of lions in hot pursuit. Also called gnus, wildebeest are a type of antelope capable of running up to 50 miles per hour. The chase scene became a moving blur as the hooves of the chased churned up the dry topsoil, creating a thick cloud of dust between them and their enemies. The long black tails of the wildebeests streamed out behind them like flags in a gale-force wind. As the racing animals thundered far into the distance, our driver shifted the vehicle back into drive. Since I hadn't wanted to witness the end of that race, I was relieved we were proceeding forward on our track.

A short while later, as the track turned back toward a sparsely wooded area, we found a herd of five elephants, two mature parents with three youngsters. They wasted no time in shuffling away from our vehicles, leaving us with only great shots of their hind ends. One of our passengers, unsatisfied with only seeing their rears, asked, "Would it be possible to get a little closer so we can see their faces?"

The driver politely declined his request, "Don't worry. We'll be seeing more elephants. Be patient." In fact, his prediction materialized as we later passed two additional herds. One group eyed us suspiciously as our vehicles slowed to a stop for a photo op directly opposite where they loitered under a large canopied tree. Seeing these enormous beasts so relaxed in their natural environment filled my heart with joy.

The dense vegetation edging the river had hidden from our view all these species that roamed the savannah. Now traveling by safari vehicle, the unobstructed views of so many species of wildlife left me breathless. In the distance, clustered herds of antelopes and gazelles wandered grazing on their favorite grasses and leaves. When our driver spotted 20 zebras bunched together about 100 yards off the track, he swerved onto the grassy plain, carefully navigating closer to offer us a better view of them.

"Anyone know what a group of zebras is called?" he asked.

"Inmates?" joked one passenger. "They're all wearing black-and-white stripes!"

The driver wobbled his head at the ridiculous answer.

"Anyone know?" he asked again. Hearing no other response, he revealed the answer. "They are a dazzle of zebras."

The description fit perfectly because, as they huddled together, it was impossible to tell where one zebra ended and another began. The collage of wavering stripes as they meandered away from our vehicle was mesmerizing in a dizzying way. I made a mental note never to buy a thousand-piece jigsaw puzzle of a dazzle of zebras because I doubted even a genius could complete it.

Amused by our lack of knowledge about the native wildlife, our driver continued his goading. "So are zebras black with white stripes or white with black stripes?"

The question prompted a lively debate. From what we could tell, a zebra was exactly half black and half white. Even their wiry manes bore stripes. A patch of black on the top of their muzzle provided a clue that turned out to be the giveaway. Smug that he'd stumped us again, our driver confirmed zebras *are* black with white stripes, a fact likely useful only for trivia buffs.

At one point where the track curved 90 degrees to the left, I gasped as three magnificent giraffes came into view. Gathered around the base of a tall acacia tree, their long necks stretched up, while their blue tongues gathered tender leaves into their mouths.

Our driver pulled to within 100 yards of where they stood, then continued schooling us about African wildlife. "Although giraffes are the biggest animal out here, they're quite timid, so we'll watch them from here. They're the tallest animal on the planet. That group over there is between 14 to 18 feet tall. You know, if you could get close to them, you could walk right under them without hitting your head on their bellies because their legs are six feet tall, but I wouldn't recommend you try that. Their necks are six feet long too."

I studied the square brown markings on the beige background covering their bodies. "The intricate mosaic design on their coats looks like an artist's masterpiece," I remarked.

"You're right," said our driver. "Their coats are like our fingerprints. The pattern on each giraffe's body is unique to that animal only. Anyone know what a group of giraffes is called?" There were no wisecracks this time, only blank stares.

"Well, it's really quite obvious," he teased. When no one hazarded a guess, he provided the answer. "They're called a tower of giraffes." This name for the group also perfectly described the trio that had captured our attention.

I surmised our driver held a special affection for these gentle giants as he continued with giraffe trivia. "How long does a giraffe sleep each day?" Again, he'd stumped us. Clearly, he enjoyed teaching us about Beho Beho's resident wildlife. "If you were out here roaming around with lions, how much time would you want to sleep?"

"Not much," said Frank.

"That's right. They take short power naps now and then, getting about 20 minutes of sleep a day."

"What do those cute little knobs on top of their heads do?" I asked.

"They are a horn made of bone covered with skin and fur called an ossicone. When males fight with each other, they use them as clubs to deliver blows to their opponent."

One of the male passengers chuckled, "Guess that's where the expression 'butting heads' comes from."

The temperature climbed steadily as the sun reached its apex overhead. The rising temperature drove the grazing wildlife from the open grasslands into the shade of wooded areas. It also drove us to return to the shade of Beho Beho's main building for lunch and naps. On the return drive along a section of the track surrounded by thick brush on one side, our driver abruptly hit his brakes to stop the vehicle. He turned to us with his index finger in front of his lips, signaling we should be silent. He slowly pointed to the right and mouthed, "Lion."

I saw a dense stand of leafy bushes, but nothing that resembled a lion. Nor was I the only one in our vehicle having trouble spotting it. Then she moved ever so slightly, and the outline of her shape and her glassy eyes revealed where she lay camouflaged under a bush. I couldn't believe our driver had even spotted her. It shocked me to think I could be just 10 yards from this lion, yet completely oblivious to her presence.

In retrospect, I realized how lucky we'd been to see the lioness sunning herself in plain sight on the riverbank a few days before, as opposed to a disappointing glimpse of a lion hidden by thick bushes or streaking through a dust cloud after a herd of wildebeests.

Back at Beho Beho, we dove into a buffet lunch of fresh breads, cheese, hard-boiled eggs, German potato salad, smoked meats, and a platter piled high with tropical fruit salad. After eating, a few of our group trekked to visit the nearby grave of the namesake of the reserve. Frederick Courteney Selous, the renowned professional elephant hunter, lost his life in 1917 while fighting the Germans in World War I at Beho Beho. His life's adventures inspired the Allan Quartermain character in Sir H. Rider Haggard's *King Solomon's Mines.*

The rest of us spent a leisurely afternoon enjoying the expansive view of verdant plains and distant hills from the raised veranda. I needed this break to relax and recharge before tackling the next part of our Tanzanian journey. I used an hour of my free time to hand-wash all my river clothes in tepid soapy water. Spread over the wooden rail in front of my hut, they dried quickly in the brisk afternoon breeze. My thoughts drifted with fondness back to the Ifakaran ladies washing their *kangas* by the side of the stream. How I wished I could send a message to my *rafiki* to let her know I had survived this wild adventure. I regretted she might wonder her whole life how her American *rafiki* made out. I hoped she would grow into a strong, smart woman, find a good man to marry, and have her own babies to nurture and love.

28

RETURN TO DAR ES SALAAM

AS SOON AS I FINISHED DINNER, I returned to my hut to repack my river bag and backpack. I hadn't gotten far when fatigue caught up with me. I relished the comfort of the indoor cot knowing how soon we'd be back to sleeping in tents on the ground. I stretched out on it and fell fast asleep.

A wakeup call delivered by elephant was not in the cards for our last morning at Beho Beho. Instead, a staccato rapping on my door let me know it was time to gather for breakfast. I loaded my plate with tasty items from their breakfast buffet and found a seat at the table just as Conrad shared a few logistical tidbits. "Listen up. You all need to eat a hearty breakfast 'cause there's nowhere to get lunch on the way back unless you want to snag something from the vendors hawking stuff outside the train windows."

I interrupted, waving my hands in front of me like a football referee signaling an incomplete pass. "Don't even think of buying mangos. They taste like kerosene. Believe me, you'll regret it."

Conrad continued, "We need to be at the train stop by 11 a.m. *If* the train is on time, we'll be boarding at noon. Unfortunately, we have no way of knowing if it will arrive on schedule, so we may be in for a wait. The good news is that since we've been heading east on the river, we're already halfway back to Dar. We've only got a train ride of 118 miles left. When we arrive in Dar, we'll check in at the hotel, then gather in the evening for our farewell banquet. Keep your fingers crossed the train will be on time."

Two growing mounds of river gear, separated by the crew, accumulated in front of the lodge. The reusable equipment we'd take with us: rafts, oars, paddles, coolers, and the kitchen setup. The rest of the gear and unused food supplies we'd leave behind to be recycled by the Beho Beho camp. Less than half of the trip gear we'd brought with us would make the return trip to Dar. The drivers jostled the Sobek gear, cramming it into every nook, cranny, and onto the rooftop of the vehicles. Afterward, we crowded into the remaining cramped space. I dreaded the claustrophobic 25-mile ride over the rutted track, ridged like an old-fashioned washboard, to reach the train station.

At 10 a.m., our safari vehicle caravan pulled away from Beho Beho, the Rufiji River, and the greatest adventure of our lives. I twisted around to watch the camp grow smaller and tried to memorize every charming detail of this tiny paradise before it disappeared from sight.

Like the herd of wildebeests chased by lions, the vehicles in our caravan billowed up a blinding cloud of thick dust. Only the clients in the lead vehicle enjoyed the drive. Our two trailing vehicles had to drop far behind so we wouldn't suffocate from sucking in the particles of airborne dirt. So much for being clean. Whether from river silt, dirt-track dust, soot from diesel exhaust, or wood smoke from cooking fires in the city, it seemed impossible to stay clean for long in Tanzania.

After an hour, our vehicles pulled up to the train station at Fuga. A few local men loitered near the cinder-block office. For a few shillings, they agreed to help unload equipment from our vehicles and, once the train arrived, to load it onboard. Conrad searched for someone in charge. When he returned, he told us the station attendant informed him the train was going to be "a little late, not much, just a little late."

"It's hard to know if this guy has any clue what time the train will arrive or if he's hedging his bet because the train usually arrives late." Conrad sarcastically mimicked the official, using the thumb and index finger of his right hand to make a gap of half an inch of space. "Just this much late."

While we waited, I took pictures of the vibrant fuchsia bougainvillea growing abundantly around the periphery of the train stop. No one tried to stop me from using my camera at this tiny station, perhaps because other than the flowers, there wasn't much worth photographing. As the midday sun climbed to a position directly overhead, the patch of shade that the walls of the office building had cast disappeared. I picked a place to sit on the ground where a trifling breeze swirled every so often. Even so, trickles of sweat dampened my t-shirt, and my face flushed from the heat.

A few local villagers meandered up to the edge of the tracks as gossip spread in the village about a large group of foreigners hanging out at the train stop with a mountain of unusual baggage. Noon came and passed without a train appearing. However, local women dressed in brightly colored *kangas* balancing fully loaded baskets of fruit on their heads began to gather at the stop. None of the locals seemed impatient. Apparently, waiting for the overdue train was their normal routine.

One of our clients suggested, "Maybe we should make bets on when the train will arrive. Closest to the actual time wins all."

No sooner than he voiced the suggestion, we heard the shrill whistle of a train far in the distance. The locals Conrad hired jumped up and stood next to our pile of gear. Five minutes later, our train rumbled into the station. A conductor hopped off the train to check tickets and assist passengers wanting to board.

Conrad spoke briefly with the conductor, who then directed the hired workers to an empty baggage compartment that slid open along the lower half of the train car. Under Conrad's watchful eye, the workers quickly stowed our possessions. Meanwhile, the conductor inspected the return tickets that Conrad handed him. Finding all in order, he directed our group to follow him as he climbed aboard the train. One by one, the group boarded, then shuffled single file down the narrow hallway toward our assigned first-class cabins. Other than Conrad, I was the last of our group to embark.

Grabbing the two metal railings, I hoisted myself up three steep steps onto the train and started down the hallway. I detected a commotion ahead as another passenger forced his way down the hallway in the opposite direction of our group. He maneuvered his way around them, hurrying in my direction.

"*Jambo*! Hello!" he exclaimed with delight.

I glanced at the face of the approaching man. He looked familiar, but from where? Then it hit me whose eyes I was peering into.

"Solomon!"

"Yes, I saw you stand outside by train tracks. I couldn't believe it, that I saw you. But when I saw all equipment, it must be!"

Staring at him in amazement, for a moment, I was speechless.

"I thought I'd never see you again," I said.

"Me too. In that wilderness, I wasn't sure you come back."

"Oh my god, Solomon, a hippo bit one of our boats, and we had to shoot another. And there were lions roaming around outside our campsite one night."

His eyes grew wide in astonishment.

"I told you was dangerous, what you want to do. Tell me this adventure."

"I will, but first let me get settled in, and then we can talk."

Conrad rushed up behind us, concerned that I was being hassled by the appearance of our animated discussion.

"What's going on here?"

"Conrad, meet Solomon. On the way out to Ifakara, he shared a compartment with Richard and me. He taught us the proper way to eat a mango and gave us lots of information. He thought we were on a suicide mission." I laughed, remembering our conversations.

Conrad's stern expression softened when he realized I knew the man. Conrad had booked three six-person compartments for the 17 of us returning to Dar, so we had one empty spot.

"Hey, Richard, look who I found." From the bench where he was sitting, Richard sized up the stranger standing next to me in the entryway. His face gave no hint that he recognized our friend.

"It's Solomon!" I said.

As it dawned on Richard that he knew Solomon, he greeted him warmly with a handshake. "Hey, man, how ya doing? How was the border?"

"Solomon wants to hear all about our trip," I said while stowing my backpack and camera case below our seat. I slid along the bench next to Richard, leaving space for Solomon to squeeze in next to me.

For more than an hour, we entertained Solomon with vivid stories of hippo attacks and stalking lions. Our encounters with the wildlife left him astonished. I described my *rafiki* and how deeply her friendship touched me. Richard added an embellishment here and there to my narrative. Even the three clients sharing our compartment joined in the storytelling. I couldn't tell if Solomon actually believed us or thought our whopper-sized tale was all bullshit.

Before he returned to his own seat, he spoke to me in a whisper.

"You still want get one of our real Tanzanian *kangas*?"

"Yes, of course," I whispered back, wondering what illegal shenanigans he had planned.

"My sister might want sell one of hers, but you no tell anyone where you got, or she get in trouble."

"I promise."

"How long you be in Dar?"

"Actually, just tonight. We have to fly out to Moshi tomorrow."

The tight deadline caused him to cringe. "Might not be enough time. I need go to my sister's house. *If* she want sell, where I meet you tonight?"

What was I doing getting involved in this type of deal? I knew I should walk away. But I wanted to have one of the beautiful *kangas*. If I didn't use Solomon's help, how else would I get a real one?

"How much do you think she'd sell one for?"

"I not know if she want to. Maybe 20 US dollars cash?"

I was worried he might tell me 100 US dollars, in which case I would pass on his deal. At 20 US dollars, she'd be making a decent profit, and I'd have my authentic *kanga*. It seemed like a win-win for both of us.

"Let me ask Conrad where our banquet will be."

Solomon followed me down the hallway as I checked in the other compartments to find Conrad. He'd arranged the farewell dinner for 7 p.m. at one of the nicer hotels in the center. Solomon knew exactly where it was.

"Can you meet me in the hotel lobby at 6:30 p.m.?" To keep this deal from being witnessed by my colleagues, I wanted to finish our exchange before we gathered for dinner.

"I be there if I can."

"Don't worry if it doesn't work out. I'm just happy that you're willing to try." I shook his hand to seal the deal. As he headed back to his compartment, I called out, "Good luck!"

29

KANGA

AFTER THE TRAIN CHUGGED into the Tazara station in Dar, Conrad wound his way among the queued taxis, waiting to provide transport for the train's passengers. He haggled vigorously with the drivers until he'd hired enough taxis to transport us and all of our gear to our hotel. He also negotiated for some of them to return to our hotel the next morning to bring those of us who were continuing on the Mount Kilimanjaro climb to the airport.

I'd expected to feel relief that we made it back to "civilization," but the constant din of the bustling city, the polluted air, and the days' old piles of trash littering the sides of the road already had me yearning for the unspoiled wilderness we'd left behind. The stark contrast of the absence of nature on land populated by humans in the city compared with the Selous's virgin land untouched by people troubled me. With the planet's population constantly trending upward, one day a sprawling metropolis of concrete, asphalt, and steel might completely cover the earth's surface. What a dreadful thought.

After checking into the hotel, I used my free time before our banquet to indulge in a rejuvenating, hot shower. Standing in the tiny, white-tiled cubicle, I lathered from head to toe with a fragrant scented soap that dissolved away the day's grit. My body relished being swathed in this spray of hot clean water, but my mind wandered back to the lush outdoor paradise of Beho Beho's hot springs at sunset. The experience of such a splendid bath made the hotel's shower seem shabby in comparison. When I was done with my shower, I donned my only piece of clothing not meant for outdoor adventuring, a lightweight cotton sundress, and went to wait for Solomon in the lobby.

A waiter from the small bar inquired if I wanted something to drink. I couldn't resist ordering. "A cold Fanta, but no ice, please." The availability of ice to add an extra chill tempted me. However, I'd be crazy to risk another bout of gastric distress immediately before climbing Africa's tallest mountain. I watched as a few of our clients left the hotel to poke around in the nearby gift shops.

Our appointed meeting time came and went with no sign of Solomon. That he'd stood me up disappointed me, but I couldn't hold it against him. Although he'd suggested the *kanga* sale, it was a long shot he'd deliver on the deal. At 7:00 p.m., our group assembled in a private dining room on the second floor of the hotel. The chintzy European décor seemed oddly out of place. A single long table tastefully draped in white linen stretched nearly the length of the room. A brightly colored floral arrangement of tropical blooms adorned the center, surrounded by 17 place settings.

Of the three entrees offered, I chose the jumbo prawn curry. How extravagant it felt to order such a gourmet dinner after days of eating fried SPAM, canned sardines, and spaghetti. Yet I had to admit that when eaten on a riverbank marveling at a magnificent sunset, fried SPAM tasted scrumptious. It proved the truth of the marketing mantra—location, location, location.

The presentation of the food on the plate placed in front of me tantalized my senses. Five silver-dollar-sized prawns, chunks of translucent onion, half-inch rounds of bright orange carrot, and one-inch-square hunks of fresh pineapple simmered in an aromatic yellow curry sauce topped a generous serving of basmati rice. Minding my manners while waiting for the others to be served only heightened my expectation of the burst of flavor I'd get from the first bite: spicy, creamy, and exotic.

Everyone complimented the food served to them. Chatter dwindled, replaced by the clinking of silverware, scooping up morsels of food. I'd eaten half my delicious meal when a waiter approached me. He leaned down and spoke in a hushed voice.

"Madam Brenda, excuse me, but there is a gentleman here who says that you are expecting him." He glanced over at the entrance to the dining room, where Solomon stood nearly out of sight just beyond the doorway. He motioned for me to join him outside the dining room. I thanked the waiter for delivering the message. Then I spoke to Richard, seated directly across the table from me.

"I'll be right back. Will you make sure they don't clear away my plate?"

As I headed toward the exit, I'm sure Richard must have thought I simply needed to use the ladies' room.

When I reached Solomon, he said, "Follow me."

"Where are we going?"

"Outside."

He trotted down the staircase leading to the lobby. I followed, noticing he had a package wrapped in newspaper tucked under his left arm. Was it his sister's *kanga*?

"We must go outside so no one see us," he explained, opening the hotel's front door.

Though a few people still passed on the street in front of the hotel, outside, it was dark. I paused. Every one of my instincts screamed, "Are you nuts? You can't just follow this man you've known for less than 10 hours out of this hotel alone. What are you thinking?"

"How far are we going?" I tried not to let my anxiety show in my voice. I needed to know more about his plan before stepping outside the safety of the hotel.

"A little way down street." He whispered, "This very dangerous for me if I get caught."

Though I didn't say it, I wanted to counter with, "And it's dangerous for me too. Maybe even more dangerous for me as a foreigner." Was I really about to break the law? Maybe I hadn't thought this through well enough. He sensed my hesitation.

"I have it. Right here." He touched the end of the package wedged between his upper arm and chest. "You really like it."

I faced a quandary. I wanted to have a real *kanga*, but would I be stupid to take this risk? Doubts clouded my mind. Could I trust Solomon? Why hadn't I asked Richard his advice before I got myself into this predicament?

"Come on. Let's just do this." The scowl on his face reflected his growing impatience.

As soon as I took a step outside, I saw his features soften with relief.

"We go down this way." He turned to the left and began walking. I followed a few steps behind. All of Conrad's warnings about Dar being dangerous at night, rife with thieves and not wanting me to be left on my own, raised dreadful scenarios in my mind. What if Solomon wasn't alone? Was I being duped? Did he see me as a naïve, vulnerable woman, an easy target to exploit, to rob, to beat up and leave for dead?

50 yards down the street, he ducked into a narrow alley between two buildings. I stood in the street peering into the pitch-black alley. My heart pounded. My mouth was parched with fright. Solomon waited about five paces inside the alley. If anyone else lingered beyond him, the dark confines of the alley provided the perfect cover.

He motioned for me to join him. "Come here off street. No one see us here."

It all came down to this moment. Do or die time—literally. Solomon, who'd been so helpful on the train, so excited to see me on the return, and so eager to find a *kanga* for me, couldn't want to hurt me, could he? Still, I trembled with uncertainty. Wasn't this whole trip teaching me to overcome my fears to reveal wonders if I found the courage to take judicious risks?

I stepped into the alley facing Solomon. I fished a $20 bill from my money belt and offered it to him.

"Thank you, Miss Brenda. Here is *kanga*. I hope you like. Think of me when you use." I took the package from him.

"No, thank *you*, Solomon. I really appreciate you did this for me. I will never forget you. Even if we hadn't met up again, I would never have forgotten you."

"Remember, no one can know where you got this."

"I promise."

"Be careful on Kilimanjaro. I go now."

"Will you walk with me back to the hotel?"

"Better not. But I watch from here. You go back in hotel. You first."

"Thank you, Solomon. I'll treasure this."

I stepped out of the alley and scurried quickly back to the hotel. Only after I entered the hotel lobby did the adrenaline rush subside. I couldn't return to the dining room with the package. Everyone would question where I went and what was in the package. Instead, I brought the precious parcel to my hotel room.

There I unwrapped the newspaper and found a pair of stunning cloth rectangles printed in yellow, maroon, and black. In the center was a geometric floral medallion, surrounded by black polka dots on a yellow background. Alternating columns of branches and hearts formed the border. I searched for the Swahili proverb on the edge. When I found it, the one word I recognized astounded me—*rafiki*!

Was it just a coincidence, or had he somehow finagled to get me a *kanga* with this word on it? I felt my eyes moisten with tears as I thought of my two *rafikis*.

My absence, much longer than a trip to the bathroom, raised a few eyebrows when I returned to the dining room. The group was halfway through dessert and coffee.

"Are you okay?" Richard asked. "I thought you might not be coming back, so they took your plate. Sorry."

"I'm fine." Because of the success of my *kanga* deal, I no longer felt hungry. I fibbed to avoid any further questions.

"I just needed to walk off a little indigestion. I'm fine."

30
OFF TO KILIMANJARO

THE NEXT MORNING after breakfast, we shared some heartfelt goodbyes with the members of our group who were not continuing on to Mount Kilimanjaro. Richard needed to fly back to California to research itineraries for future adventures. Michael would continue with his research on mountain gorillas back in Uganda. A few of our clients had opted out of doing the Kilimanjaro climb. Conrad helped arrange transportation to the airport for all the different departure times needed and settled up with the hotel for our meals and lodging charges.

For those of us heading to the mountain, the flight took an hour to Kilimanjaro International Airport. As we flew north, I spotted snow-capped Mount Kilimanjaro majestically rising from the flat plains surrounding it. The mountain looked immense from my aerial view. It seemed impossible that we could climb to the top of it.

Because it's not part of a mountain range, Mount Kilimanjaro is unique as the earth's tallest freestanding mountain at 19,341 feet. As the highest peak in Africa, it is one of the Seven Summits, sharing the distinction of being the highest peak on each continent. Considered the easiest of the seven summits to climb, reaching its top requires no technical climbing skills or mountaineering equipment. Yet its relative lack of difficulty makes it one of the most dangerous, as climbers often underestimate the impact on their bodies of being at such a high altitude.

At the airport, Conrad negotiated with two vans to transport our slimmed-down group of just 10 to Moshi, the largest town on the south side of the mountain. After an hour-long ride, skirting the heavily vegetated lower flanks of Kilimanjaro, we arrived at a small hotel in the bustling center of town. At the conclusion of the climb, the Sobek crew would spend a couple of nights at this hotel, so Conrad had convinced them to store our rafting equipment while we were on the mountain.

With that errand completed, we headed to the headquarters of the Kilimanjaro National Park, where we had reservations in their hostel that night. As we turned off the heavily trafficked two-lane road from Moshi onto a narrower road that sloped steeply uphill, I felt that familiar fear of the unknown trying to take over. After surviving our Selous adventure, at least the excitement of a new adventure helped to temper the panic.

My ears popped twice during the mile-and-a-half drive up to the park headquarters, located 6,000 feet above sea level. However, it was still at the foot of the mountain we were about to tackle. The highest mountain I'd ever climbed was Mount Washington, the highest of New Hampshire's White Mountains range with a summit of 6,288 feet above sea level, and an iconic point on the Appalachian Trail. I couldn't believe that we were already close to the highest point I'd ever hiked to. This first night would allow our bodies to adjust to being 6,000 feet higher than where we'd started our day.

The headquarter complex's modern design housed administrative offices, charged with overseeing both conservation of the park's natural resources and the needs of the tourists who came to climb the mountain. Just beyond the entrance and reception desk, stretching out to the rear, was a wing of sparsely decorated rooms offering basic lodging that catered to climbers who needed a cheap, clean room for one night before or after their climb. A small restaurant with a limited menu extended onto a garden terrace from the lobby. After checking in and receiving our room keys, Conrad returned to where we waited with our duffels and backpacks.

"I've got your room assignments, so go ahead, find your rooms, and get settled in." He circulated among us, handing out keys. "We're gonna eat an early dinner tonight because you all need to hit the sack and get as much sleep as you can. The climb tomorrow will use up your energy sooner than you think. For the last 12 days, you've been using your arms to paddle the rafts. Starting tomorrow, your legs are going to get a real workout."

For hours each day on the river, we *had* been sitting on our butts, building strength in our arms, shoulders, and core. This climb would be a further test of our endurance and stamina. Slade, who was an accomplished technical climber, told us that success on this mountain was not only a test of our physical fitness, but our mental toughness. I wasn't convinced I had that required toughness, but once again, I stood at the base of this monstrous mountain with no option but to take on the challenge and give it the best I had.

AS WE ATE OUR DINNERS, Conrad told us about the logistics for the next day.

Scorn tinged his voice as he spoke. "Most of the climbers that come here choose the traditional Marangu route, otherwise known as the tourist route, because it's like following the damn 'yellow brick road.' You use the same wide, smooth track both up and back." With a roll of his eyes and a flick of his hand, he dismissed that climbing route.

"To make things more interesting, *we're* going to hike the Mach-ame Trail. Less than 100 climbers use this trail each year, so it's still in its natural condition. It's more scenic, but with alternating uphill and downhill sections, definitely a more strenuous route. It's a loop trail, so except for the ascent day, there's no backtracking."

Back in Angels Camp, I had read about the climb in Sobek's brochure. "Seeking as always the less-trammeled ways, we have chosen a route rarely traversed but safe and of only slightly greater difficulty. The Machame Trail, winding underneath the spectacular glaciers of the southern face, requires an extra day of ascent, allowing additional time for acclimatization. The climb is taxing and the altitude restrictive."

Before our group scattered back to our rooms, Conrad announced, "Breakfast is at eight o'clock. Get packed up and bring your gear with you. We'll be leaving right after we're done eating. I hope you all sleep well."

Lying on my bed that night, I thought about the 37-mile hike with a 13,000-foot elevation gain we faced over the next seven days. I wondered if I would succeed at yet another intimidating challenge. Somewhere I had read that altitude sickness bothered younger people more than older climbers, contrary to what one might expect. Being the youngest member of our group. I didn't relish the thought of being the first one to experience the dizziness and vomiting symptoms of this condition that frequently stopped trekkers in their tracks.

What would I do if it turned out to be more than I could handle? I tried to fight the fear that told me I wasn't good enough to climb this mountain. At least I'd collected a few badges for courage while on the river. What I'd accomplished already on this trip gave me reasons to believe that I probably could do it. At least I had to try.

AFTER A SUBSTANTIAL BREAKFAST to fill our stomachs and extra liquids to hydrate our bodies, we turned in our keys at the reception desk and brought our personal gear to the courtyard in front of the building, where a group of 20 men and boys sat on the grass.

One of the park managers introduced Conrad to the man in charge of our entourage of guides and porters.

After a quick conversation, Conrad turned and gave us instructions. "I need all of you to put your duffel bags and backpacks together in small piles in a straight line along the cobblestone walkway." He placed his own possessions at the end of the line, and we followed his example.

"Each of you will be assigned a personal guide who will carry your gear for you and stay with you as we climb. The other porters will carry our food and general provisions."

The process to match us with our guide was nothing short of unruly madness. The man in charge uttered a word in Swahili, which must have meant "go." Instantly, the waiting guides raced forward to the line of baggage and began hoisting the duffel bags in a chaotic attempt to find which bag weighed the least, to claim as their own.

Since my gear was among the lightest of the piled mounds, two of the guides, each with a hand on my duffel, argued animatedly and engaged in tugging my bag back and forth between them. After a minute of struggling, the younger man released his grip on my bag, and I went forward to meet my guide. His name was Kulmansini.

The weathered, wrinkled brown skin on his face marked him as one whose life had been lived primarily outdoors. He wore tattered clothes and burrowed his feet into a pair of cheap plastic shoes, likely scored from a past client. He belonged to the local Chagga tribe, known for their sense of enterprise and strong work ethic. There was an intensity in his eyes as he pantomimed how seriously he took the job of transporting and guarding my possessions with his life for the next week. I shook his hand to seal the deal, though it occurred to me that perhaps I was better equipped to be carrying my own things up the mountain.

Shortly afterward, our trekking party crammed into the vehicles that would take us to the Machame Gate, the departure point for our next adventure. As we drove west, we passed small plots of land cultivated for growing coffee. Beyond the coffee trees was a dense green band of cloud rainforest. Nearly all the rainfall in the Kilimanjaro region fell on this side of the mountain. Afternoon showers occurred frequently, thus the inclusion of a Gore-Tex rain suit on Sobek's equipment packing list.

An hour and a half later, we arrived at the entrance to the Machame Trail. With blue skies overhead and the sun shining brightly, our porters unloaded the expedition gear they would carry in the open, grassy clearing. Then they were off in a flash. To give them a bit of a head start, we scarfed down some cheese sandwiches prepared earlier that morning. Everyone made sure they topped off their water bottles. Our local guides suggested we all use a walking stick on the climb to help us stay balanced and upright on the uneven and, in places, muddy trail. Over the years, they had collected a supply of discarded walking sticks of various lengths that they regularly recycled.

Conrad whistled to get our attention. "We're going to head out now. If you like to hike at a fast pace, come up to the head of the line. Everyone else fall in behind."

Starting out at 6,000 feet of altitude, I'd remain content bringing up the rear. Yikes! My heart pounded before I'd even taken my first step. Was it warning me to be cautious? Diehard doubts wanted to muscle their negative influences into my mind. What if I hiked for half an hour and became too exhausted to continue? Sure, I was young and relatively fit, at sea level, but what if up here, I turned out to be the slowest person in the group? How terrible would altitude sickness make me feel, though I hoped to avoid that for at least a couple of days? The dreadful "what ifs" were weighing me down. That had to stop.

Slade, an experienced mountaineer, took off with the group of fast hikers and their guides. Bart meshed into the middle of the hiker caravan, while Conrad, who had several Kilimanjaro summits under his belt, stayed with our slower group. Immediately, the path narrowed to barely a foot-wide rut that funneled us into single file. The forest engulfed us with green lushness. Moss and lichen dangled from tree branches, vines wove an ornate web between the branches, and giant ferns splayed their fronds, forcing us to push them out of our way to stay on course. At times, it seemed we weren't following *any* path, that our bushwhacking through overgrown foliage had led us astray to an impenetrable dead end. Without our Tanzanian guides, who I prayed to God knew this obscure route, we would have gotten hopelessly lost as we traversed across one jungle ravine after another.

As difficult as the portage through the Selous jungle had been, the tangled mess of vines and trees in this tropical zone on Kilimanjaro seemed denser and more grueling. The air inside this fairytale forest was cool but humid. Mist swirled around the tops of the trees, and while it didn't actually rain, our clothing became damp and clammy from the excess moisture saturating the air. The exertion of the hike itself caused us to sweat as we inched higher with every step.

Kulmansini walked in front of me, demanding I follow in his footsteps. In places that concealed hidden hazards, he'd stop, gesturing down at partially hidden exposed tree roots or camouflaged fallen tree branches that could easily cause someone to trip. Conrad harped at us to keep drinking whenever we stopped for a brief rest. It took us nearly six hours to cover the first three miles. The higher we climbed, the more the forest thinned.

When we emerged from the forest at dusk into a vast open plain of highland heather, we got our first glimpse of the summit of Kilimanjaro since starting out that morning. Skirting the bushy clumps of heather proved so much easier than bushwhacking through the forest. With another half mile of hiking, we reached the Machame Hut, our first night's campsite.

Our porters already had a fire going in a pit outside a small, round, steel-sided hut, topped with a conical-shaped, corrugated-steel top. The aromas drifting from their kitchen area caused my mouth to water. The energy I'd spent in making the 4,000 foot gain in altitude left me bushed. I felt immensely grateful to see our porters occupied with their kitchen chores because if it had been up to me, even though I was ravenous, I would have opted to skip dinner and snuggle into my sleeping bag for a long night's slumber. Kulmansini helped me pick a patch of level ground for my tent. In no time, we had it set up and my gear stored inside.

As darkness fell, the air temperature plummeted. The cooler air didn't seem to bother our guides and porters, but it sent all of us scurrying to don fleece jackets and hats. Snuggled in close to the fire's warmth, we ate our supper of stew and boiled white rice. Too tired to even care about what ingredients the stew contained, I shoveled spoonfuls of it into my mouth. Its hearty spiciness tasted delicious. I noticed that the rarified air at 10,000 feet seemed to make every task slightly more difficult. Even my spoon felt heavier.

Slade, who still seemed to have plenty of energy in reserve, goaded us. "So, how's everyone feeling?" We answered his question with groans and rolling eyeballs. "Come on now, this is just day one. We've got 9,000 feet to go!"

"Don't remind us please, all right?" one exhausted client snapped. "All I want to do is sleep."

A chorus of voices echoed, "Me too."

Slade's demeanor changed to concern. "Seriously, how are your feet? Anyone got blisters?" No one admitted to any problem.

"You've got to stay on top of any sign of irritation on your feet. We've got moleskin and band-aids if anyone needs them. I really don't relish the thought of hauling any of you back down this mountain because you can't walk."

Bart added, "At this altitude, some of you may get a headache. We've got Tylenol for that, but everyone needs to keep drinking water so you don't get dehydrated."

From the weary looks on people's faces, it was obvious no one wanted to linger for a fireside chat. On this mountain, nighttime would be reserved for sleeping and healing sore, tired muscles.

After climbing into my tent, I pulled off my hiking boots and socks to study my aching feet. Rubbing all the surfaces of both feet, I found one spot on the back of my right heel that was mildly irritated from a seam high on the inside back of my hiking boot. I'd get a piece of moleskin from the first aid kit in the morning. Sleep, mercifully, robbed all thoughts from my brain. I conked out like a boxer taking a knockout blow.

31
HIGHER WE GO

BY THE TIME MY EYES fluttered open, bright sunlight warmed the nylon walls of my tent, evaporating small dots of condensation clinging to its inner surfaces. Outside, the clanking of pots and pans held the promise of a tasty breakfast whipped up by our porters. Boy, did I feel stiff. I expected the soreness in my muscles would magically disappear overnight, but the soreness had turned into stiffness.

I threw on a few layers of clothes and crawled out of my tent. Yawning in a deep breath of the cool fresh air, I pulled myself into a standing position and gently stretched my arms high over my head. A night of restful sleep had worked wonders to restore my energy and prepare me to tackle new challenges. I gazed up to see the sun's rays sparkling off the snow-capped peak towering over us. Still, over 18 miles separated us from our ultimate goal.

I devoured a bowl of hot mushy oatmeal with local honey melted over the top and peeled a hard-boiled egg for breakfast. While Conrad laid out the day's plan, I sipped bitter black tea that provided warmth and got my heart pumping.

"We'll cover another three-and-a-half miles and gain another 2,700 feet in altitude to reach the Shira Plateau. It'll probably take us five to six hours. Today we'll be hiking through moorlands all day, and I think you'll find the going easier. We'll see lots of tussocks of tall grasses and heather bushes, but they're short enough that we'll have a clear view of the top of the mountain all day, as long as it doesn't cloud up. Tank up on water now. You're going to sweat a lot today. And make sure your canteens are full before we leave."

Before we broke camp, I cut a piece from the sheet of velvety moleskin in the first aid kit to cover the sore spot on my heel. Overnight, the pain had lessened, but I was determined to prevent a blister from forming as we climbed higher.

Kulmansini lingered by my tent, anxiously waiting to claim possession of his load. I hustled to stuff my sleeping bag back in its sack and gather my scattered belongings, to cram them back into my duffel bag. We took down my tent, carefully folding it around the aluminum support poles and shoved it into its bag. Kulmansini heaved my duffel up to rest on his head, adjusting it to balance perfectly. I hoisted my lightweight day pack, filled only with sunscreen, my camera, and an extra jacket, onto my back.

Their fine-tuned system for cleaning and packing equipment gave our porters a sizable head start on our second day. Just as the rest of us finished packing, Conrad, displaying the skills of a veteran camel herder, barked at us, "Ready to go? We're outta here."

I obediently trailed Kulmansini, as we took our places in the single-file procession, leaving the Machame Camp. I found the open moorland easier to navigate, though the trail steepened, especially during the morning. I got winded more easily and needed to stop for a minute-long break more often. Kulmansini seemed annoyed each time I stopped, but he patiently waited until I felt ready to continue. At least three climbers lagged behind me, and even with my minibreaks, I stayed ahead of them.

By midmorning, puffy white clouds accumulated high on the mountain, draping their mist over its summit. Up here, dry air pleasantly drafted up the slope. Except for the flapping wings of birds, rousted from their perches in the shrubs by our unexpected approach, the only sounds were of breezes rustling through the tall grasses and the crunch of our own plodding footsteps on the gravel surface, one after another. We stopped for lunch just as we arrived at the edge of a vast flat highland called the Shira Plateau.

Mount Kilimanjaro formed as the result of the rifting of two continental plates, which allowed the initial eruption of a volcano 2.5 million years ago. Over time, three separate cones erupted. First, the Shira Cone erupted for 250,000 years. When those eruptions ceased, the Shira Cone collapsed in on itself.

Later, eruptions led to the formation of Kilimanjaro's two other major cones. Those eruptions helped to fill in Shira's crater to create a level plateau. Mawenzi, also extinct, was the second cone to erupt. Our climb would be up Kibo, the third and highest cone, which first erupted one million years ago. Classified as dormant, it last erupted around 150,000 years ago.

A stop for lunch provided a much-needed chance to get off our feet for a while. Already, I felt as wasted as I had been at the end of our first day's climb. We had already gained 2,000 feet in altitude in a little over four hours. This lunch spot, at almost 12,000 feet, rested at more than twice the altitude of the city of Denver, Colorado, and over two miles above sea level. The feeling of being winded while I was climbing did not abate while we paused for lunch. With involuntary yawns, my body tried its best to suck in more oxygen.

The spectacular view from the bench-shaped stone I'd claimed humbled me. My presence in this magnificent vista amounted to a mere speck. The miniature round steel roofs of the Machame Camp, far below, visibly proved the extent of our day's progress. Below the camp lay the indistinct band of green rainforest, and far beyond, the fertile plains of the Rift Valley fanned out endlessly.

"Everyone doing OK?" Slade made his rounds checking on our group.

"I could do with a little more oxygen," Frank quipped. "Got any of that handy?"

"Can't help you with that. But the good news is we've only got about another hour to go after lunch, and it's pretty flat compared to what we've been climbing."

"Thank God!" Kate raised her arms to the heavens.

Because we'd stopped burning energy during the hiatus for lunch, suddenly, the air temperature felt cooler. I fetched my windbreaker from my daypack and donned it. Our local guides handed out cups of hot broth and thick slices of yeasty-flavored bread that reminded me of sourdough, crunchy on the outside and chewy soft on the inside. I knew I should be hungry, but oddly, my stomach was not growling for food. The soup tasted good and warmed me up. I ate the slice of bread and almost immediately regretted it as a wave of queasiness rose in my stomach.

I closed my eyes and concentrated on taking deep breaths. Eventually, it passed. Our break lasted for an hour. Then we were on our feet again, trudging out across the Shira Plateau to our next camp. Slade had not lied. The slope of the trail was now gently upward. I still felt winded, but not enough to require the breaks I had taken during the morning's ascent. The vegetation on the plateau grew much shorter and sparser than in the moorlands. Rather than dodging bushes, we wove our way around rocks and boulders.

Before I knew it, we had arrived at the Shira Campsite still a few hours before dark. The campsite, situated well above the tree line, offered us stunning views in every direction. While the porters worked on preparing dinner, we Sobek climbers gathered to relax and share stories of the progress we'd made. Conrad pointed up at the summit shrouded in clouds.

"This is the best spot on the mountain for getting spectacular sunset pictures. When the sun sets, all these clouds are going to come alive with vivid colors. Looks like it will be a beauty tonight. Get your cameras ready."

A porter brought over a plate of crackers, cheese, and nuts for us to nibble on. I asked for a cup of tea. Our porters used a new filtering device that eliminated the need for iodine drops. They made our beverages with water purified both by boiling and using this new filter. My hot tea tasted like tea!

Our front-row seats in this outdoor dinner theater were free. With heaped plates resting on our laps, the show began. Billowy white clouds took on pale hues, previewing the coming attraction. Cameras clicked. We sat mesmerized as the feature presentation began. The slowly drifting clouds deepened to vibrant hues of gold, peach, and amber. The star of the enfolding drama floated with brilliance arcing down toward the horizon.

Oohs and aahs and more camera clicks punctuated the silence. The performance was worthy of a rousing John Williams soundtrack to accompany the scenery as it deepened to amethyst and sapphire. But I was content to be hearing Vivaldi's "Four Seasons" in my imagination. From our upper balcony seats, we looked *down* on the sun's exit from the stage as it dissolved below the distant horizon. A few of us applauded as the show's curtain fell dark. It left us hoping for an encore presentation the following night.

If animals prowled around on the plateau that night, I didn't hear them. My cozy sleeping bag encased me in warmth, protecting me against the chilly night air. Weariness made falling asleep easy and sound slumber even easier. Sleep never felt soooo good.

OUR THIRD DAY allowed us extra time to acclimatize. We'd camp for a second night right where we were. By staying at 12,600 feet for another day, it would help our bodies adjust to the lower amount of oxygen in the air we breathed. Each day as we climbed higher, the effect of oxygen deprivation would increase, and our every effort would become more difficult.

Since there was no need of an early start for trekking, we happily took advantage of an extra hour of restorative sleep. In the late morning, we took a short hike to explore the Shira Plateau. I took pictures of a few alpine flowers, including one with orange blossoms that reminded me of a columbine plant. Otherwise, various odd-shaped, ancient lava rock formations captured our attention. We spent the rest of the day catching up on journaling, reading, and writing postcards purchased the first day at the park headquarters. It felt great to just relax and drink liquids.

During the afternoon, clouds formed that obscured the top of the mountain and stealthily crept down its slopes, leveling off just a few hundred feet above our camp. How odd it felt to be so close to the bottom edge of the cloud bank. I wondered what it would feel like to reach up and touch a cloud. That evening, the local guides told Conrad that we might be in for a patch of damper weather over the next few days, so he warned us to keep our rain gear easily accessible. Uncooperative weather resulted in the cancellation of the magical sunset show that night, as the clouds in the valley below shrouded the day's last rays of sun.

OUR FOURTH DAY on the mountain started gray and dreary. We had several miles of hiking ahead, during which we would gain another 2,500 feet in altitude before descending to our next camp at 13,000 feet. This would give us all a preview of what the last two days of climbing would entail. I drank a few cups of hot tea at breakfast and considered skipping solid food, fearing the queasiness I had felt on the second day might return. Yet I knew I needed energy from food to give my body the strength to keep climbing. Oatmeal seemed too glutinous and heavy, so I went with some toast and peanut butter, hoping for the best.

By 8 a.m., we sought stable footing for our hiking boots with every step forward on the uneven ground as we trudged up the steepest section of trail we'd yet to confront. The rocky surface at this altitude had little vegetation. Every time I looked up to peek at the scenery, I caught myself stumbling, so I spent most of the day staring down at the ground. This became quite tedious after about an hour. I struggled to breathe and felt constantly winded.

I needed more frequent breaks to catch my breath. I took 10 steps uphill, stopped, and counted slowly to 10, then repeated this routine. My sluggish but deliberate progress resembled that of a tortoise. But my stop-and-go approach irritated Kulmansini. He summoned over one of the local guides who spoke English to give me and a few other nearby Sobek trekkers a pep talk.

"This isn't the way to climb this mountain. You mustn't stop."

He had to be kidding, I thought as I sat on a rock panting for air. Were we just supposed to pass out from lack of oxygen and sheer exhaustion?

"If you get tired, slow your pace down, but don't stop. Keep going, no matter what. Just keep going. We have an expression, *'polepole.'* It means slowly, slowly. Starting and stopping, over and over, it isn't good for your heart. First, your heart pumps hard, then it stops for a bit, then it pumps hard again. You want to keep your heart pumping at a steady pace, even if it means going slower. Try to do that, okay?"

Heads nodded. The English-speaking guide nodded back. "Good then. *Polepole.*"

His explanation made sense, so what the heck? I'd try his approach. I looked at Kulmansini and smiled meekly, to let him know I understood what he wanted me to do. We exchanged places. He no longer led but walked behind me, while I followed in the footsteps of the hiker in front of me. He started a soft chant of "*polepole*." It reminded me of the steady, rhythmic beat of a metronome. Before the pep talk, the length of my steps had been about 10 inches. Afterward, they were barely five inches. Instead of walking, I shuffled. But I kept going, still trying to suck in as much air as I could.

The fit, middle-aged male client in front of me kept up a slightly faster pace than mine. Gradually, the distance between us grew. But 20 minutes later, I spotted him doubled over just off the side of the path. When I caught up to him, I realized he was retching from waves of nausea, an unpleasant symptom of acute mountain sickness (AMS).

Kulmansini motioned for me to keep going, while he diverted from our route to offer the sick client's guide help. I wasted no time scurrying up the path, in part because stopping wasn't encouraged, but more so because the man's guttural gagging started to make me feel nauseous. If I didn't get out of earshot quickly, there'd be two of us off the side of the path. In the sweep position, Conrad, who carried our first aid supplies, stopped to evaluate our client's condition.

When I could no longer hear the conversations of people huddled around the sick man, I sat and relaxed on a comfortable rock to wait for Kulmansini to rejoin me. I recalled Slade had told us about the health risks of trekking at high altitude. When climbers reach altitudes above 8,200 feet, medical studies estimate 70 percent of them suffer from AMS symptoms including headache, nausea, diarrhea, vomiting, or loss of appetite. For the majority suffering AMS, the symptoms are temporary and resolve as soon as the climber descends. However, for an unlucky few, AMS can progress to pulmonary or cerebral edema, fluid in the lungs or brain, a potentially lethal condition.

I pondered whether the queasiness and lack of appetite that had bothered me that morning could be signs of a mild case of AMS. It wasn't long before the huddle below me broke up. Once our client recovered enough to continue moving, Kulmansini came trudging up to my resting spot. His thumbs-up gesture meant our trek was back on track. We slogged on for a few hours before reaching our midday rest stop at 15,000 feet, the highest point we would reach that day. Above us, thick gray clouds blanketed the mountain's peak, but a breathtaking vista spread out in front of us. We couldn't have dreamed of a more picturesque picnic spot.

At this height, the temperature was downright cold. Even dressing in layers with an outer jacket couldn't completely fend off the chill once I stopped moving. We had entered the Highland alpine desert zone where hats and gloves were essential. A few people stretched out prone on the ground for a catnap. Others complained about pounding headaches and swallowed paracetamol tablets with lunch.

What bothered me was an odd feeling of spaciness. Both my physical movements and thoughts muddling about in my brain moved in three-quarter time slow motion. But damn, why would I expect to feel "normal" in a place three times as high as Denver? At this altitude, an airplane could fly safely over Denver's tallest skyscrapers. The rocky outcrop where I rested equated to sitting on a wing of a jet flying three miles high! Up here in the realm of the clouds, my mind drifted along like one.

The still-nauseated client refused to eat any solid food, though he sipped at a mug of steaming broth. Despite the energy we'd already depleted that morning, food held little appeal for any of us. We quickly nibbled on bread, cheese, and fruit, washed down with broth, knowing that as soon as we finished eating, we could escape the penetrating wind that swept across the exposed ridge.

To deal with the group's restlessness, Conrad delivered some spirit-lifting news. "We're gonna get going. Once we're walking again, we'll warm up. Gravity will be on your side for the rest of the afternoon. Breathing will get easier, but watch your step for loose rocks that might buckle when you step on them. We don't need any sprained ankles."

Initially, it seemed much easier going downhill, but the loose scree was treacherous. Because of the unstable footing, my boots frequently took turns sliding three to six inches farther downhill than I expected. Staying safe still depended on going slowly. The tips of my toes shifted slightly forward with each step, ramming the sturdy inside front of my hiking boots, causing my feet to cramp and my toe joints to ache. The pain intensified as the afternoon proceeded. Even tightening my shoelaces to prevent my foot from shifting forward offered no relief.

By midafternoon, a veil of mist swirled around us like the fog that creeps in from the ocean to engulf beach walkers. Amazingly, we were trekking directly into the middle of a cloud. Visibility decreased as the saturated air became denser and damper. We stopped to don rain jackets and pants. Minutes later, the heavy mist formed into droplets. Rain poured down on us inside the cloud. The ground muddied, hindering our descent as we cautiously wound our way down the slippery slope.

Without warning, a crinkling flash of light and thunderous clap exploded all around us and reverberated through the air. I froze in place, gripped with fear. Holy shit! Countless hazards had threatened this expedition from the start. But really? Now we stood trembling *inside* a thunderstorm!

As a child, thunderstorms topped the list of things that petrified me. Lightning had struck the lightning rod on our roof, the rock ledge behind, and several trees surrounding our house during my youth. Dad, a master electrician, had a blasé attitude toward the thunder boomers. But Mom insisted my brother and I put on rubber-soled sneakers, while we huddled on the edge of the sofa, ready to flee to God knows where if a bolt struck, setting the house on fire. One night, we heard the rumblings of a thunderstorm approaching. After one nearby heart-stopping strike, we raced into the bedroom where Dad lay reading in bed.

My mother barely finished saying, "Boy, that was a close one!" when the hair on my father's head stood straight up. A second later, a bolt struck a cedar tree only 15 feet from the corner of the house. The brilliant flash and electric crackle terrified me. I shrieked. Sobbing, I leaped onto the bed and wiggled into the protective curve of my father's arm. Countless times he'd explained to me that as long as I took cover inside a house or car, I didn't have to worry about being hit by lightning.

Now, standing in the middle of a wide-open expanse of rocky terrain, *inside* a thunderhead, I wanted to scream out loud, "Dad, what do I do now?"

The unexpected lightning bolt startled all of us. Even the local guides seemed spooked by it.

"Let's keep going." Conrad figured reaching the lower altitude of our next camp and changing into warm, dry clothes would make us feel better. As we slogged on, I prayed the lightning bolts wouldn't follow us. Luckily, the storm only conjured that one rogue lightning bolt. The rainfall persisted, gathering in rivulets skirting lava rocks jutting from the ground, gravity pulling their swelling flows downhill.

As if the altitude, slippery loose scree, and pelting rain weren't challenging enough, soon the rain switched to small ice pellets that stung our exposed cheeks. The icy hail coating the ground's surface further increased the danger that one of us might slip and break a bone. Honestly, this maniacal obstacle course could have been used to determine if any of us had that 'special something' to gain admittance to an elite Special Ops team.

Miserable didn't even begin to describe my state that afternoon. Shivering, my shirt wet from condensation inside my fleece jacket enclosed inside my rain gear, sore feet, suffering from exhaustion, and struggling to breathe, the current conditions made the long portage hike through the jungle look like a piece of cake. I wondered why people would willingly put themselves through this torture to get to the top of this mountain. But there was no turning back now. *Polepole.*

The hail storm lasted for an interminable 15 minutes, after which the precipitation stopped, though we continued to be blanketed in mist. About an hour later, we arrived at the Barranco Campsite. At 13,000 feet, we were only a few hundred feet higher than our last camp. Coming down into our campsite, I noticed a plentiful number of strange plants that stood guard around the perimeter. After getting my tent set up and my wet clothes replaced with dry ones, I went outside to inspect these unusual, unfamiliar plants.

"Conrad, what do you call these things growing all around here?"

"Those plants?" He pointed toward several five-footers that stood close by.

"Yeah, they look like old cactus trunks crowned with a shaggy pineapple with a spiked-up green hairdo."

He laughed at my description. "I guess that's one way to describe them. They're called giant *senecios*. Those there are babies. They grow to 10 feet or more. Keep your eye out for even taller ones tomorrow."

"But how can they grow in the freezing weather way up here above the tree line?"

Conrad shrugged. "The only places in the world they grow are here on Kilimanjaro and a few other East African peaks."

How the heck could any vegetation survive and grow in the cracks between the volcanic rocks in this harsh environment? I wandered over to one plant that was not quite my height. I could see that the green "crown" was a large symmetrical rosette of thick, short, pointed leaves.

The mist obscured any sort of view from this location, and the grayness hastened nightfall. Every night, we climbed into our tents earlier, falling asleep more quickly, and sleeping for more hours. This high on the mountain, we didn't have to worry about any animal attacks. However, the key to comfortable sleep required finding a level spot, a nearly impossible task on the sloping mountainside. The spot also had to have a wide enough gap between the sharp-edged lava rocks for the footprint of the tent.

Then, when laying out my sleeping bag inside the tent, I had to position it so my head would rest at the higher end of the tent. Otherwise, I'd wake in an uncomfortable position with a splitting headache from gravity pooling blood in my brain. Believe me, I only made that mistake once.

32

THE BARRANCO WALL

I COULD TELL the weather must have cleared up overnight because a bright sun shone outside my tent when I woke up. Yet with the air still chilly, I could see my breath when I exhaled. Already I could hear clanging pots and smell the aroma of coffee and hot chocolate as the porters prepared hot drinks and whipped up oatmeal in the camp kitchen. Those sensory triggers spurred me to get dressed so I could join my colleagues by the firepit.

I climbed out of my tent and stood up to stretch. My jaw dropped at the site in front of me. Towering over our campsite loomed a massive, sheer-faced cliff, which the mist of the previous evening had completely concealed. I felt insignificant standing at the base of this immense rock wall.

Bart noticed how dumbstruck I looked, gazing at the distinctive landmark. He couldn't resist a wisecrack. "Meet Barranco. It's ready and waiting for you."

This 834-foot-high cliff, nearly the height of three football fields stacked end to end, straight up, stupefied me. Knowing we were going to have to get to the top of it, I scanned the wall for any sign of a trail or path on its face. From what I could see, none existed. Perhaps an extremely nimble mountain goat would have some success, but I couldn't see any sign of a single living creature anywhere on that wall.

The seemingly hopeless task of going 800 feet straight up this cliff suggested to me that humans probably shouldn't be attempting to scale it. My constant companion, fear, taunted me. "You can't possibly even consider doing this ridiculously risky climb."

I snapped back, "Just shut up! It's not like I have any choice."

Conrad's sharp whistle brought us scurrying to the center of camp. He stood in front of us, holding a steaming cup of black coffee while the ominous-looking sheer cliff behind him loomed over us.

"So that cliff behind me is the Barranco Wall. Today we're going to climb it."

"How?" asked Martha. Oddly, her astonished expression comforted me.

"Exactly! How the hell are we going to get up that wall without ropes and technical climbing gear? The brochure said this trip didn't require rock-climbing skills." Perturbed, Jay threw his hands in the air in exasperation.

Conrad extended his arms forward with wrists bent upward to ward off further complaints. "Look, I know it appears impossible from where we are right here, but trust me, I've climbed it before, and there's a decent trail that zigzags diagonally to the top of the wall. I won't kid you. It's a steep climb, and it will tax you, but it is *not* a technical climb."

Not fully persuaded, Martha cut straight to the chase. "How many people have fallen off this wall?"

Conrad grinned and made a circle with his right thumb and index finger and held it up. "Zero. I've never heard that anyone has fallen here. Look, let's just get packed up and head over to the base. Then you can see for yourselves this is completely doable."

OVER THE EONS, pieces of rock had slid off the cliff face, accumulating at the bottom in a sloped mound that served as a ramp leading up to the wall. We scrambled over the fallen debris, and at the top, a four-foot wide, mostly level path waited for us.

Conrad stretched out his arms, standing a few paces onto the path, to show the ample clearance we'd have. "As you can see, the trail is plenty wide enough for one person, a little iffy for two people side by side. From here on, it's single-file only. Don't try passing anyone. Just try to keep a slow but consistent pace."

Polepole, I thought. Just put one foot in front of the other and don't look down. We allowed our porters to head up the trail first. Quickly, they disappeared from sight. I felt safer having Kulmansini go in front of me so I could mimic his foot placement whenever we had to scramble over a group of larger stones partially blocking our way. I also intently watched where he held onto the side of the wall with his hands to stay balanced when the trail narrowed in places.

After we'd been on the trail for half an hour, we admitted Conrad had told us the truth. No one would plunge off the side of Barranco unless they intended to do so. Some parts of the trail were moderately steep, but in certain sections, the path resembled stairs, each step being a good six inches higher than the last. The quads and calf muscles in my legs tightened and started to burn.

I swore Kulmansini could read my mind, because every time I thought it might be time for a quick break, he'd look back over his shoulder and command me to continue, "*Polepole.*" Once I felt relatively confident about walking along this trail, absent of any railings or guardrails, I occasionally dared to look out over the Barranco Valley below and saw our previous night's camp growing smaller the higher we went.

I won't lie. Toward the end of two exhausting hours on the wall, my legs had turned to jelly. I felt I'd been conscripted to serve as a laborer in a Siberian work camp, even though Kulmansini covered the heavy lifting.

Polepole. With each painful step taken, I needed one less step to put this wall behind me. Each step sapped a bit more of my waning energy. Did I have enough left in my tank to make it to the top? Pole. Goddamn pole. When the perfectly vertical wall on our left began to tilt away from the inside edge of the trail, exposing more blue sky, I figured we'd nearly completed our climb up to heaven.

A short time later, we stood atop a ridge just under 14,000 feet, where we reveled in a clear view of the majestic snow-capped Kibo Crater. From this closer perspective, its height seemed more daunting than ever. We rested just long enough for a hydration break before Conrad had our caravan marching again. He told us, "We still have miles to go before we can sleep." I wondered if he'd intentionally paraphrased Robert Frost. Our route didn't take us through a snowy wood, but our way to the summit weaved up a snow-capped mountain.

I decided a better literary reference for our situation might be the proverb, "There's no rest for the weary." It didn't matter how tired we felt; we just had to keep going. For the rest of the day, we crossed an alpine desert devoid of anything but rocks and soil. It reminded me of a lunar landscape. At this altitude, night temperatures nearly always dipped below freezing, resulting in the ground beneath the exposed topsoil layer being permanently saturated with ice crystals.

Polepole. The mantra took on a whole new meaning this high on the mountain. No longer did it matter that I tried to pace myself so I wouldn't be breathless. Now my body worked on only one speed. Pick up one foot, move it forward a few inches, plant it securely on the trail, shift weight forward, inhale, lift trailing foot, and repeat.

At best, drudgery described our exhausting progress. It became torture when waves of nausea hit. Most of our group experienced some degree of stomach upset or headache. Three of the men threw up that afternoon. Still, we kept going and going, one tiny step after another, wondering when this agony would ever end.

I would have enjoyed gazing out over the vast Tanzanian plains as I plodded along, but I couldn't afford that luxury, since the loose scree beneath my feet demanded my full attention. I considered it a useful step taken if I didn't slide backward when my weight shifted forward. Too many of my steps that should have measured eight inches of forward progress took me only four inches as the rocks gave way under my boots. But with every step I took, I broke the record for the highest altitude I'd ever reached.

All of us too winded to hold any conversation, our caravan proceeded in silence, broken only by the monotonous din of trudging boots crunching on the ragged-edged lava rocks heavily carpeting the ground. The grinding sound of the rocks scraping against each other, as they found new positions under my feet, also ground on my nerves. The slowness of our pace, the gasping for breath, and the overwhelming fatigue were taking their toll.

At one point, I'd had it with the motherfucking loose rocks. I wanted to pick up a few and hurl them with all my might out into the air, certain it would make me feel better. As much as I wanted to, I didn't act on my brilliant idea. If I started chucking rocks, I worried our group might question my sanity, though I secretly suspected every one of them would spontaneously join my rock-flinging revenge. Anyway, I didn't have enough energy to hurl one rock, let alone several. *Polepole.* Take deep breaths to get more oxygen.

Throughout the later afternoon, gray clouds accumulated above us, concealing Kilimanjaro's summit. The wind picked up and chilled any exposed part of our bodies. I hoped it didn't signal the approach of bad weather. Finally, about an hour before sunset, we reached the Barafu Hut, perched on an exposed ridge, the highest hut on the mountain at 15,330 feet. As the porters worked on setting up the kitchen, I collapsed onto a contoured, bean-bag-shaped boulder to celebrate the relief of being off my feet.

"Hey, listen up. Don't put your tents up yet," Conrad told us as he approached the local guides. During their lively discussion, they inspected the inside of a round hut made of steel. As we waited for instructions, we felt falling sprinkles of rain and ice crystals, plunging our spirits likewise. Everyone scrambled to pull out rain gear. The porters hustled to erect a portable tarp over the cooking area.

Returning to where we waited, Conrad delivered the bad news. "The Kili guides think that we're in for a stormy night. I think the 10 of us can fit inside the round hut here, but we'll be really cozy."

Based on the hut's physical size, I estimated that six people would overcrowd the capacity of the interior space. We'd been traveling together for many days in the rough and still liked each other. But coexisting, even for one night in this tiny space, would test the strength of our camaraderie and goodwill.

Conrad continued, "At least the hut will keep us and our things dry, and the heat from our bodies will help to keep us from freezing. Inside the hut is a wooden sleeping platform raised about eight inches off the ground. Clients get the platform. The rest of us will spread out on the ground around the platform. Everyone okay with being real cozy tonight?"

"Dry versus sopping wet in freezing temps really isn't a choice." Kate perfectly summed up the position of the entire group. Everyone immediately accepted Conrad's proposal. I thanked God we wouldn't have to sleep outside.

"Great, we have a plan! Clients, you can claim your spaces on the platform and get your sleeping bags laid out. Once you're settled, the Sobek crew will fill in around you. Another thing, the cooks are making us a hearty stew, but you'll want to eat lightly tonight, or you'll pay for it tomorrow. By that, I mean our day tomorrow will start with a wakeup call at midnight, and we'll start our ascent at 12:30."

"Why so early?" asked one of the men. "I'm so exhausted I could sleep for about 10 hours."

Every one of us felt like hibernating for a long sleep, but for us to make it to the summit would require making the climb within a brief window of time.

Patiently, Conrad explained to our clients, "If we're going to be successful, the climb has to start at night while the scree is frozen solid, because once the sun comes up, it melts the frozen crystals, and the scree becomes treacherously loose. Leaving here by 12:30 gives us the best chance of making it to the top. So that means we are going to get to sleep as soon as we finish dinner."

There were plenty of groans and sighs about the thought of heading out on the final 4,000-foot ascent in the middle of the night, but after all, we'd all come here to reach the summit.

"Get your clothes for the climb ready tonight, so you can dress quickly in the morning. Don't forget your rain gear."

"Do they think it will rain tomorrow?" A look of concern reflected Kate's serious doubt about tackling the peak in adverse conditions.

"The only thing I can tell you is the weather on this mountain is fickle and can turn on a dime. Whatever weather we wake up to at midnight is what we'll have, and we'll make the best of whatever Mother Nature throws at us."

My Sobek colleagues and I gathered under a second tarp the local guides put up next to the tarp over the cooking area. Protected from the rain, we drank tea, grateful for its warmth, and stayed out of the way of our clients as they dragged gear into the hut and jockeyed themselves into place on the platform. Once they finished, we swapped our place under the tarp with them and stowed our gear inside the hut, though we wouldn't lay out our sleeping bags until our clients had settled themselves in for the night.

Outside, the freezing rain continued, soaking everything not under cover. Just after darkness fell, our porters brought us steaming bowls of stew and hot beverages, which we ate enthusiastically while huddled together, sitting along the three sides of the raised platform inside the hut.

"Before we get settled in, I'd highly recommend that you visit the latrine. Take your headlamps with you so you can see where you're going. If you wait until later, you'll probably wake everyone up to get out of the hut, so do your duty now." Conrad gave directions to the nearby rudimentary latrine. One by one, we took advantage of the pit toilet, clients first, handing off our roll of toilet paper to the next needy latrine user as if it were the baton in some bizarre relay race.

As the clients returned, they removed their rain jackets and boots and snuggled down into their sleeping bags. The six of them barely fit on the platform. The sight tickled me. "If one of you turns over tonight, I think everyone will have to turn, like one domino starting a chain reaction," I joked. I figured we'd find out who snored and, God forbid, who farted when they were sleeping.

I had no desire to be part of the human club sandwich plated on that platform. I thought I'd gotten the primo spot, horizontally stretched out between the foot of the platform and in front of the door of the hut. Only after I had laid down my sleeping pad and bag and climbed into it did the two drawbacks of that spot present themselves.

The ground in front of the platform sloped gently down toward the door. Not much, but enough that I felt the undeniable force of gravity pulling me toward the door. I finally stopped fighting gravity and just let my body lean against the door and wall of the hut. Even with the door closed, an inch gap between the bottom of the door and the ground allowed a frigid draft to pummel the side of my sleeping bag. When one side of my body got cold, I'd roll over 180 degrees to give the cold side a chance to warm up.

As people gradually nodded off to sleep, restless bodies ceased squirming, becoming motionless. But I remained wide awake. I listened to raindrops beating on the hut's steel roof, trying to use the constant pattering as a lullaby to help me drift off to sleep. After quite a while, the patter turned to a crisp pinging as hard pellets ricocheted noisily off the metal roof like someone playing the drums.

It frustrated me that my sleeping pad didn't prevent the chill of the permafrost in the ground below from penetrating through it and my sleeping bag straight into my flesh. My whole body grew colder and colder as each minute ticked by. Still, sleep remained elusive. My mind reluctantly turned to evaluating whether I would be in any shape to tackle the monstrous climb just a few hours away.

How could I back out at this point? I'd come here to climb this frigging ice-covered volcano. I desperately wanted to succeed. But trudging through the hail, sleet, or rain in the middle of the night, with no sleep and already half frozen before even starting out, seemed a rather idiotic idea. Once I started shivering and could feel that the chill had invaded my aching bones, I knew I wouldn't sleep at all, no matter how much I wanted to.

I lay there as miserable and uncomfortable as I had ever felt in my life, the constant shivering wasting the small reserve of energy I had left. If precipitation continued at midnight, I sadly decided I wouldn't attempt the climb. Even if I tried, I strongly suspected I wouldn't make it because I felt completely drained. I shivered uncontrollably for what seemed like an eternity, wondering if even time itself had frozen in place. Finally, I heard the alarm on Conrad's watch go off.

33

TIME TO CLIMB (OR NOT)

"UP AND AT 'EM, everyone," Conrad grumbled.

Sleet continued to pelt the roof. Still nestled in my sleeping bag, I sat up and scooched out of the way so Conrad could go rouse the guides and porters if they weren't already up. As he pushed the door open, a blast of arctic air stole what little warmth there had been inside the hut. Everyone else climbed out of their warm cocoons and began adding layers of clothing.

Conrad quickly reappeared with a promise that the porters would have hot drinks ready soon. "The weather out there is nasty. You need to dress really warmly and wear your rain gear. You should bring extra headlamp batteries with you. Otherwise, bring only what you absolutely need. This will be a really strenuous climb in these conditions. The first peak, Stella Point, will take about six hours to reach, and the highest elevation, Uhuru Peak, is another hour or more beyond. If, at any point along the way, you feel you don't want to continue, you are free to turn around. Your safety and health are the top priority for us. Is everyone game for this?"

I felt guilty about backing out with all the others readying themselves for the attempt. After pulling on an extra jacket and my boots, I followed Conrad outside.

I fidgeted for a moment before confessing my decision. "Hey, Conrad, I think I'm not going to go up with the group."

My statement came as a surprise. "You sure?"

"Yeah, I didn't sleep at all last night, and lying on the ground with a draft coming in under the door got me really chilled. I shivered all night."

"I know what you mean. I felt the cold from the ground too, but got in a few hours of sleep."

"I know I'm giving up an enormous opportunity, but I *need* to sleep. Maybe I'll be able to if I'm on the raised platform while all of you are gone."

"With this rotten weather, I'm doubtful how many of this group are going to make it to the top. I suspect many of them will make it part of the way and then turn back." Suddenly, he smiled as a thought occurred to him. "You know, Brenda, having you stay behind might work out really well."

How could my staying behind be a good thing, I wondered, when I saw it as a personal failure?

"Look, with you here, when others arrive back cold, wet, and exhausted, you can make sure they get into dry clothes and get them to drink hot liquids. Make sure they don't get hypothermic. The porters will get food prepared that people can eat when they arrive back. But we've also got to camp at a lower altitude tonight. I'm guessing I won't be back down before the porters need to get packed up and move to the lower camp. I'm going to put you in charge of seeing that gets done so the rest of us can relax and enjoy a hearty dinner tonight." Conrad beamed with pride at his new plan.

I knew one or two of the local guides could speak English, but they would be on the mountain with the clients and our Sobek guides. "Do any of the porters speak English?" I asked. How did Conrad expect me to be in charge if I couldn't even speak to our porters? But at least now I had a reason to feel better about my decision to stay behind. I might actually be useful to others in the group.

Just before they departed, Conrad brought one of the younger porters over to me for an introduction.

"Brenda, this is Williness Julius. He speaks some English and will help you with whatever needs to get done."

I FELT A SURGE of regret as I watched the group plod slowly away from the Barafu Camp. Before long, they disappeared into the sleet-filled darkness. Fortunately, my decision left me with a sense of peace. My body was screaming for help, and I was listening.

I returned to the hut, intending to make space for my sleeping pad and bag on top of the platform. Once inside, when I saw all the empty sleeping bags, I had a flash of genius. I fully unzipped the three sleeping bags on the right-hand side of the platform and stacked them in order on top of each other. Then I placed my sleeping bag on top of those three. I also unzipped the other three sleeping bags from the left-hand side of the platform. Once I had climbed into my sleeping bag, I covered myself with those three bags. What luxury! Sandwiched in the middle of seven sleeping bags on top of the platform, I felt toasty in no time and gratefully fell into a deep, restful sleep.

When I woke up, dull daylight shone through the gap under the door. I felt as warm and cozy as I would have been sleeping in my bed at home, despite the freezing temperature inside the hut. The extra warmth provided by the nest I'd made of sleeping bags had helped to restore a life force in me I'd nearly lost six hours before.

Feeling more refreshed than I dreamed possible, I dressed in layers, booted up, and emerged into a misty gray morning. The porters had built a firepit under the second tarp and squatted around it, seeking the warmth it offered. I wandered in their direction. Williness Julius saw me approaching and waved me over to join them.

"Are you hungry, Miss Brenda?" he asked.

"Actually, I'm starving," I answered.

Williness spoke in his local dialect to the other porters, who jumped up and began preparing food. I looked around for Kulmansini, assuming that he'd been spared the effort of the ascent since I'd opted to stay in camp.

"Where's Kulmansini?"

"On mountain. Mr. Conrad ask him help out. Since bad weather. If anybody get hurt."

I thanked Williness for the information. By using Kulmansini, Conrad had made a practical decision that would enhance the safety of the group. The porters presented me with a bowl of oatmeal, a hard-boiled egg, and a cup of pinkish-orange juice. I had no idea what type of fruit they used, but I enjoyed its citrusy, sweet flavor. Just as I finished my last spoonful of oatmeal, one porter pointed up at the mountain.

"Miss Brenda, one group coming," said Williness.

My eyes followed where his extended arm pointed. I could make out two people, one of our clients and their guide, slowly picking their way back down the trail. Still too far enough away, I couldn't tell which client had turned back. I hustled back into the hut and zipped and rearranged the sleeping bags as I had found them on the platform. I also stuffed my sleeping bag back into its sack and into the end of my duffel bag. Then I went back out to wait for the first two of our group to arrive.

As they came closer, I could see her guide struggling to hold Kate upright.

"She not good shape, Miss Brenda," Williness worried, shaking his head.

I stood and walked up the trail toward them. The woman's hair looked like she'd just stepped out of a shower, and several strands hung limply in front of her face. She sobbed in obvious pain.

When I reached them, she burst out crying. "It was horrible, so steep, and we couldn't even see where we were going. You were so smart to stay here. I should never have tried to do this." She continued to cry.

"Let's get you into the hut. It's going to be okay. You're safe now," I tried to reassure her. Williness went to get some hot tea or broth ready. I followed her into the hut, where she collapsed onto the platform. She pointed at her duffel, where I found her some dry clothes. I helped her pull off the soaked clothes she had on so she could put on dry ones. I tucked her into her sleeping bag just as the hot tea arrived. She stopped sobbing and gratefully sipped at the tea.

"I went as far as I could, but the sleet was awful, and my feet got cold, and the sleet was blowing and getting inside the hood of my jacket. It was so awful. The others got way ahead, and I finally just gave up."

"Well, I think you made a smart decision to come back. You can rest and get warmed up. When you feel better, we'll get you something to eat. Rest now because we have more hiking to do this afternoon."

Probably the last thing she wanted to hear were the words "more hiking." Imagining what it had been like to be in her shoes only reinforced that I'd made a wise decision. I returned to sit with the porters by the fire. The rain had stopped, and the mist gradually lifted. Williness eagerly shared his knowledge of the mountain.

"Very top, we call Uhuru Peak. Means Freedom in Swahili. It's highest part of Kibo. Kibo is easy, but Mawenzi," he pointed eastward out across a saddle of land at another nearby peak, hidden by mist until now, "not is possible. Not high like Kibo, but very steep, very dangerous."

Williness's description of Kibo as "easy" shocked me. No doubt he climbed this mountain several times a year, but what about it did he think was easy? The other peak, Mawenzi, was the third cone of the Kilimanjaro volcano. Its sides looked nearly vertical and its crown a series of jagged spires that topped out at 16,893 feet. Based on Williness's assessment of Mawenzi, I made a mental note not to make the mistake of signing up for any climbing trips on it in the future.

"They say Europeans got to top of Kibo first. I don't believe. Our people knew the mountain. They got to top first." He made the statement with such pride and confidence that I believed him.

What I had read about Mount Kilimanjaro before our trip gave bragging rights for the first ascent to a German geology professor, Hans Meyer, and an Austrian mountaineer, Ludwig Purtscheller, who reached the summit on their third attempt in October 1889. But a Chagga guide led them, along with two local tribal leaders, nine porters, and a cook. So much for giving credit where credit was due.

While we were talking, a porter pointed out another person stumbling his way back toward our camp. He, too, looked miserable. It shocked me to see his mustache caked with ice. Shivering, he headed straight into the hut. I told him to get into dry clothes and into his sleeping bag to get warm. After he'd had time to change, I brought him some hot broth. He took the cup and wrapped his stiff hands around its surface, and hoisted it to just under his nose where he breathed in the vapors, melting the remaining ice from his mustache.

"The mountain beat me. I gave it my best, but it's a beast. Some others were getting discouraged, too. I came back because I was getting so cold and tired and wondered if I even had the energy to make it back here. By the way, thanks for the broth."

Our conversation wakened Kate from her nap.

"How are you feeling?" I asked, noticing her improved appearance.

"I thought I'd never be warm again, but I feel so much better to be dry, and it really helped to snuggle inside the sleeping bag. I have feeling back in my feet."

"That's what *you* need to do now," I said to the still-shivering man. "Get some rest and warm up. The porters will have some food ready for us soon. We've got a pleasant fire going in a pit outside, so come join us when you're ready."

Seven of our group were still up there somewhere in the clouds swirling around the top of Kilimanjaro. They had been gone for nine hours, enough time to reach the top and return to camp in optimal weather. Meanwhile, I enjoyed my chat with the porters as they worked in the kitchen. Williness served as interpreter as they asked about our river trip and many questions about America. I asked why they enjoyed being porters and how long they'd been doing their jobs.

About an hour and a half later, Martha and Jay hobbled back into camp, each supporting the other. They were so utterly exhausted I didn't think Martha could stand on her own, and Jay hadn't fared much better. Their guide explained to Williness that they had almost made it to Stella Point, but Martha had problems breathing, so they aborted the mission.

The first two back to camp, now both in much better shape than when they'd arrived, layered up and vacated the hut so the married couple could change clothes in privacy. I repeated the same advice about resting and getting warm. Shortly afterward, I entered the hut with cups of hot broth. They had buried themselves in their sleeping bags and were curled up around each other like two cozy cats getting warmth from each other. They thanked me for the hot drinks I'd brought them.

The group surrounding the firepit grew as the guides reunited with their porter buddies. By 11 a.m., the porters served us plates of boiled white rice slathered with a thick stew of potatoes, carrots, onions, and peppers. The porters kept the pots of food warm so those who hadn't returned yet could dig into a hot meal. About noon, Frank and another man from our group, with their guides, shuffled back into camp. Kulmansini accompanied this group. I waved when I saw him, glad he'd be with us when we left for the lower camp.

These guys were tired, but had weathered the climb more easily than the others. Both had reached Stella Point and were high on their accomplishment. They went straight to the firepit and gulped down some hot broth and devoured the plates of food handed to them in celebration of their hard-fought accomplishment.

One of them explained, "Five of our group reached Stella Point. We had to decide whether to continue. The two of us felt that hiking uphill in icy slush for two more hours to reach Uhuru wasn't worth it. But Conrad, Slade, and Bart went for it. Oh, and Conrad told us to tell you that if they aren't back by 2 p.m., you need to move the porters and us down to the lower camp."

I did a rough calculation in my head that left me in doubt they'd be back by 2 p.m. Everyone, other than the three still on the mountain, had returned safely. Once Martha and Jay ate lunch, the porters could get the kitchen packed up and ready to roll.

"Williness, Conrad wants us to leave this camp by no later than 2 p.m. If the couple in the hut sleep for another hour, will it give the porters time to clean and pack up?"

"No problem. We leave pots next to firepit. One porter stay behind with food for Mr. Conrad and people. We start packing now."

"Great, let's do it!"

He seemed so eager to please. I guessed his youth put him at the low end of seniority among the porters. My rank among the Sobek crew also occupied the bottom rung. Only because Williness could communicate with me, we'd both gotten temporary promotions. Unwittingly, we found ourselves "in charge." How'd that happen? He shouted out a few orders that set the porters in action. Soon, we'd be on our way back down the mountain.

At 1 p.m., I woke Martha and Jay. While they ate lunch outside with the rest of us, I explained the plan for the afternoon to our clients. They had an hour to pack up their belongings, drink another hot beverage, and warm themselves by the fire before our trek started at 2 p.m. I repacked my own belongings into my duffel and handed it off to Kulmansini. I'm not sure whether people just wanted to leave Barafu behind or if they were ready to be at a lower altitude where oxygen was more plentiful. But clients, porters, and guides were ready to march by 1:45 p.m.

"Williness, does everyone know where we're headed? It's the Mweka hut, right?"

"Yes, Miss Brenda, no problem."

"Okay, then the porters can head out. Please get things set up as soon as you reach the camp."

He whistled, said something that included Mweka, and the procession headed off. Soon after, our clients and their local guides and I followed in their footsteps. I felt a little sorry for leaving one porter behind, but he'd keep watch over the personal gear still in the hut and the cooked food and broth. Maybe he'd enjoy a bit of peace and quiet.

FOR MANY REASONS, the descent was easier than climbing. The trail was steep, so we lost altitude quickly. There wasn't the constant gasping to breathe in enough oxygen or strain of our hearts furiously pumping with the exertion that was needed to climb uphill. Gravity had become our friend. Even our view had changed. No longer were we constantly staring at a barren rocky slope that led into the clouds. The weather had cleared on the lower flanks of the mountain and the view looking out over the rainforest. The checkerboard of coffee plantations beyond and the high plains in the distance looked like a carefully pieced medal-winning quilt.

It wasn't long before I could feel the tips of my toes jamming into the front of my hiking boots again with each step I took. Although I had broken my boots in well wearing them on several long hikes back in California, I'd never felt my feet slide forward inside my boots like this. I stopped and tried again to tighten the lacings, hoping that change might hold my feet more securely in place. And again, it didn't make much of a difference. I alternated walking straight ahead with side-stepping down the trail to buy short periods of relief for my toes that were cramping. My unusual dance steps amused Kulmansini, but I figured he knew exactly what my problem was.

This became yet another reason to be grateful I had not tried for the summit earlier that morning. The others were all doing fine. Moods improved with some light banter as we continued downhill. With the extra energy we gained, the lower we went, our speed increased. It took us about two-and-a-half hours to reach Mweka. I couldn't believe how much better I felt when we arrived back at 10,000 feet at the upper fringes of the rainforest belt.

Williness and the porters had their kitchen area already set up. In the firepit, crackling red dancing flames set the mood for a celebration. I got my tent set up and took off my hiking boots to find two blisters on the ends of my big toes. I used a few band-aids from my personal first aid kit to cover the blisters and put on a clean pair of socks. Our clients and I sat around the campfire, enjoying its radiant warmth. The two climbers who made it to Stella Point described their journey and answered questions from the others.

With the weather clearing, I stared up at the partially exposed top of Kilimanjaro. I tried to imagine how many other humans over the eons had gazed in wonderment, or trepidation, at this snow-capped giant, just as I was doing at that moment. It stood practically on the equator, yet glaciers and snow covered its peaks, while tropical rainforests surrounded its lower flanks. There was no place else on Earth quite like it. No wonder the locals and early explorers believed that Mount Kilimanjaro possessed magical powers.

Formed by violent explosions of an erupting volcano, then shaped and reshaped by later eruptions, cracked apart during ice ages and whittled down over the millennia by winds and erosion, the ancient mountain had endured. It beckoned adventurers to challenge its frontiers, and doubtless would continue to draw the curious to experience its marvels. RESPECT is what I felt for Kilimanjaro.

Sure, I felt disappointed that I failed to reach the summit, but I learned something important from this magnificent mountain. As long as I gave it my best effort, it didn't matter if I stood atop its peak. I had climbed three times higher than I had ever climbed before. In sacrificing the summit attempt, it positioned me perfectly to provide comfort to my weary fellow climbers as they returned to Barafu. I could help orchestrate a timely and smooth transition to our last camp. Sometimes I believe fate puts you exactly in the place you're meant to be.

Just as I started to worry about our missing Sobek guides, Conrad, Slade, Bart, and their three guides and the porter we'd left behind hauled themselves into Mweka Camp. If anyone could reach the summit, it would be these three. Failure did not exist in their vocabulary. Exhaustion racked their bodies. Conrad surveyed the camp, noting all of our group were together relaxing by the fire, nibbling on crackers and nuts. He sniffed the scents of dinner being cooked. Then he swung his daypack to the ground and flopped down beside me.

"Brenda, I can't thank you enough for getting everything moved down to this camp. And for taking care of everyone. What a tremendous help that was."

"Don't forget to thank Williness. He did as much, if not more, than I did."

Conrad called Williness over and praised his efforts in Swahili. His face lit up with pride. He bowed his head at the compliment. "*Asante sana*, Mr. Conrad. *Asante sana*."

Everyone seemed to be energized by the hearty dinner the porters served just as darkness fell. Spirited conversation ensued for another hour, with accounts of the attempts and successes our group members experienced. It sounded to me as though everyone had achieved a personal best of one sort or another. My feet hurt, but I wasn't nearly as exhausted as the others, who, one by one, peeled away to their tents for a well-deserved deep sleep.

There would be no restless rolling around during the night unless I wanted to because I'd found a truly level spot for my tent. I got cozy inside, grateful that I would *not* freeze to death tonight. And I *would* sleep soundly. These minor comforts, blithely taken for granted in the past, now brought such sweet relief.

34

THE END OF THE TRAIL

ALL THAT REMAINED of the expedition was a leisurely three-hour hike, dropping another 4,000 feet in altitude back through the rainforest, then a short walk on a level dirt road to the spot where our pickup vehicles would be waiting. Pure elation filled my heart. I'd survived some of the roughest conditions found in the natural world. Before our launch 19 days ago, I honestly doubted I'd complete this expedition unscathed, but amazingly, the only damages, I had to show for my adventures were two toe blisters! I yearned for a hot shower, a soft bed, and to tell my friends about all the wondrous things I'd seen.

I had tasted only a tiny bite of the feast our planet could serve up to those hungry to partake of adventure. Richard had hooked me on a wanderlust high without my even realizing what he'd done. When I got back to California, I didn't know if I'd chastise him for pushing me into the realm of adventure junkies or whether I'd hug him gratefully, to thank him for infecting me with his chronic malady. How soon, I wondered, would it be until I could get out on another expedition?

The porters didn't need to hurry off ahead of us today, no more camps to get set up, firepits to build, or dinners for them to cook. For them, payday had arrived, to be supplemented with generous tips once we were safely off the mountain. Williness hung back so we could walk the lower part of the trail together, with Kulmansini trailing a few yards behind. We passed time chatting about his family and the village where he lived. He wanted to know more about the United States and what my home looked like.

Black-and-white colobus monkeys performed daredevil flying tricks in the branches and vines above our heads and yammered at us for intruding into their territory. We watched a few of them using flexible branches like trampolines to give them the lift to make enormous leaps of at least 10 yards. Some used loose vines to swing Tarzan-style from one landing spot to another. Amused by their antics, I could have watched their acrobatic show for hours.

Williness hefted two bags, a well-worn gray backpack whose contents bulged heavily against his lower back, and a newer blue duffel that was heavily loaded. Neither the weight nor awkward shape of the duffel bothered him. As we walked, he alternated balancing it perfectly on top of his head, or with his head bent forward against his neck and the top of his shoulders.

Once we emerged onto the dirt track, we encountered a young girl of not more than five years old. She wore a shabby, diamond-patterned, sleeveless dress and balanced an oversized woven basket loaded with green bananas on her head.

We called out "*Jambo*" to her. She gave us a shy smile. Williness agreed to pose with her while I took their picture. Her small body barely reached to his waist. Both looked relaxed despite the loads their heads and necks supported. The sight of the bananas got Williness excited.

"Miss Brenda, have you ever tried *banana bieri?*"

I didn't know what that was, but I could guess.

"Is it beer made from bananas?"

"No beer, but like it. Makes you drunk. We drink *banana bieri* when come from mountain. To celebrate." He waggled his head, causing the duffel to dip and sway. "Is very good. You like try some?"

I didn't know what to say. I didn't want to offend him, but I sure didn't want to drink something that would make me sick either. He sensed my hesitation. "I give you some. You try it, you like it! Is very good."

At last, a group of vehicles came into view farther down the road. Two transport trucks would bring the locals back to their work base, and two vans would deliver our Sobek group to the Marangu hostel where we'd spend the night. When we were all assembled by the vehicles, our Tanzanian guides "officially" handed us back the personal gear bags they had lugged up and down Kili for us. We, in turn, "officially" presented them with their well-deserved tips. Kulmansini shook my hand, grateful for the reward he'd received. Each of us also chipped in money toward a generous pooled tip for the porters.

The celebration started when the local members of the expedition raised their voices to sing the traditional "*Buana Jambo*" song and engaged in a joyful ritual of dancing and clapping they performed after every successful climb. When they finished the song, the *banana bieri* flowed. The locals passed out cups of the celebratory beverage among themselves. Williness took an extra cup and found me standing next to a van.

"Miss Brenda, for you, *banana bieri*!"

"*Asante*, Williness."

Inside, a dense, brown, foaming, fermented mash filled the cup. It looked gross and smelled dreadful, yet our guides and porters were eagerly gulping it down, enjoying every drop. I needed to find Conrad to figure out what to do with it, so I tactfully said, "Williness, please don't let me hold you up from the party. Thank you for all your help."

I waved my hand toward his colleagues, so he knew he had my blessing to return to the celebration. I know he really wanted to hear how I liked the drink he had handed me, but realized from my cautious reaction I might not taste it after all. He shrugged and answered simply, "My pleasure." Then he spun around and rejoined the celebration.

I found Conrad and showed him my cup of *banana bieri*.

"Ah, that stuff is deadly. It's got twice the alcohol content of regular beer and tastes like mucus."

"You've tried it?"

"Yeah, once. Never again."

I didn't want to be seen discarding the liquid, so I carried the cup with me into the van. As we pulled away, I raised my cup in a toast to Williness, as proof I still had his offering. Returning my salute, he raised his cup in a last farewell.

THAT AFTERNOON, we had free time for showering, washing clothes, napping, and writing postcards purchased at the hostel. I must have stood for half an hour under the showerhead, shampooing and lathering up my body twice before I deemed myself clean. The water wasn't hot, but warm enough to enjoy. Sadly, the only piece of semi-clean clothing left in my bag was the sundress I'd worn at the end-of-trip dinner in Dar. It didn't smell too rank from being stuffed in with my dirty clothes. I pulled the dress over my head and went to find a plastic bucket I could use to give my clothes a good scrubbing.

Our clients would depart from the Kilimanjaro International Airport on an 8:05 a.m. flight the following morning. They busied themselves making sure they had everything packed and ready to leave the hostel by 6 a.m. Later that morning, our Sobek crew would move over to the hotel Conrad booked in Moshi to clean and pack up our permanent rafting equipment for storage at the home of Conrad's expat friends in Arusha, a quaint colonial village not far from Kilimanjaro.

We celebrated our last hurrah with a group dinner that evening. Everyone looked so spiffy we hardly recognized each other. We resembled normal jeep safari tourists, not survivors who had battled hippos and crocs, adverse weather, stifling portages, and altitude sickness on

Africa's highest mountain. We dined on grilled meat with a variety of roasted potatoes and vegetables until our stomachs ached with fullness.

As we shared stories of our favorite moments, I sensed we had become invisibly bonded for our lifetimes by sharing one hell of an intense experience together. How could we ever describe to our friends and families what we had seen, heard, smelled, or felt on this unique expedition? How could I ever describe to someone the chilling sound of a wild lion roaring? A hippo snorting its indignation? The terror of being inside a thunderstorm?

All these moments I'd shared with our eclectic group of adventurers. Since our first introduction in Ifakara, we'd grown together, each of us enriched by the knowledge we'd gained and each of us gaining confidence with every experience we'd survived. Now, like the mature seed head of a dandelion blossom, each of us would take to the air, being carried back to distant homes. We hugged and wished each other well and expressed a hope that we might one day travel together on another expedition. Yet, sadly, I suspected I would never again see them.

Africa had changed me: changed my beliefs about her precious people, animals, and lands; changed my perception of the world at large and my place in it; but, *most* importantly, changed how I saw myself. I likely always would have some fear of the unknown, but instead of dreading and avoiding it, spending this time in Africa had given me courage and curiosity to step toward the unknown with an open mind and heart.

35

TRAVELING SOLO

I HAD LOOKED FORWARD to a laid-back day hanging out
with the Sobek guys. But, when Conrad assigned tasks to the crew that
morning, the job he had for me was not at all what I expected. He could
make better use of the male crew members' brawn to repair and clean
the equipment, so he decided my mission would be to return a few
pieces of equipment we'd rented for the trip from the Park Headquarters.

"Brenda, I need you to take a local bus back to Marangu and walk
up to the headquarters to return the rental stuff."

Aghast at the idea that he would even consider putting me on
a local bus all by myself and sending me on an unfamiliar journey, I
said, "I don't think that's a good idea. You know I can't speak Swahili.
I don't know where to go. How would I even know where to get off
the bus? What if I get lost?"

I'd faced plenty of harrowing moments on this journey, but I'd
always had a safety net in the other guides and fellow travelers. Asking
me to travel solo on this mission with *no* frigging safety net would
be the most challenging test I'd faced yet. A million things could go
wrong. What he expected me to do petrified me. Warm tears welled
up in my eyes.

"I'll go down to the bus station with you, buy your tickets, and get you on the bus. You'll be fine."

In silent panic, I thought, "No, you won't!" If ever there was a mission impossible, this was it. Conrad saw how stressed out I was, but adamantly insisted that I could manage this assignment.

"I know you can do this. I wouldn't be asking you to do it if I thought you couldn't. It's a simple bus ride out and back and a hike up to the place we stayed the first night. Actually, you've got the easiest job of all of us today." His pep talk did nothing to convince me I would have it easy, only that my side of the argument held no merit for him.

After breakfast, walking down the busy road to the central bus station, Conrad prepped me. "Your bus is going to take you 25 miles to Marangu. When you get off at that stop, you'll need to walk up the road on the left for about a mile and a half until you get to the headquarters office. You remember the park headquarters where we stayed the first night with the passengers?"

"Conrad, I really don't remember the road to the headquarters or where we turned. From the back of the van, I couldn't see anything. Why would I ever think I'd be going back there?"

I knew from previous rides in our rental van to and from the park that the local buses made stops in every small village or crossroad along its route. "What if I get off at the wrong stop?"

For a moment, I thought he might be rethinking the idea of sending me off into oblivion with no way to communicate with the crew if I got in trouble.

"I'll tell the bus driver that he needs to be sure you get off at the right stop. There's one really important thing you need to know. The last bus back to Moshi comes through Marangu at 4:30 p.m. You need to be on that bus. If you miss *that* bus, you'll have to go back up to the park headquarters and stay overnight in their hostel."

I sighed. This just kept getting more complicated.

"You can use the equipment deposit refund to pay for the hostel if you need to. Buy yourself some lunch with it too, but don't forget to get a receipt," he said, chuckling. "You know those people back in Headquarters are sticklers for getting receipts for everything." His inappropriate joke stung given the situation...

At the bus station, Conrad located the colorfully decorated bus that would travel the route through Marangu and beyond. I climbed aboard ahead of Conrad. Tanzanians loaded down with baskets of food, supplies, and even live chickens already packed the bus. All eyes turned to size me up: a terrified white foreign woman whose obvious discomfort they all could feel. Men and women already occupied every seat on the bus, I discovered as I squeezed my way down the crowded aisle to the last extra-wide bench seat at the back of the bus. Conrad spoke in Swahili to the locals already seated there, and they shifted their bodies to open a space just wide enough into which I uncomfortably wedged myself. Using both arms, I hugged my backpack and canteen tightly on top of my lap and secured the walking poles between my legs.

"You can do this," Conrad said. "It's 11 a.m. now. It'll take 'til about 12:30 to get to Marangu. Then you should make it up to the headquarters in less than an hour. Return the gear to the office, get some lunch in the café, and head back down to the bus stop by no later than 4:00 p.m. Got it?"

I glared at him, willing myself to be anywhere but on this noisy, smelly, overpacked bus.

"You should be back before dinner. Good luck!" He turned and jostled his way around passengers who were still boarding the bus. I saw him stop and speak to the driver, who nodded his head. And then my only lifeline disappeared. I felt as helpless as I had ever felt in my life.

Five minutes later, with the aisle of the bus packed so tightly I could no longer even see the driver, the bus slowly rolled out of the dusty parking lot onto the asphalt road. About half of the windows, the ones that still functioned, were open, affording at least some air circulation. I found it impossible to see anything ahead of us. My only clear view came from looking out the two side windows of the last row. I begged God to spare us from getting in an accident. There would be no way to escape should the bus overturn.

The bus ground to a halt frequently. At each stop an intricate tango ensued, of people trying to exit the bus pushing toward the door while those wanting to be sure they could get on the bus were pushing their way up the steps. Whoever stood in the aisle closest to a seat vacated by a departing passenger quickly claimed the space. The people standing in the aisle shifted further back to make room for the newcomers.

At one stop about 40 minutes into the journey, a group of three white men in their mid-20s, heavily loaded with camping gear, boarded the bus. Tall, with rugged physiques, they wore folded bandanas tied around their heads to hold back long curly hair and keep salty sweat from stinging their eyes. These trekkers, I assumed, had to be on their way to Kilimanjaro. I felt a sense of relief that I wouldn't have to find the Marangu stop on my own, now that these guys would get off there as well.

I heard them speaking among themselves in what sounded like German. Standing a good head taller than the other passengers surrounding them in the aisle, I'd easily be able to see when they got off the bus. One of them spotted me in back of the bus and threw a nod my way. He must have realized we shared the same destination. It appeared they were as much out of their element as was I.

The bus rattled on through more small villages for a while longer before reaching Marangu. Pulling off on the side of the road, the driver shouted, "Kilimanjaro!" I rose and began the obstacle course of climbing around and over people to get off the bus. The Germans ahead of me did an admirable job of clearing a pathway. As soon as we had tumbled out of the bus, the driver steered back onto the roadway, picked up speed, and disappeared around a bend in the road.

The Germans spoke in heavily accented English. "Are you going to Mount Kilimanjaro?"

I graciously replied, "Yes, just follow me." I pointed across the road. "The Park Headquarters is a mile and a half up that road." We started out walking together up the moderately inclined roadway. Despite their heavy loads, they had youthful stamina and walked much faster than I did. I wished them a successful climb before they charged on ahead.

Conrad had given me an accurate estimate of an hour's walk to reach the headquarters. I inquired of the gentleman sitting in the reception area where I should return the rental equipment. He picked up a phone, dialed a number, and spoke into it. After receiving an answer, he replaced the phone on its cradle.

"Just have a seat please, madam." He pointed to a wooden chair against the wall. A little while later, a nicely dressed man arrived and, in perfect English, invited me into his office. I explained I needed to return the equipment that Sobek had rented. He offered me a drink, chai or soda, and asked if I had enjoyed the climb. He seemed impressed by my account of our climb on the less traveled Machame route.

I asked him to direct me to the café so I could eat some lunch. I also confided in him my anxiety about getting back down to the bus stop, so I wouldn't miss my ride back to Moshi. After directing me to their café, he generously offered to give me a lift in his car to the bus stop, as he had to pass by it on his way home. He told me to meet him in the reception area when he got off work at 4:00 p.m.

I couldn't believe my luck. I'd be able to skip that mile-and-a-half hike back down the road, which meant I'd have time to enjoy my lunch and maybe even spend some time updating my journal. My anxieties about completing this assignment were waning. Maybe Conrad was right. I might just pull this task off after all.

AT 3:45 P.M., I went back to the reception area to wait for my ride. 4:00 p.m. came with no sign of the official. After another 10 minutes passed, I wondered if he had forgotten his offer and had already left for the day. I figured that even if I ran all the way down the hill, I would be too late to catch the bus. I began to panic, and I felt nauseous. Each minute that passed seemed like an hour, and my mood sank as I realized I was going to be stuck here for the night. Finally, at 4:25 p.m., the official showed up and told me to come along.

"I'm afraid that by the time we get to the bus stop, it will have already come and gone," I told the official.

He laughed. "Don't you worry," he reassured me, "that bus is never on time. You'll catch your bus."

It was already past 4:30 p.m. when he pulled up to the bus stop. A crowd of 20 people milled about, waiting for the bus. Again, my heart sank. With this many people already waiting, I wondered if I had any chance of getting on the last bus back to Moshi. The official parked on the side of the road, and we both got out of his car. He walked toward the crowd and spoke in a loud voice. After a brief back and forth with the crowd, he led me to a young black couple.

"These people are missionaries, also headed to Moshi. They speak English and will take care of you to make sure you get there safely." I thanked the official and wished him a pleasant evening.

"My name is Brenda," I said, introducing myself. The woman told me her name was Mariam, and the man with her was her husband.

"When the bus arrives, you need to stay very close to us because it can be very difficult to get onto the bus."

Five minutes later, the bus came rattling around the bend. As it slowed, Mariam grabbed my arm, tugging me through the throng, rushing along beside the bus. We were lucky because her husband had planted himself almost directly in front of the door when the bus squealed to a stop. I could tell he had nailed this game of anticipating precisely where the door would end up.

Only a few people exited the bus, already packed beyond its capacity. We dove up the stairs onto the bus and shoved our way toward the back. We made it about halfway down the aisle. The metal grab bars on the back of the seats to help people stay upright were unnecessary, given how tightly wedged in we were. It would be impossible to fall.

Shouting volleyed between the driver and the handful of people who could not fit onto the bus. Finally, with no arms or legs trapped, the driver's assistant forced the door to close, and the driver pulled onto the roadway. What a relief it was to be on the bus to Moshi. I'd be back with the crew soon.

A third of the way through our journey, the bus driver pulled over to the side of the road. He and his assistant started screaming instructions in Swahili. The passengers shouted back, unhappy about what was happening.

"We've come to a police checkpoint." Mariam explained. "They are going to come onto the bus and check for contraband goods. Everyone has to get off the bus now so they can do their search. We have to get off."

I stayed close to Mariam and her husband. Slowly, the bus emptied of passengers as military police, armed with rifles, inspected everyone as they came off the bus. Mariam explained to them I didn't speak Swahili and translated their demand.

"You need to open your backpack. They want to look inside."

I unzipped my pack so they could see the contents: my camera, some film, my journal, some postcards, and the crumpled-up poncho I carried in case the weather turned bad. I suspected they would have been overjoyed to find some illegal item they could seize, but they didn't. Once satisfied, they ordered us to move five yards into a field next to the bus.

The police climbed aboard the empty bus and began searching through the baskets and *kanga* cloth sacks that people had been told to leave on the bus. 15 of the male passengers moved further out into the field, forming a straight line facing away from the rest of us. Almost as if choreographed, they watered the dry earth and weeds in front of them, achieving the immense relief of an emptied bladder.

The search didn't take long, and from what I could see, the police did not find any evidence of smuggled goods. They left the bus empty-handed. There seemed to be an unspoken rule for reboarding, that everyone would return to the exact seat or place in the aisle where they had been before the bus stopped. In the reverse order that we had filed out, we reentered the bus.

One more time, the bus pulled onto the roadway. We swayed and teetered each time the bus braked or sped up as it made stops along the route. Every passing minute meant I was that much closer to a reunion with my colleagues. My legs and back muscles ached from being locked in an immovable stance that prevented any way of stretching.

Suddenly, the driver barked out a command. Everyone standing dropped into a squat, leaving only me still standing in the aisle. The driver's assistant launched into a tirade directed at me, but whatever he was yelling meant nothing to me.

Mariam tugged at my shirt. "Sit down. Now!"

"Where?" I asked. I couldn't move even an inch. The others in the aisle had claimed every bit of space. Even trying to squat meant my butt would rest on someone else's head or back. The assistant continued to scream at me.

"Just get down. Now!" Mariam's voice was terse.

Sitting in the seat next to where I stood sat a portly, well-endowed woman colorfully wrapped in a *kanga*, a complete stranger to me. Her lap was empty, so spontaneously, I did the rudest thing possible. I collapsed onto her lap. At least I was down. I glanced at her face, expecting to see outrage at my unwelcome intrusion. Instead, she sported a grin from ear to ear, and her brown eyes twinkled with amusement. I'm sure this was her first and last time to be sat upon by a white tourist. She would have a story to tell her family that evening.

The driver again pulled over to the side of the road, but not at a bus stop. We were in the middle of nowhere. The driver and his assistant carried on a brief animated conversation. Then the assistant started yelling again. Most of the people on the bus yelled back at him. A loud, intense debate carried on.

"What's going on?" I asked Mariam.

"The bus is overcrowded. The driver is afraid he'll get fined. There's a roadway checkpoint about half a mile up the road. He wanted to see if the bus would still look crowded, if we all crouched down, but he thinks that it still looks too crowded. They want some people to volunteer to get off the bus."

Why would anyone agree to that request, given it was the last bus of the day back to Moshi? I knew for sure that I was not budging from the comfortable seat I'd just gained. The tone of the argument escalated.

"The bus driver says that he won't go any farther until at least 10 people agree to get off the bus."

The yelling and screaming continued, no doubt colored with profanities. As the confrontation grew in intensity, the mood shifted. A few men got up and left the bus.

Mariam said to me, "We're going to get off the bus."

My eyes must have conveyed my fear. "But this is the last bus to Moshi today," I argued. "How can we get home from here?"

"Trust me. We'll be fine. Come on, let's go."

Her husband stood and cleared the way for Mariam and me to follow. I felt so conflicted about this odd change of plans. The mission-aries spoke the only language I knew, and they had taken care of me as they promised the park official. But not knowing what we would face if we left the bus frightened me. Reluctantly, I decided to trust Mariam. After we got off, a few more people also got off the bus. The assistant handed each of us a few shillings. Then the bus departed, leaving us on the side of the road.

I watched with deep misgivings as the bus passed through the checkpoint. The driver paid a toll and continued down the road. A half mile past the checkpoint, the bus suddenly stopped and pulled off the road. That was strange.

Several trucks and automobiles passed by as we waited by the edge of the roadway. The other Tanzanians, who joined us, stared down the roadway in the direction we'd come and waited. When a flashy, slightly lopsided minivan came into sight, they excitedly waved their hands above their heads. As it drew closer, they hollered, as if more commotion would coax the van to stop.

One of the men explained our plight to the van driver, but he shook his head and pointed back toward the already crowded interior of his vehicle. I could see there wasn't any chance of catching a ride in this van. With a shrug of his shoulders, as if to say he was sorry he couldn't help, he pulled away from us and back onto the road. Secretly, I was glad that we didn't have to climb into that vehicle. It looked less than roadworthy, listing severely enough that if the van hit a pothole just right, it might have tipped over.

The next minivan that the men waved down had some room, so five of our group handed the driver their shillings and wedged their way into the back. Off they went toward Moshi. A few minutes later, another minivan came along, and Mariam's husband signaled that one to stop. The driver had room for three people, so, after a quick discussion, Mariam, her husband, and I claimed that van. Mariam pushed me into the one space on a seat while she and her husband sat on the step just inside the side door. Once again, we set off toward Moshi.

Our van cleared the checkpoint with no problem. As we approached the Moshi bus on the side of the road, the van slowed and pulled off the road just ahead of the bus. Mariam and her husband opened the van door and hopped out. "Come now, Brenda," Mariam directed as we walked back to the bus and climbed aboard it. The unspoken rule still applied, so we wriggled our way back down the aisle to assume our original places.

When the big black lady, whose lap I reluctantly occupied during the overcrowding frenzy, saw me back on the bus, she gave me a genuine smile. Luckily for her and me, with the checkpoint in the rearview mirror, I could stand for the rest of the trip. We waited a bit longer for the last two of our passengers to be shuttled past through the checkpoint.

All the pieces fit together now. If the bus driver had pulled up to the checkpoint with an overcrowded bus, he'd have incurred a fine for each person over the maximum capacity. Though more inconvenient for his passengers, it was cheaper for him to pay a small fare to a minivan to shuttle his extra passengers through the checkpoint. Farcically, the whole drama played out within sight of the checkpoint. The scam fooled no one.

Dusk had turned to night when the bus pulled into the Moshi bus station. With the bus running late to begin with and the timeouts for the police contraband search and the overcrowding fiasco, we arrived at nearly 7 p.m. When we stepped outside, I gave Mariam a quick hug and thanked her for all her help. She wished me well.

I heard Conrad's voice behind me saying, "You're just making friends all over this country."

"Oh, Conrad, you won't believe what a day I've had! Have I got a story to tell."

"I'm glad you made it back. I was wondering why the bus was so late. But the crew is waiting for us for dinner, so let's head back to the hotel. You can tell me on the way."

I reflected on the complexity of the task I had just somehow completed while I spilled out the details of my adventure. So many things could have gone wrong, but they didn't. I just plodded along and asked for and accepted help to make it to the finish line. Though I had just spent 12 days in one of Africa's remotest wildernesses, and seven days climbing higher on a mountain than I had ever climbed before, it had been in the company of well-seasoned experts.

On this day, I had traveled solo, just me against the world. A joyfulness filled me, when I realized that I *could* travel by myself in a foreign country and manage the task assigned to me, even while experiencing a rollercoaster ride of emotions. With a newfound confidence, I celebrated the possibility that I could accomplish anything I set out to do! Just like Jimmy Buffet.

36
42 YEARS LATER

THE ONLY CONSTANT IN LIFE is change. The discoveries I experienced on my maiden adventure affected me profoundly and radically changed the path I have followed since then. I have always wondered if Richard Bangs had any inkling, when he informed me we were going to Tanzania, how much that trip would transform me from a fearful introvert into a curious, emboldened woman who would fall hopelessly in love with our wonderfully diverse world. He fed me just a spoonful of mouthwatering wanderlust, then encouraged me to fly on my own journeys of discovery. What a gift that was!

The following year, I spent 30 days rafting on the Omo River in Ethiopia, with Conrad Hirsh as our trip leader. The last seven days of that trip, we navigated territory that Sobek had never explored. In 1983, I journeyed to Pakistan with Sobek on a jeep safari that traversed parts of the Karakorum, Hindu Kush, and Himalayan mountain ranges. My growing hunger to immerse myself more fully in other cultures led me to search for opportunities to live and work overseas.

Shortly after returning from my jeep trip to Pakistan, I was hired by the United States Agency for International Development (USAID). Ironically, my first four-year-long posting turned out to be in Islamabad,

Pakistan! With a heavy heart, I bade farewell to my Sobek colleagues and took flight to a strange new world halfway around the globe. That move launched my lifetime of traveling, living, learning, and working in countries around the world.

Since I boarded that plane to Africa 42 years ago, the changes in our world have been profound. The Tanzania of 1981 that so charmed me is nearly unrecognizable today. Underlying forces of "human progress," money, and greed have transformed the land in unthinkable ways. The population of Tanzania in 1980 was 18.5 million. By 2021, it had more than tripled to 61.1 million. Its commercial hub, Dar es Salaam, has exploded from 905,000 residents when I wandered its streets to 7 million people in 2021. And in Ifakara, the tiny speck of a village where our trip started, the population has grown to 49,500 people. Online pictures show modern multistory buildings, a two-lane paved road through its center, and a bridge that was constructed over the Kilombero River just outside Ifakara in 2012.

This level of population growth has demanded enormous increases in infrastructure, particularly expanded production of electricity. In 2018, Tanzania approved the construction of the Julius Nyerere Hydroelectric Dam and Power Station across the Rufiji River at Stiegler's Gorge, a move that has had tragic consequences for the Selous's wildlife and ecosystem.

Construction on the $3.6 billion arched concrete and rock-filled dam started on July 26, 2019. The structure is 440 feet high and 2,300 feet wide. Behind the dam, a catchment area will hold 34.3 billion square meters of water covering 460 square miles. When full, the reservoir measures a maximum length of 62 miles, drowning one-third of the length of the river system we traveled down.

Significant populations of savannah-dwelling giraffe, wildebeest, and zebra, along with countless river and forest-dwelling species, have been forced from their habitats. By eliminating the natural cycle of seasonal floods during the rainy seasons in the Rufiji's delta below the dam, there will be a negative impact on agriculture and fisheries as well.

At a ceremony to lay the foundation stone for the power station in 2019, the president of Tanzania directed the government to divide the Selous Game Reserve, making two-thirds of the northern Selous into Nyerere National Park covering 12,000 square miles, with the goal of promoting tourism. The remaining third remains as a game reserve for use by big-game hunters. They also carved out an area of 155 square miles from the southwest section of the reserve to allow for private exploitation of uranium reserves. Mining in this area is likely to leach heavy metals and arsenic into the area's water systems.

At the time of our trip, only the small lodge at Stiegler's Gorge, from which Richard and Slade did their aerial reconnaissance, and the Beho Beho camp inland from the river, where we stayed at the end of the river trip, provided accommodations in the area. Both were rustic with few amenities, no radios, no maintained tracks, and no means of contact with the outside world.

Now, tour operators routinely send adventure seekers to 11 high-end safari camps in the northern section of the park near the Rufiji River and surrounding lakes. In 2004, Beho Beho underwent a complete renovation and upgrade to a high-end luxury resort. Enthusiastic safari clients now eagerly fork over $1,450–$2,450-plus per person per night, as of 2022. Every room now features an open wall for optimum wildlife viewing, a large-screen TV, Wi-Fi connections, and individual plunge pools. There is also a swimming pool and treehouse accommodation on the property.

With the beauty of the hills surrounding Beho Beho marred by huge transmission lines carrying power to urban centers and paved with new roads carrying trucks to and from the power station, it remains to be seen whether tourists will continue to see the value in paying for these new, less-than-magnificent views.

While the dam poses an existential threat to the Selous's wildlife now and in the future, African wildlife conservation efforts have also suffered devastating blows over the last several decades.

Until 1800, an estimated *26 million* elephants lived on the continent of Africa. Since then, collectors in the United States and European and Asian countries have grown insatiable appetites for elephant tusk ivory to make combs, pool balls, brush handles, piano keys, trinkets, delicate pieces of art, chopsticks, and religious symbols. In 1979, Ian Douglas Hamilton, founder of Save the Elephants, conducted the first Pan African Elephant survey and found only *1.3 million* elephants remained in Africa. The Great Elephant Census, completed in 2016, undertaken and funded by the Paul G. Allen Foundation along with the International Union for Conservation of Nature, found that illegal poaching for ivory remains the major reason for the ongoing loss of wild elephants. They reported the African elephant population had dropped to *415,000*.

Poaching of elephants within the Selous game reserve has been equally devastating. At the time of our trip in 1981, an estimated *100,000* elephants roamed within its borders. By 2014, an elephant census revealed only *15,217* elephants remained in the reserve. That same year, UNESCO placed the Selous on its List of World Heritage in Danger based on the severity of elephant poaching. Since then, the Selous elephant population has stabilized, suggesting government efforts to curb poaching may be working.

Unfortunately, other species are also being decimated. In 1980, rhinoceros conservation organizations estimated there were *5,000* black rhinos in all of Tanzania. By 1995 poachers had killed all but *100* of the critically endangered species. Since that low point, numbers of black rhinos have edged up only slightly.

In 1940, *450,000* lions roamed the earth. At the time of our trip in 1981, there were *200,000* lions worldwide, mostly found in Africa.

By 2012, the number of wild lions had decreased to *32,000*, with half of them found in Tanzania. Every year, tourists on African trophy hunts kill roughly *600* lions after paying trophy fees ranging from $10,000 for a lioness to $35,000 for a black male lion.

On a lighter note, I am happy to confirm we did not encounter any dragons in the Selous. Scientists and local Tanzanians hold the belief that these creatures are in no danger of extinction, but remain alive and well as colorful figments of an overactive imagination.

Over the past four decades, Mount Kilimanjaro also has seen many changes. Per park records in 1965, they registered barely 1,000 visitors. By 2003, the park hosted 28,000 visitors. In recent years, it's estimated that annually 50,000 people visit the park, of which 35,000 attempt to climb the mountain, with 66 percent of them successfully reaching the summit.

I hold myself in good company with a few American celebrities foiled by their summit attempts. Ann Curry, then a *Today Show* host, fell victim to altitude sickness. Super Bowl champion linebacker Ray Lewis suffered a foot injury. And superstar tennis player Martina Navratilova experienced pulmonary edema.

Currently, there are seven climbing routes up the mountain. Most of the 20-plus tour guide services operating on the mountain offer the option to climb via the highly scenic Machame route, the one we climbed on our trip. It has become so popular that it has overtaken the Marangu route as the most heavily trafficked climbing route on Kilimanjaro.

In December 2020, the Tanzanian government approved installing a cable car aerial suspension system that will run above the Machame route from the Machame Gate to the Shira Plateau at 12,000 ft. It will be the fourth-highest cable car in the world, carrying 15 cars with a capacity of six people per car on a one-way ride that will take 20 minutes. And in August 2022, the Tanzanian Tourist Department launched broadband internet service on Kilimanjaro.

Besides the growing number of visitors crowding the trails, littering, and trampling vegetation, global climate changes are taking their own toll on the habitat. A study by the European Geosciences Union reports that between 1912, when there were 4.4 square miles of ice cap, and 2011, when the ice cap had shrunk to .68 square miles, 85 percent of glacial ice has disappeared. Some climatologists fear that Mount Kilimanjaro may lose all of its glaciers and snow by 2033.

In 1989, the organizing committee of the 100-year celebration of the first ascent of Mount Kilimanjaro awarded posthumous certificates to the African porter-guides who accompanied Meyer and Purtscheller in a long-overdue acknowledgement of the essential roles they played in the ascent.

There have been many changes in the lives of my fellow Sobek crew members who made this magical journey with me, many amazing and some sad. Tragically, Conrad Hirsh waged a courageous battle with brain cancer and, despite aggressive treatments in London, passed away in October 1999. After his death, a fund in his name to support elephant conservation and research was established.

His friend Cynthia Moss wrote about Conrad, "He loved Africa, its landscapes, rivers, people, animals, cities, and towns, but he was not selfish or possessive in that love. He enjoyed sharing it with others by taking them on glorious adventures. In doing so, he changed the lives of those visitors, and Africa touched their souls." After our Selous adventure, Conrad founded his own company based in Madagascar, called Remote Rivers Expeditions. It continues his legacy by offering safari and river trips to East Africa.

Richard Bangs continued his prolific career in adventure travel, logging 35 first descents of wild rivers. He has authored 19 books and over 1,000 magazine articles about his adventures, winning a National Outdoor Book Award and a Lowell Thomas Award. He founded and served as editor-in-chief of Mungo Park, Microsoft's pioneering travel publishing effort, and later became a founding executive of Expedia. com, serving as its editor-at-large. He also founded the Well Traveled feature for Slate magazine and was founding editor and executive producer of Great Escapes for MSNBC.com.

In the movie world, he has been a stuntman, director, and executive producer of four adventure-based films. He has launched two successful television series traveling through 10 countries called *Richard Bangs Adventures with Purpose* and *Richard Bangs Quests*, for which he has won numerous media awards, including two Emmys. His current project is Steller.com, a digital storytelling platform dedicated to travel.

In 2022, ExplorersWeb named Richard as one of the 100 great explorers of the last 100 years alongside the likes of Thor Heyerdahl, Reinhold Messner, Edmund Hillary, and Neil Armstrong. His latest book, *The Art of Living Dangerously*, chronicles 50 years of Sobek's explorations, adventures, and misadventures. He lives in Venice, California with his wife, Laura Hubber of the BBC, and his younger son, Jasper. His other son, Walker, lives in Bellevue, Washington. Richard recently led the first commercial adventure to Eastern Angola and has a list of things to do that never shrinks.

During Jim Slade's 44-year career as a professional adventure travel guide, he never stopped exploring the remotest nooks and crannies on all seven continents of our planet, searching for the wildest rivers and highest peaks to conquer. Sobek first hired Jim as a guide for its inaugural commercial trip on the Omo River in 1974. Jim stayed and led many of the trips that followed. Of those early Omo trips, he recalls, "It was a very special experience for me. It was truly wild back then. I was in my second year as a guide leading month-long trips in remote parts of Ethiopia. Crazy!"

Favorite highlights from his life of adventure include the following: a six-week-long, first, and only, crossing of the long axis of Borneo, including the first descent of the Kayan River; six expeditions to

Antarctica, three by ship and three land-based trips; the first descent of a section of the Yangtze River and the first descent of the Yarkland River, both in China; and attempting Mount Everest from the Tibetan side. Having reached the summits of dozens of mountains, several of them multiple times, Jim is sure of one thing: "Mount Kenya is, and has been, my favorite mountain in the world."

Other favorite expeditions he's proud of include planning and leading a motorcycle trip through southern Chile for Lyle Lovett and renowned motorcyclist Malcolm Smith, and joining Bart Henderson in 2014 as they replicated the 1000-mile journey made by John Wesley Powell in 1869 to explore the Green and Colorado rivers. In 1990, during a break in his travels, he met his wife, Barbara, a successful lifelong equestrian. Jim retired at the end of his 80th and last commercial Colorado River trip in 2017. He and Barbara continue to explore the world with recent trips to Vietnam, Colombia, Namibia, and Portugal.

In 1976, Bart Henderson fell in love with Alaska's wilderness while exploring the Tatshenshini River with its sweeping glaciers, icebergs, grizzly bears, and moose. After our Tanzanian expedition, Bart returned to his job managing all of Sobek's Alaskan operations. At the same time, he started a small guiding business of his own to demonstrate how wilderness tourism could provide the region with a robust economic alternative to logging and mining. Over the next three decades, Chilkat Guides effectively proved that point, by leading more than 1.5 million awe-struck passengers into the wilds of Alaska.

Whenever Sobek planned an expedition on an unexplored river, Bart jumped at the opportunity to be one of the guides. Reflecting on his river guiding career, during which he has logged over 50,000 river miles, he is grateful for "all the exploring, all the adventure, all the challenges of figuring out how to get clients out into such exotic and amazing places. No longer is it feasible for a single guiding company to hold on to an elite group of guides traveling to the far corners of the globe to set up new adventures on virgin rivers. We were so lucky!"

In 2022, during his 53rd year of guiding, Bart completed his 111th trip on the Colorado River through the Grand Canyon and says, "I still love every mile." He is especially proud of his 29-year-old twin sons, one of whom has followed in his footsteps as a year-round guide in the Arctic. The other is a lawyer in Juneau.

Michael Ghiglieri's love of the natural world and rivers has resulted in his logging over 47,600 miles on rivers during 47 years of guiding. He claims working on 700 commercial whitewater trips has endowed him "with a highly varied set of challenges, experiences, and arcane problem-solving." He is the proud author of nine books ranging from his studies of chimpanzees in Uganda, to his river-guiding exploits, to the multitude of ways one might die in the Grand Canyon and Yosemite.

Michael and I worked together again in the late 1990s for the School for Field Studies, where he directed overseas semester-abroad programs for sustainable resource management in five locations in Kenya, Turks and Caicos, Palau, Australia, and Vancouver Island. He still occasionally guides rafting trips in the Grand Canyon for OARS, Inc., often with one of his sons, and continues to write books.

Seven years after our trip, while reading a gardening column in the *Miami Herald*, I saw a subscriber had written to ask, "Are kerosene mangos edible, and, if so, how can the kerosene flavor be removed?" To learn that such a thing actually existed was sweet vindication. The paper's expert botanist replied, "This species of mango does indeed have the distinctive flavor of kerosene, and, for this reason, it is not considered edible, though it does not cause any physical harm if eaten. The kerosene taste cannot be removed."

Writing about my little elephant friend, Kitubinissa, made me curious to know more about her life. I discovered that when the Stone Zoo closed (temporarily) in 1990, they sent Kitubinissa to a zoo in Syracuse, NY. While there, she became pregnant, but sadly died in childbirth.

Our world continues to change. There are still wonders to be found, and the chance to experience wilderness in its pristine state, but the time to do so is growing short. I would never have dreamed that in just 42 years, the Selous Game Reserve and Mount Kilimanjaro could undergo such massive transformations. Because of the changes I've just described, it would be impossible to replicate my life-changing journey now. It is heartbreaking to know that others will never experience the same magic I felt on my trip through Tanzania. The best I can do is share my story about the time that Richard and I indeed found lions roaming in terra incognita.

While there is still time, run to the wilderness. I can guarantee you won't encounter any dragons. But I hope you *will* get to hear the magnificent, mighty roar of a wild lion.

AUTHOR'S NOTE

IF I HAD TRIED to write this story soon after I returned from Tanzania in 1981, it would have lacked a mature awareness of how the world has transformed, as well as the beneficial value of the insights I gained from this life-changing journey. I owe more gratitude than I could ever express to Lois Simes, who always welcomed me home from my travels and listened eagerly to my tales of adventure over a shared cup of tea. Trip after trip for 40 years, she urged me, "You must write these stories so people can learn about the world you've discovered, especially the places where they might never venture." For 40 years, I told her I would, but never found the time. Just before she died, she made me promise I would grant her wish. This book is the fulfillment of that promise.

Richard Bangs, you inspire me today as much as you did in Sobek's early fledgling days. I am grateful for our 42 years of friendship. In response to a holiday card I sent you years ago, I've kept the letter you sent back in which you wrote, "I think you are a wonderfully evocative writer with a knack for storytelling and a unique treasure chest of experiences. I smell a bestseller!" What optimism you've always had in my abilities. I find myself once again being urged by you, a journalistic giant, to take a scary leap into another type of adventure—being an author. I can hardly wait to see if your prediction comes true.

Along with Richard, I have great respect and affection for Sobek guides Jim Slade, Bart Henderson, Michael Ghiglieri, and Conrad Hirsh (deceased), who navigated us safely through the African wilderness on our Tanzania trip. Based on your intelligence, survival skills, and teamwork, I'd trust all of you with my life anywhere in the world. Collectively, you have the entire planet covered.

I will always have a special place in my heart for Solomon and my little Ifakaran *rafiki*. I have often dreamed of one day revisiting Tanzania to find my *rafiki* to discover the woman she has become. This journey opened my eyes to the people's kindness, the magnificent wildlife, and the spectacular landscapes of Tanzania, which will live forever in my heart.

Thanks to OARS, Inc. and Sobek Expeditions, I realized my dream of becoming a skilled river guide, taking hundreds of rafters on whitewater trips where I could share with them my appreciation of natural wild places, some of which have disappeared and others that are endangered.

My "bestie," Monica Langley, has read this manuscript almost as many times as me. I have always appreciated your comments, astute attention to details, and straight-up honesty. Your feedback has made the story better. A big thank you goes to Michael Ghiglieri, with whom I fact-checked details of our trip. An accomplished writer with many titles to his credit, Michael, I am grateful for your influence and perseverance in schooling me on writing skills I needed to improve. Many friends, writing partners, memoir classmates, and instructors read and commented on pieces of my writing as the story came to life. A part of each of you is in this book.

In hindsight, I respect and appreciate my parents' efforts to protect me when I was young. I'm equally grateful they knew when to set me free, so I could own my successes and failures, while exploring far corners of the world. I returned home years later, a wiser, braver woman.

When I decided to self-publish this book, I chose Paper Raven Books to guide me through the mysterious and convoluted maze of turning a completed manuscript into the product you are holding today. The specialized skills and support I got from each person on their team has taught me so much. Thanks for holding my hand and leading me through the publishing process.

I don't think Lois Simes expected me to stop at just one story. I have many to tell. I write every day, often into the wee hours of the night, with the hope I can bring you with me on more of my magical adventures. You can follow my progress on www.eyeopenerpress.com. But right now, I invite you to read *Shattered*, for a sneak peek of the stories I'm working on for my upcoming book, *Wanderlust Snippets*.

FREE GIFT

If you enjoyed *BECOMING FEARLESS*
Brenda's adventures continue in

GET A FREE SNEAK PEEK BY SCANNING THE QR CODE

Ascension Theology

DOUGLAS FARROW

t&t clark

Published by T&T Clark International
A Continuum Imprint
The Tower Building, 11 York Road, London SE1 7NX
80 Maiden Lane, Suite 704, New York, NY 10038

www.continuumbooks.com

British Library Cataloguing-in-Publication Data
A catalogue record for this book is available from the British Library

ISBN13 : 978-0-567-14405-8 (Hardback)
978-0-567-35357-3 (Paperback)

Typeset by Pindar NZ, Auckland, New Zealand
Printed and bound in India

In memory of
my mother, Kathleen Farrow
my father-in-law, Vincent Whelan
my Doktorvater, Colin Gunton
and Fr Richard John Neuhaus,
to whom I had also hoped to show this book.

'then we who are alive and remain shall be
caught up together with them'

Contents

Preface xi

1 The Upward Call 1
. . . in which the ascension is located in its biblical context,
allowing us to see how it forms the natural outcome of the
story of Jesus.

2 Re-imaginings 15
. . . in which is traced the rewriting of the ascension story
as a movement of the mind, a view pioneered by the early
Alexandrians and newly developed in the Enlightenment.

3 Raising the Stakes 33
. . . in which an alternative with stronger roots in scripture,
reckoning with the implications of ascension in the flesh, is
discovered in St Irenaeus and supported from the later tradition.

4 A Question of Identity 51
. . . in which is clarified the fundamental issue at stake in the
doctrine of the ascension, *viz.*, the identity of the risen Christ and
hence also the identity of the church.

5 Presence in Absence 63
. . . in which the rite that sustains the church during the absence
of Christ is considered, with reference to the problem of the real
presence.

6   The Politics of the Eucharist   89
      . . . in which the church's eucharistic relation to the world is
      explored in connection with the mystery of lawlessness that
      shadows the heavenly session of Christ.

7   Ascension and Atonement   121
      . . . in which is discussed the cleansing of heavenly things and
      the glorification of earthly things, and the great transition
      to glory.

Epilogue   153

A Summary of the Anaphoric Work of Christ   158

Prayers for Ascensiontide   159

Bibliography   163

List of Images   169

Index   171

Preface

THE CREED, like the scriptures, forms from its several parts a unified whole. At a particular juncture in history, however, this or that credal assertion may become especially important for our grasp of the whole. It seems to me that, in the West, we are presently at such a juncture where the witness to the ascension is concerned. Yet a great many western Christians are quite unaware of this. For them that witness has been obscured. The doctrine of the ascension has become an enigma, if not an embarrassment. The corresponding liturgical feast, once one of the church's great feasts, is poorly celebrated. The Rogation Days that preceded it have disappeared and, whether marked on Thursday or on Sunday, Ascension pales beside Pentecost. The latter's dependence on the former is seldom noticed, which helps to explain why it too is fading in many places, along with the ecclesial confidence and communal bonds that all these feasts formerly nourished.

Reviving the feast of the ascension will require a theological renewal, but as I argued in *Ascension and Ecclesia* that renewal will require a choice between theologies. From the turn of the third century there have been two main streams of interpretation of the ascension, streams that until modern times have flowed side by side in the level landscape of a broadly Christian culture. Our culture is in upheaval, however, and it has become more apparent that these streams are running in opposite directions. Their headwaters are different, and so are their destinations. Simply put, our choice is between a doctrine of the ascension that truly affirms our humanity in Christ and one that secretly or openly denies it.

In the preface to *Ascension and Ecclesia* I presented it as the precursor to a more substantial discussion of the church; that is, of the community in which – thanks to the Spirit who is sent from the ascended Lord – authentic humanity becomes possible. The present much shorter book is not an attempt to fulfill that rather rash promise, but, in making a more modest contribution, to provide at the same time a more accessible sketch of the relevance of the doctrine of the ascension and of its relation to other key Christian doctrines. While I have tried to take into account some of the scholarly responses elicited by *Ascension and Ecclesia* and to give fellow theologians something further to ponder, it is also my business here to reach out to the less-specialized reader, including the adventurous undergraduate, the seminarian and, more importantly if only indirectly, the person in the pew who prays that the preacher would not, on this particular feast, ascend the pulpit in vain.

No doubt some of the criticisms that have been made of the earlier work will have even greater force here. I intend to present the aforementioned choice rather starkly, even if this leaves the unwanted impression that the theological world is populated only by heroes and villains, rather than by people like ourselves who stand in a mixture of light and shadow. Nor can I help being even more selective in this shorter work, overlooking a great many things that ought to be considered and being content to allude to others in a rather cryptic fashion that may tease or puzzle. Yet it seems right to carry the discussion and debate forward into a broader forum. And though I am only recently a Catholic, and as such not ready to attempt the ecclesiology at which I was aiming, it seems right also to say what I have to say as a Catholic – with important differences that will please some and displease others – even though the book was better than half-written when I was received into communion with Rome.[1]

Some years ago my wife, Anna, and I were having dinner one evening in Montreal with some august company, including Fr Richard John Neuhaus, who observed that in chapter four of *Ascension and Ecclesia* I had been rather hard on the Latin church (he might have said the same about the eastern churches) and on its Mariology especially. I pointed out that the chapter in question acknowledged that it had, for a good purpose, indulged in caricature, and that I had always seen it as an exercise in clearing the ground both for the more damning criticism of modern Protestantism pursued in the penultimate chapter ('Where is Jesus?') and for a more constructive approach to Mariology that I hoped some day to offer. I repeat this here not only because I had hoped to seek his counsel about what I have written but also to help two groups of readers: those who might otherwise not see any continuity between chapter four of that book and chapter five of this; and those who, having started with this one, might be scandalized by what they later find in that one. Of course, the former may conclude that I simply lost my way, the latter that I finally discovered it – or am at least looking like I might discover it. God helping me, and the intercessions of those who have gone before, I will.

With the exception of chapters five to seven, which are the most difficult and require some additional support – indeed, much further thought – I have tried to keep the notes relatively brief. I would like to think that they will be consulted on at least a few points. The many biblical and patristic references they contain should not be regarded as proof texts but as invitations to more important reading.[2] Among the fathers, I continue to recommend Irenaeus especially,

---

1 I have told the story of my entry into that communion in 'Are you Catholic?' (Conrad Black *et al.*, *Canadian Converts*, 59-109).

2 With patristic or other classical texts I have, for the most part, indicated only the primary source and not the secondary. When quoting I have generally employed standard works (such as *The Ante-Nicene Fathers*, e.g.) that are listed in *Ascension and Ecclesia*, while taking the liberty on occasion of modifying capitalization or punctuation for the sake of clarity.

whose name ought to have featured prominently in the Litany of the Saints, at least during the Minor Rogations. For he it was who saw most clearly how the assertion of the humanity of Christ, even and especially in his ascension, made good on all the promises contained in the assertion of his deity – that is, of his identity as the very Word 'through whom the wood fructifies and the fountains gush forth, and the earth gives "first the blade, then the ear, then the full corn in the ear."'[3] St Irenaeus, *ora pro nobis.*

Rogation Sunday 2010

---

3   *Against Heresies* 4.17.4. Irenaeus, I suspect, would have enjoyed a mediaeval Rogation procession along the lines described in the Golden Legend, or at all events as expounded in Dom Guéranger's *The Liturgical Year*. The latter's section on the ascension would have excited him much more, of course.

CHAPTER ONE
The Upward Call

HE ASCENDED INTO HEAVEN AND IS SEATED AT THE RIGHT HAND OF THE FATHER. So say the baptized, so says the church; and when they do, they say something no one else has ever said about anyone. Ascension stories abound. One has heard of Enoch, for example, who walked with God and then vanished, 'because God took him'. One has heard also of Elijah, whom Yahweh caught up to heaven in a whirlwind. Pagan tales of apotheosis, gnostic redeemer myths, Muhammad's rumoured night journey – all sorts of examples can be found. One has even heard quite recently of the ascent of Man. But 'he ascended into heaven and is seated at the right hand of the Father' – that is something which, taken in context, is without parallel. By context I mean the sequence of events in which the ascension is lodged in the Christian creed, *viz.*, crucifixion, descent to the dead, resurrection, ascension, heavenly session, parousia and judgement. This combination of claims about Jesus makes his story utterly unique, in each of its parts as well as in the whole. But if unique, I hasten to add, by no means unanticipated.

To begin to understand the story of Jesus we can and should go back as far as the legend of Enoch, and even further. Indeed we must go all the way back to Eden, to that cosmic mountain on which was planted the garden where the very first humans are said to have walked and talked with God. For in Eden begins the pattern of descent and ascent that provides the main story-line of the holy scriptures, a story-line that reaches its climax in Jesus and his ascension into heaven.

Now according to Genesis the first man and woman were not, as we are, in bondage to death. The open horizon of divine blessing lay before them. For their disobedience, however, they were cast out of paradise and lived with their progeny in the plains below, east of Eden 'in the land of sepulture' (as Irenaeus somewhere describes this world of ours). Their way back to the garden was blocked by the flashing swords of the cherubim, for through them nature itself had become twisted and distorted, unfit for the presence of God, unready for his blessing. When the wickedness of the descendants of Cain, who had settled in Nod, reached its apogee, and even the sons of Seth had become unfruitful and corrupt, God sent a great flood to carry them all away into the abyss. But as the waters of judgement rose up and up, a single righteous man – one who like Enoch still walked with God – was carried in the ark 'high above the earth'. To Noah and his household new horizons were granted. Yet the mountain where the ark landed immediately became the witness to a new act of faithlessness and disobedience. As Noah's own descendants spread out in the plains below and to the east of Ararat, they too increased in wickedness. One expression of this was their attempt to construct in Shinar an eternal city, with a tower that would reach to heaven: a new Eden of human design, which as it climbed felt the heat of the cherubim, and came to nought in confusion and strife.

This pattern of descent and ascent, ascent and descent, continues as we pass from cosmopolitan history (couched initially, as need be, in the language of myth) to salvation history. Out of the new dispersion which follows the débâcle at Babel, Abram is called up from Ur of the Chaldees into the hills of Canaan, where the promise of divine blessing upon man is renewed.[1] But the heirs of the promise, the offspring of the miracle child Isaac, soon go down to the plains of the Nile to suffer their centuries of servitude in Egypt. Brought up again by the mighty hand of God, having passed through the deep on dry ground (that is, through the waters of judgement that swallow up Pharaoh's army), they make their rendezvous at Sinai. There Aaron and the seventy elders are invited to dine in Yahweh's presence while Moses ascends to the peak of the mountain, lost from view for forty days in the glory cloud which rests upon it. Down below the people dance round their golden fertility god, and waste away in the desert. Not until Yahweh establishes on Mount Zion the throne of David, and David the tabernacle of Yahweh, does it begin to seem that the cherubim might at last drop their guard, and Eden reappear: 'Lift up your heads, O you gates; be lifted up, you ancient doors, that the King of glory may come in.'[2]

The story takes another downward turn, however, beginning with the fall of David, who from the vantage point of the palace's rooftop garden covets Bathsheba. It ends with the sack of Jerusalem by the Babylonians and a return to the floodplains of Shinar.[3] The sequel follows suit. Hoping to encourage the unhappy exiles, the prophet Isaiah predicts the demise of the Babylonian pretender – no King of glory he!

---

1 Lot foolishly chooses the Edenic illusion provided by the plain of the Jordan 'towards the East', which was well watered 'like the garden of Yahweh' (Gen. 13:10ff.).

2 Psalm 24 (New International Version), q.v. in full.

3 See W. Gage, The Gospel of Genesis, 68ff.

> You said in your heart, 'I will ascend to heaven;
> I will raise my throne above the stars of God.
> I will sit enthroned on the mount of assembly,
> on the utmost heights of the sacred mountain.
> I will ascend above the tops of the clouds;
> I will make myself like the Most High.'
> But you are brought down to the grave,
> to the depths of the pit.[4]

Afterwards, under a more benevolent Persian regime, a tiny pilgrimage led by Nehemiah ascends Zion once more to restore the walls of Jerusalem and its temple to Yahweh. Then (under the wretched Antiochus?[5]) a seer sees a vision of 'one like a son of man', standing tall among the beast-nations which rend and devour God's people. Like Daniel himself, this figure is a hero personifying that people. He represents the new Adam, the new humanity, that God is wanting to mould out of the stubborn and resistant material of Israel. He is called upwards into the divine presence to receive what was promised long ago. He is seen 'coming with the clouds of heaven', approaching the Ancient of Days to receive glory and sovereignty over the earth, a dominion that will never pass away.[6] But the new Man comes bearing the standards of the invading Roman army, and after a lengthy occupation the legions of Titus tear down again what Nehemiah had raised up.

Descent and ascent, ascent and descent. It is worth noticing that Israel's most sacred liturgy – that of Yom Kippur, the Day of Atonement – answered to this pattern. The high priest, by an elaborate series of steps, would ascend annually into the sanctuary of God, into the holy of holies, to present the blood of atonement at the mercy seat. From there, if all went well, he would carry the divine blessing to the people waiting outside.[7] Descent and ascent, ascent and descent: the pattern had, even in Israel, the closed horizon of the circle – almost, we might say, the monotony of the wheel of fate. For the Hebrew cultus effectively perpetuated the standoff that existed with Yahweh. Year after year Israel's representative was

---

4 14:13f. (NIV)
5 Or, as now seems more likely, sometime before Antiochus.
6 Dan. 7:13f.; cf. Ps. 110:1 and see also Isaiah 11f.
7 The nature and pattern of the priestly ministry is established in Leviticus 9f.; cf. chap. 16.

received into intimacy with God and granted an audience; year after year he was put out again. Not that the task of the cultus was merely to sanctify the yawning gap between promise and reality. In its own way it was oriented to the eschaton; it pointed ahead to the closure of that gap in the coming kingdom of God. Were that not the truth of the matter, were expectations of arrival at the destiny foreseen by Daniel not constantly rekindled in the life and worship of Israel, we should have no occasion to speak of *salvation* history at all. Indeed, we should be tempted to revert to the history-rejecting beliefs and rituals of the Gentiles, for whom time was strictly a cyclical affair and fate absolute. Yet the gap itself remained, and with it the need for this priestly descent and ascent.

Israel's sense of expectation was rooted, not in any abstract conviction that history itself had a purpose or goal, but in the covenant faithfulness of Yahweh. Naturally enough, then, the form which that sense of expectation took was increasingly messianic, in keeping with the specific terms of the promises made to King David, who represented the true high point of covenantal history.[8] Embodied in the prayers and hymns of Israel, as in the visions of its prophets and seers, was an anticipation that one day would come that very special person of destiny who, by the grace of God, would fully redeem those promises and fulfill the Davidic dream:

> The One enthroned in heaven laughs;
> the Lord scoffs at [the rulers of the nations].
> Then he rebukes them in his wrath, saying,
> 'I have installed my king on Zion, my holy hill.'
> I will proclaim [to them] the decree of Yahweh:
> He said to me, 'You are my son;
> today I have become your Father.
> Ask of me, and I will make the nations your inheritance,
> the ends of the earth your possession.'[9]

Of his kingdom there would be no end, and in his time the marriage of heaven and earth, which had never yet been consummated, would

---

8   See 2 Sam. 5–8, especially chap. 7; cf. Rev. 22:16.

9   Ps 2:4-8 (adapted from the NIV) takes up the language of 2 Sam. 7:14. Surely Josephus' attempt to find in Vespasian this person of destiny can only be regarded as a quite astonishing betrayal of the Jewish hope (cf. N. T. Wright, *The New Testament and the People of God*, 376).

take place, with Zion standing in for Eden.[10] Psalm 110, which later
became the touchstone of Christian messianism, went so far as to
predict that this person would have a priestly as well as a kingly role.
Like the mysterious king of Salem in days of yore, he would unite the
functions of the two historic institutions by which the divine blessings
were mediated to Israelite society.

> Yahweh said to my lord,
> 'Sit at my right hand until I make your enemies your footstool.'
> Yahweh commits to you the sceptre of your power:
> reign from Zion in the midst of your enemies.
> Noble are you from the day of your birth upon the holy hill,
> radiant are you even from the womb in the morning dew of your youth.
> Yahweh has sworn and will not turn back:
> 'You are a priest forever after the order of Melchizedek.'[11]

Since the throne to which he would ascend would be as the heavenly
throne itself – since he would sit with God, so to speak, presiding over
the union of heaven and earth – he would surpass and render obso-
lete the Aaronic priesthood, the priesthood of the gap. And having
ascended he would no more descend, at least not in the usual way.[12]

AGAINST THIS BACKDROP, or something like it (obviously I have
sketched in bold lines just a few of its contours), the story of Jesus was
consciously told. That is everywhere apparent in the New Testament,
not least in those documents most attuned to the gentile mission.
Let us take Luke–Acts, for example, by far the largest work. Luke
appears to structure his entire contribution so as to draw out a parallel with David. Already in the birth narratives an analogy is set up
between the stories of Samuel and David and those of the Baptist

---

10 The walls of the temple 'were adorned with figures of the guardian cherubim, palm
    trees and flowers', graphically portraying the intention that the temple in Zion should
    replace the garden sanctuary as the centre of the universe (W. J. Dumbrell, *The End of
    the Beginning*, 52). But this aim, as the prophets soon realized, would never be fulfilled
    by Solomon's temple. It would require a completely new temple established by Yahweh
    himself, one isolated from the corrupted structures of the monarchical period (ibid., 59).
    Indeed, it would require a new creation.
11 Ps. 110:1-4 (adapted from the Liturgical Psalter).
12 Cf. Ps. 16:8ff. On Melchizedek, see Genesis 14 and Hebrews 7–8; see also Farrow,
    'Melchizedek and Modernity', *The Epistle to the Hebrews and Christian Theology*, ed. R. J.
    Bauckham *et al.*, 281–301.

and Jesus. Then, after Jesus is anointed with the Spirit for his messianic role, we find him wandering like David on the borders of the land, skirmishing with the various enemies that oppress the people of God. As 'the time for his ascension' approaches, he resolutely sets out to conquer the stronghold of Jerusalem.[13] He too ascends Zion triumphantly and is enthroned there. But his throne is a cross; his ascent is already a descent.

That is the great twist in the plot, in the Christian telling of Israel's story, and it is immediately followed by another. Jesus rises from the dead on the third day – not, like David or the rest of the righteous, at the end of the age – and ascends into heaven itself. His throne is established at God's right hand. As Peter's sermon on the day of Pentecost is designed to show, that establishment is just what David dreamed of but never achieved.[14] That it has implications both for the people of God and for the whole human race is the burden of Luke's second volume, which is hinged to the first by the twice-told story of the ascension.[15] The book of the Acts of the Apostles describes the progress of Jesus' servants throughout the Roman Empire and among all the nations, just as David's loyal men once carved out for him a secure kingdom between the Jordan and the sea.

It may be added that in each of the synoptic Gospels the crucial episodes of baptism and the transfiguration already point to the fundamental pattern of descent and ascent,[16] long before we actually come to the passion story and to the miraculous events that follow. Jesus' entire ministry takes place between the abyss and the mountaintop. Thus are we slowly but carefully prepared for its climax – the cross, the resurrection, and finally the ascension, which is the proper

---

13 Lk. 9:51; cf. Wright, *The New Testament and the People of God*, 379f.

14 'For David did not ascend into heaven', says Peter (Acts 2:34), quoting Psalm 110. Behind this lies 2 Samuel 7 and the question as to who would establish whose house: David, Yahweh's or Yahweh, David's? In the event – the event of one who makes *himself* the temple of the Spirit, and is ultimately brought by the Spirit into the Father's presence – both are true.

15 In the first account (Lk. 24:50ff.) the overtones are priestly; in the second (Acts 1:1-11) they are kingly or magisterial. Cf. J. G. Davies, *He Ascended into Heaven*, 24ff., and M. Parsons, *The Departure of Jesus in Luke–Acts*. M. Sleeman, who in *Geography and the Ascension Narrative in Acts* builds on Parsons, expands on the importance of the ascension motif but does not take up the David analogy, as he might have. (While happy to discover that Sleeman has built on *Ascension and Ecclesia* as well, I am not persuaded that theologians are quite so 'insensitive to narrative position' as he seems to think; cf. 6, 18ff.)

16 See Lk. 9:28ff.

outcome of his messianic career. To be sure, in Matthew and Mark
the ascension does not feature so prominently as in Luke–Acts, and
this is often taken by modern scholars as a reason for not allowing it
much attention. The doctrine of the ascension, it is sometimes said,
is just another way of talking about the resurrection. But to take such
a line is not only to marginalize Luke but also to ignore both the general witness of the scriptures and the logic of the Gospels in question.
Since Matthew and Mark have not themselves taken up the task of
writing about what Jesus continued 'to do and to teach',[17] there are
different literary and theological pressures on their concluding sections. Yet the ascension is by no means passed over, nor is it conflated
with the resurrection. Rather, appropriate variations on the theme
offer a rich texture of complementary perspectives.

Matthew works not with David, but with Moses, as his main
christological type. Not surprisingly, then, his version of the story
concludes with Jesus on a mountain somewhere on the fringe of the
promised land, blessing and commissioning his disciples as Moses
once blessed Joshua and the leaders of the tribes. As in the case of
Moses, whose last act was to ascend Mount Nebo, Jesus' final resting place is known only to God, who, we may assume, carries him
thither. But that resting place is no secret grave, as (*pace* Josephus
and the Jewish tradition) it was for Moses. For Jesus has already conquered death, the last enemy of God's people.[18] The implication of
Matthew's ending is clear enough, and we will return to it when we
come in a moment to the epistle to the Hebrews. Matthew does not
spoil the rhetorical effect by trying to spell it out.

Mark is rather more pointed. He introduces the language and
imagery of Daniel into the synoptic tradition, with ascension imagery
being especially prominent. We may take as the best example the following excerpt from Jesus' trial, as he is questioned by the high priest
on the eve of his awful descent into the abyss of God-forsakenness:
'"Are you the Christ, the son of the Blessed One?" "I am", said Jesus,
"and you shall see the son of man seated at the right hand of the
Power [on high] and coming with the clouds of heaven." '[19] The
primary reference here, as some have suggested, may well be to the

---

17   Cf. Acts 1:1f.

18   Hence the mission of the disciples is in the opposite direction to that of Joshua – out of
     the land and into the wide world beyond it. Cf. Mt. 28:16ff. and Deuteronomy 32ff. (espe-
     cially 34:9); for Josephus' account of Moses' alleged ascension, see *Antiquities* 4.8.48.

19   Mk 14:62; cf. 8:38, 13:26f. See also Davies, *He Ascended into Heaven*, 34ff.

judgement on Jerusalem foreseen by Jesus and fulfilled in AD 70, but there can be no doubt that the prediction itself hinges on the actual deliverance and vindication of Jesus; that is, on his resurrection and ascent to the Father. We do not need the unreliable and somewhat anticlimactic longer ending, which makes the allusion explicit,[20] to recognize the import of this regular Markan refrain.

Of all the evangelists, however, it is John who makes the most of the descent–ascent motif, and of the ascension itself. That may at first glance seem an odd claim, since the fourth Gospel tells neither the story of the ascension nor those of the baptism and transfiguration. Like the institution of the eucharist (a mystery joined to that of the ascension in 6:62) all these events are simply assumed by John, who takes it as his task to interpret them through the medium of dialogues and monologues. Early on, Jesus is introduced to us quite openly as the one who has gone up into heaven, and who indeed (in another sense) came down from heaven; as the one who is essentially defined by his place 'at the Father's side'.[21] To this a priestly dimension is added when his body is spoken of as the temple in which an altogether decisive liturgy is to be conducted.[22] He is even portrayed, like Eden itself, as an oasis in the wilderness where the glory of God can be found, a garden on a mountain from which the whole earth can be watered:

> On the last day, the great day of the feast, Jesus stood and cried out, saying, 'If anyone is thirsty, let him come to me and drink. He who believes in me, as the scripture said, from his belly will flow rivers of living water.' This he said concerning the Spirit, whom those believing in him were about to receive; for the Spirit was not yet, since Jesus was not yet glorified.[23]

To be that oasis Jesus must first 'sanctify' himself for our sake, stooping so low as to allow himself to be lifted up on a cross by sinners; but in and through this mock ascent he will also be lifted up by God as the beginning of a new creation. The divine lifting up begins with his resurrection, naturally, but its true goal is a homecoming – reception

---

20 Mk 16:19
21 Literally, 'in the Father's bosom' (1:18; cf. 3:13, 13:3).
22 Jn 2:13-22; cf. the allusion to Jacob's ladder in 1:59.
23 Jn 7:37-39. It is a pity that Northrup Frye missed this vital key to biblical symbolism in his fine book, *The Great Code*.

in the Father's house – as the famous farewell discourse in chapters 14ff. makes plain. The point is powerfully reinforced in chapter 20: 'Stop holding me', says the risen one when embraced by the astonished and delighted Mary, 'for I have not yet ascended to my Father; go rather and say to my brothers, "I ascend to my Father and your Father, my God and your God!"'[24]

What John emphasizes is that Jesus' destiny is our destiny; or rather that, in reaching our destiny, he has reached it not only for himself but also for us. Long before the Gospels were in circulation, of course, St Paul had already got hold of that point, just as he had also proclaimed the good news of Jesus' enthronement and the coming subjugation of his enemies in every corner of the cosmos.[25] He communicated it to his friends in Philippi in the following words: 'Forgetting the things behind, and stretching towards the things ahead, I press on according to the goal for the prize of the upward call of God in Christ Jesus.'[26] Now the prize of the upward call is just what Israel had been denied at Sinai, the mediation of Moses notwithstanding:

> Yahweh descended to the top of Mount Sinai and called Moses to the top of the mountain. So Moses went up and Yahweh said to him, 'Go down and warn the people so that they do not force their way through to see Yahweh and many of them perish . . .' Moses said to Yahweh, 'The people cannot come up Mount Sinai, because you yourself warned us, "Put limits around the mountain and set it apart as holy."'[27]

But in connection with Jesus the upward call is freely extended to all who are near and all who are far off. That is because the one who has ascended is in this case one who 'also descended into the lower parts of the earth',[28] effecting a complete sanctification. And having reached the goal which his predecessors did not – having been raised from the

---

24 Verse 17; cf. 1:11-13, 14:1ff., 20:22.

25 See Phil. 2:9ff.; cf. Eph. 1:19ff., 1 Pet. 3:22.

26 Phil. 3:13f. The prize in question may be the ανω κλησις or upward call itself, or that which awaits the one who receives the call; it makes little difference to the sense.

27 Exod. 19:20ff. (adapted from the *NIV*); cf. Acts 5:31.

28 That is, who 'was crucified, died and was buried' and 'descended to the dead', as the creed has it. See Eph. 4:9f., a passage to which we will later return in another connection.

dead and welcomed into the Father's house – he is able to arrange there a welcome for others. 'I go to prepare a place for you.'[29]

Just here we also have the guiding idea of the epistle to the Hebrews, whose author must be ranked together with John and Paul among the great theologians of the early church. It has been suggested (on the basis of the synagogue lectionary) that Hebrews is a Pentecost sermon, and that may well be the case, but it would certainly do just as nicely as a sermon for Ascension Day.[30] A quick glance at its rhetorical structure reveals that Hebrews is more interested in the achievement on which Pentecost rests; that is, on Jesus' passage through the veil that separates a holy God from his sinful creatures:

We have
a high priest who
sat down at the right hand (8:1)

Draw near (4:16)		Draw near (10:22)
Sharers of a heavenly calling, concentrate on Jesus (3:1)	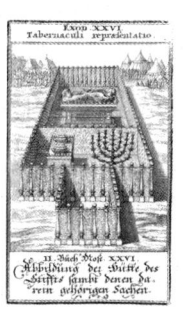	Let us run . . . fixing our eyes on Jesus (12:1f.)
He sat down at the right hand of the Majesty in heaven (1:3)		Since we are receiving a kingdom . . . let us be grateful (12:28)

By the author's own testimony in the pivotal passage, the main point he wants to make is that 'we do have a high priest such as this, who sat down at the right hand of the throne of the Majesty in heaven, minister of the sanctuary and of the true tabernacle, which the Lord built not man.' On the basis of the unique ministry of the ascended one we too may be assured of coming safely to the place to which neither Moses nor Aaron nor Joshua nor David could ever lead us.[31]

In other words, the writer of Hebrews picks up more or less where

---

29 Jn 14:2; cf. Rom. 8:34, Col. 3:1ff.

30 Ascension Day is of course a later development, as is the rest of the Christian calendar; but it is ontologically and theologically more fundamental (cf. Eph. 4:7ff.).

31 See chapters 3–4.

Matthew, quite self-consciously, leaves off. Recognizing that Jesus has gone where no other has yet been able to go – to the promised and longed-for rest, to the Paradise that exists in God's own presence – he develops a pastoral theology for the people of the new exodus, based on Jesus' heavenly ministry. For this the typologies of Matthew and Luke are inadequate. Taking up the Melchizedek motif from Psalm 110, he shows how Jesus actively sustains the link between his own destiny and ours from his place at the Father's side, that God may succeed at 'bringing many sons to glory'.[32] In doing so, he accentuates the priestly office of the messiah in a way unmatched elsewhere. It is worth noticing that the heavenly liturgy of Jesus continues to be construed by way of analogy with Yom Kippur, with one decisive difference: that, when our high priest re-emerges from the sanctuary 'not built with hands', he will do so not as one who is himself put out of the sanctuary, but as one who will exorcize from the whole creation all that is not fit to be brought into that sanctuary. In his reappearance he will bring with him the fullness of God's blessing, and of God's judgement. This then will be a descent of a very different kind, for in Jesus the stalemate between God and his people has been broken.[33]

AS A COLLECTION of literature deeply rooted in the Jewish scriptures and their presuppositions, the New Testament tells and interprets the story of Jesus in a perfectly natural way as a story that reaches its climax – though not of course its conclusion – with his ascension into heaven. This is the same story that the church regularly repeats to herself, before going out to share it with the world, when she comes together in eucharistic assembly. For those who gather round the Lord's table do so precisely in order to anticipate 'the upward call of God in Christ Jesus'. Nowhere is that privilege captured more succinctly than in Hebrews 12, with its sharp contrasts and expansive imagery: 'You have not come to a mountain that can be touched and that is burning with fire; to darkness, gloom and storm . . . But you have come to Mount Zion, to the heavenly Jerusalem, the city of the living God. You have come to thousands upon thousands of angels in joyful assembly, to the church of the firstborn, whose names are written in heaven. You have come to God the judge of all men, to

---

32   See 2:5ff. (esp. 9f.), 4:14ff.
33   See 6:19ff., 9:11ff., 12:14ff. (esp. 25ff.); cf. Acts 3:17ff., 1 Thess. 4:13ff.

the spirits of righteous men made perfect, to Jesus the mediator of a new covenant, and to the sprinkled blood that speaks a better word than the blood of Abel.'[34]

Alongside Hebrews 12, however, we might well set the opening chapters of the Apocalypse, which are fairly bursting with eucharistic meaning and with the promise of the ascension. 'On the Lord's day,' says John, 'I was in the Spirit, and I heard behind me a great voice like a trumpet.'[35] Turning to discover the source of that voice, he is confronted by the Melchizedekian figure of Jesus, in a vision that completely overwhelms him and leaves him as one dead. Restored to his senses by a touch from the priestly hand, however, John receives that peculiar upward call of the prophet or seer:

> After this I looked, and there before me was a door standing open in heaven. And the voice I had first heard speaking to me like a trumpet said, 'Come up here, and I will show you what must take place after this.' At once I was in the Spirit, and there before me was a throne in heaven with someone sitting on it . . .[36]

Near to the one sitting on the throne the same figure appears in a different aspect, as the Lamb slain – slain, yet standing very much alive in the presence of God and in the centre of all creation. This is the one whom the dragon waited to devour, but who was 'snatched up to God and to his throne'.[37] This is the one who alone in all the universe can open the seals on the scroll of destiny. This is the one who cries out to his faithful witnesses, the martyrs, 'Come up here!' And they too stand on their feet and ascend to heaven in a cloud, safe from the clutches of death and of what men mistakenly call the eternal city.[38]

With evocative scenes such as these fresh in our minds, and with the wider testimony of scripture in view, we may join the faithful down through the ages in celebrating, with unaffected enthusiasm,

---

34 Heb. 12:18ff. (*NIV*)
35 Rev. 1:10 appears to contradict Heb. 12:19 by drawing on Sinai in a more positive way, but the conflict is superficial. The voice heard at the Lord's table is not a different voice than Sinai's, which the people could not bear and 'begged that no further word be spoken to them'. It is the same voice sounding in and through Jesus Christ.
36 Rev. 4:1f. (*NIV*); cf. Exod. 24:9-11, where eating, drinking and seeing God are the prize of the upward call.
37 Rev. 12:5 (*NIV*), a text powerfully portrayed by Benjamin West in a painting that hangs in the Princeton University Art Museum.
38 Rev. 11:12; cf. Acts 7:54ff.

Christ's 'mighty resurrection and glorious ascension'.[39] But what exactly are we celebrating? That is not immediately clear from the biblical material, and certainly not from mysterious works like the Apocalypse. Nor is it something about which there has always been agreement.

---

39 Prayer of consecration, *Book of Common Prayer*.

CHAPTER TWO

Re-imaginings

IN JERUSALEM, ON THE MOUNT OF OLIVES, stands a small domed shrine. It is of course the Chapel of the Ascension. Within its dark and cramped interior one is directed to a footprint-like depression in the rock, said to be the exact point from which Christ parted from his disciples and from our world – as if he sprang into the heavens with such vigour that the very rock underneath his feet was compressed in the act![1] If this conjures up an absurdly literal interpretation of the ascension, it may be remarked that such have been with us from relatively early days. The New Testament itself shows remarkable reserve, to be sure. Even Luke's insistence that 'he was taken up before their very eyes, and a cloud hid him from their sight' is quite circumspect when compared with the so-called Epistle of the Apostles, for example, or with certain strands of sacred art.[2] But while we might make allowances for the artist, or indeed for the pre-scientific commentator, it must be admitted that the doctrine of the ascension, if construed along Lukan lines at all, is something of an embarrassment in the age of the telescope and the space probe; indeed, of evolutionary theory and faith in global progress. Truth be told, it is an embarrassment also in the age of pluralism and post-colonialism, of feminism and ecofeminism, and other reactionary movements of late modernity.[3]

Rudolf Bultmann's famous claim that a man with a wireless and a scientific world view could no longer take seriously Jesus' ascension into heaven – or rather, that in order to take it seriously he must first allow it, like much else in the Gospels, to be demythologized – was one attempt to face that embarrassment squarely and, at the same time, to point towards a more spiritually authentic reading of the story of Jesus.[4] *Pace* Bultmann, however, neither the difficulty nor the solution is peculiarly modern. Long before the radio or even the telescope, a very un-Lukan view of the ascension was already emerging, partly because of the fact that a tangible body rising up into the

---

1   The tired camel which at my visit was kneeling outside, as if in prayer that it not be burdened by tourists on a hot day, provided a suitable contrast – it looked as if it could not be raised with a crane!

2   Acts 1:9 (*NIV*). *Epistula Apostolorum* (cf. 21, 51) is a second-century anti-gnostic document.

3   The pluralist may object that the doctrine of the ascension is triumphalist, though surely it is hardly more so than the liberal doctrine of progress; the feminist may object that it is just another example of sexism (Grace Jantzen, 'Ascension and the Road to Hell', *Theology*, May–June 1991, 164f.).

4   See *Kerygma and Myth* (ed. H. W. Bartsch), chap. 1.

heavens was no more congenial to the ancient world view (literalists notwithstanding) than to the modern.

Early in the third century the brilliant Alexandrian theologian, Origen, offered the following advice, which would have been especially welcome to anyone troubled by the thought of a corporeal substance passing into a realm that by definition was for spiritual substances only. Let us seek, he says, 'to understand in a mystical sense the words at the end of the Gospel according to John, "Touch me not; for I am not yet ascended to my Father", thinking of the ascension of the Son to the Father in a manner more befitting his divinity, with sanctified perspicuity, as an ascension of the mind rather than of the body.'[5] Origen here sets in motion a long tradition in ascension theology, of which Bultmann is arguably a representative. 'Ascension of the mind' is one of the two basic paradigms used down through the centuries to interpret the outcome of Jesus' history, and it is a quite flexible one, capable of absorbing (or adjusting to) all sorts of scientific advances. It is a view we need to look into, beginning with the context in which it arose. Though that will mean traversing quickly some difficult terrain, the journey will prove profitable later on.

DESCENT AND ASCENT SCHEMES are hardly the special property of the Hebrew and Christian scriptures. They are the very stuff of religious speculation and mythology of all kinds, reflecting and interpreting (as they generally do) the patterns or cycles of nature and of human experience.[6] For our purposes the two most notable such schemes of Origen's day were that employed by the gnostic sects, on the one hand, and the more philosophical variety of his teacher, Ammonius Saccas, on the other – *viz.*, the emerging Neoplatonism best represented by one of Saccas' younger pupils, the famous Plotinus.[7] Origen sought to articulate the Christian scheme in

---

5 *On Prayer* 23.2. It is worth noticing that the context *is* a treatise on prayer, not on cosmology. The two are not unrelated, however, and in a moment I will try to show how this proposal fits into Origen's wider theological world view.

6 Catherine Playoust expands on a few of these patterns in her Harvard dissertation, 'Lifted up from the earth: The Ascension of Jesus and the Heavenly Ascents of Early Christians'. While I do not agree with her reading of certain biblical and patristic texts, or share her views on their relation to texts of gnostic provenance, her work helps to make the point that Jesus' ascension was often assimilated to existing cosmological schemes.

7 There is no need to draw a sharp line between Middle Platonism (an eclectic movement drawing on Pythagoras, Plato, Aristotle, Zeno, etc., to which Origen was already exposed by Clement) and Neoplatonism; Saccas himself was apparently a transitional figure. Nor

dispute with the former variety and in dialogue with the latter, which inevitably led to some interesting alterations to the biblical picture.

The schemes just mentioned were at once cosmological and psychological; which is to say, they correlated the dynamic of the universe with the drama of the human soul. For the gnostics that drama was played out on the stage set for it by the occurrence of a fault in the realm of divine being, through which this evil world of ours – the realm of mere becoming, of decay, suffering and death – had come into existence, or rather had fallen *from* existence.[8] From our world the spirits of select individuals were destined to escape, being freed from its toils through the aid of a saviour who would teach them how to return to the heavenly Fullness which was their true origin and home. This descending and ascending saviour they often identified as the Christ. By this they did not mean Jesus of Nazareth, however, who himself was a mere mortal, but the invisible emissary who spoke through him.[9] For the gnostics, what was important about Jesus was the secret teaching that he passed on to his disciples, not the man himself or his cross, resurrection and ascension. These latter events they indeed denied, since they did not believe that the body (or the temporal–material realm as such) was redeemable. Ascending was not something one did or could do in the *body*. It was strictly an affair of the inner man, to be anticipated in this life by those initiated into the secrets, but to be fulfilled only in death.

The popularity of gnosticism may be difficult for a modern to understand, until he or she recalls the parasitical forms of vaguely Christian belief which are widely available today through the same sort of cross-fertilization between eastern and western religious ideas that took place during the *pax Romana*. There were of course at that time those who sought something less quirky and more philosophically sophisticated (though the gnosticism of Basilides and Valentinus could certainly be the latter). For them the evolving ideas of the

---

for present purposes do we need to distinguish sharply between the various gnostic gurus and sects, though it may be observed that Basilides (who had also lectured in Alexandria) offered a quite distinct and moderate view.

8   On some accounts, this fault occurred in the female aspect of the divine pleroma and was portrayed in sexual (and sexist) stories. Underlying all such stories, however, was the old metaphysical opposition between being and becoming, between the one and the many.

9   We should be careful here about an over-literal reading of the gnostic myths. In one sense the emissary *is* the myth, which properly told and heard has its redemptive effect on the inner man.

Platonists offered a more attractive option. The new Platonists did not trouble themselves with Jesus.[10] Nor did they share the gnostic pessimism or fatalism. The world we inhabit, the world of multiplicity and of changing phenomena, was regarded as the natural product of descending emanations from the divine – that is, from the absolute simplicity of pure being – which they called 'the One'. The individual human comprised a sort of microcosm in which body, soul and mind represented advancing degrees of being, and hence of potential for union with the One, training for which was man's moral and intellectual vocation. Here too ascent was a matter for the inner self, for the soul and for the mind, in which the traces of divinity were to be found; but it was at least allowed that the phenomenal world might to some extent reflect the mind's advance.

In such ideas Origen found some of the ingredients for an antidote to gnosticism, and for a cultured Christianity that might make better headway among the educated classes. On the other hand, he recognized that the Platonists did not take the problem of evil seriously enough – even if they were to be commended for not making the gnostic mistake of attributing our world to a defective deity![11] Without blaming its Creator, the world's inequalities and injustices and suffering had to be accounted for. Origen himself argued, then, that the temporal–material world was not meant to be. It came about as the result of the breakdown of an original creation in which the souls of all living creatures had once been completely absorbed in the contemplation of God. Between them, since they had but one goal, there was no distinction in kind or in form; but falling away from God in differing degrees they took on differing conditions of bodily existence, some finer (the angels), some grosser (men, animals, etc.). That they did so, however, was an expression of the benevolence of God, who brought the present creation into existence as a school for the reformation of souls. Through their various trials in the body its pupils were invited to turn their attention upwards again to God, and so to ascend gradually through the ranks of creation to the situation from which they fell.[12]

Origen's alteration of the biblical story, and of the descent–ascent

---

10   Saccas himself is thought by some to have been a lapsed Christian.

11   That is, a corrupt demiurge. This view should not be attributed to Basilides, whose teachings about providence, and about the connection between suffering and sin, to some extent anticipated Origen's.

12   The integrity of the Creator, the responsibility of the creature – these are what Origen

motif that runs through it, thus begins with a radically different
reading of the opening chapters of Genesis. This re-imagining of
the narrative about human beginnings (putting the fall prior to
creation) is eventually answered by his 'sanctified' reading of the
Christ-narrative and its implications for human endings.[13] Who is
Jesus Christ? Where did he come from and where did he go? How
will we follow him? According to Origen, Jesus is the human embodiment of the one soul that did *not* fall away, a soul eternally united
to the divine Logos, a soul who descended freely in pursuit of the
fallen in order to show them the way back home. Against both the
gnostics and the philosophers, then, Origen bravely insists on faith
in the incarnate and crucified one as the necessary foundation for
our own upward call.[14] But that call – on this point all are in agreement – is to ascension of the mind, since only the mind is capable
of participation in the Logos. We must understand ascension not so
much as a change of place as of state, he suggests, to which there
must correspond a series of transformations of the body in which
its material and temporal dimensions are gradually stripped away,
having fulfilled their purpose. That is what happened to Jesus and
what will also happen to us if we imitate him. For ascent is simply
the reversal of descent, nothing more; and in that reversal there is
also a reversal of the material creation. There will be, remarks Origen
wryly, no hairdressers in heaven.[15]

Once again a modern person (even a modern Christian) might be
excused for finding some of this a little bewildering, but we cannot
pause to consider its finer points. We need only observe that according to Origen there is not one gospel, but two: the temporal gospel,
for the simple, which treats of Christ's work in the flesh; and the

---

sought to underscore as he steered a middle course between the heretics and the
philosophers.

13 *On First Principles* explains all the above, as well as the hermeneutical theory by which he
justifies such readings; among other things he maintains (4.9.2) that scripture includes
things which did not actually happen in order to signal the reader to seek a higher, mystical sense.

14 Bravely, I say, yet not so bravely as more orthodox theologians, since he wants to talk
about the incarnation of the soul united to the Logos rather than of the Logos *per se* (*First
Principles* 2.6). Still, he has decisively rejected gnostic views of the saviour, while putting
it to the philosophers as well that it is not the spark of divinity in our own soul – the god
within us – but the Christ whom we ought to regard as our true Teacher.

15 See J. Trigg, *Origen*, 114; cf. *First Principles* 2.3, 2.10, 3.6, 4.4. But see also J. N. D. Kelly,
Early Christian Doctrines, 469ff.

eternal gospel, for the wise, which treats of his invisible heavenly glory. Only as we progress beyond thinking about the saviour κατα σαρκα, as Paul said – according to the flesh rather than the spirit – do we begin to grasp the eternal gospel and so to commence in earnest our own ascension. Origen thus invites us to conceive of the ascended one, not in terms of his human particularity, as we would the crucified one, but in terms of his godlike ability 'to fill the universe', to run through the whole of creation restoring all to its original condition.[16] Later on we shall have to ask whether this is really Paul's meaning, and whether he would have been happy with the distinction between a temporal and an eternal gospel. But for now we must continue to follow the path marked out for us by Origen, which not only extends into the modern period but eventually broadens out into something much more familiar.

THE CRITICS OF ORIGEN have always been numerous and vocal, as have his defenders. St Augustine of Hippo, whose stature as a theologian exceeds even that of Origen, may be put down among the former. In his last great work, *The City of God against the Pagans*, he rebukes the Alexandrian for some of the philosophical assumptions he has retained from Hellenic thought, and especially for the notion of the pre-existence of souls that is so fundamental to his whole scheme of things. He insists on Christ's ascension in the flesh – that other, more primitive, paradigm which we will explore in chapter three – and pokes fun at the 'ponderous' arguments of the 'little coterie of skeptics' who still want to deny that a corporeal body could rise to the highest heavens, or that those heavens could open to receive it.[17] But Augustine is not an altogether effective opponent. Indeed, to understand the progress of the Origenist option in the Middle Ages and beyond, it is necessary to consider his somewhat ambiguous contribution.

It would not be unfair to comment that Augustine, who like Origen stands as an enormous pillar of learning at the crossroads of the late classical and the Christian era, was slow to shed some of his own Platonic assumptions; or to notice that his remarks on the ascension are sometimes more witty than weighty. What most needs

---

16 Cf. *First Principles* 2.6.7, 2 Cor. 5:16; *Princ.* 2.11.6, Ephesians 4:11; *Princ.* 4.3.13 and editor's note 7.

17 See *City of God*, books 11 and 21ff.; cf. *Faith and the Creed* §13.

to be said, however, is that what Augustine insists upon he also to a certain extent undermines, at least in his earlier writings. We are to believe, he says, in bodily ascension. But what was the point of that marvellous event? To remove from the disciples the stumbling block of his humanity, in order that they might come to a robust faith in his divinity. Labouring in the wake of the great Nicene controversy, Augustine sometimes offers a very different reading of Jn 20:17 than that which we were assuming in chapter one. In certain places there is little thought (though it can be found elsewhere in Augustine) of Jesus our brother opening up for us the true destiny of man. The thought is rather that we must not cling to him as a man, but allow him to appear to us as God. 'Through the man Christ you go to the God Christ', says Augustine, and only thus to the Father.[18] The ascension, in other words, enables us to look past the human Jesus to the eternal Logos.

Already we are in Origen's territory. Indeed, from a different starting point and for different reasons, we have come to the same conclusion, simply by shifting our focus from the objective pole to the subjective pole. The ascension happens to Jesus in the body that it may happen to us in the mind. It is something that happens for us only in so far as it also happens in us: 'Therefore he has ascended for us, when we rightly understand him. At that time he ascended once only but now he ascends every day.'[19] Augustine's earlier treatments of the ascension thus contribute both to the internalizing or subjectification of the doctrine and to the marginalization of the humanity of the risen Lord, which together characterize Origen's approach. The bishop assures us that the Lord himself expects us 'to press on, and instead of weakly clinging to temporal things, even though these have been put on and worn by him for our salvation, to pass over them quickly and struggle to attain to himself, who has freed our nature from the bondage of temporal things, and has set it down at the right hand of his Father.'[20]

---

18 'Where you are to abide, He is God; on your way thither, He is man' (Sermon 261; cf. Sermon 264). Christ's human nature, and faith in that human nature, is necessary only 'for our weakness'.

19 Sermon 246

20 *Christian Doctrine* 1.38 (see W. Marrevee, *The Ascension of Christ in the Works of St. Augustine*, 99ff.). Much depends here, obviously, on what is meant by 'the bondage of temporal things'; clearly the mature Augustine does not mean what Origen meant by such statements, as we shall see later.

Now Augustine is in good company here – whatever we make of Origen – for something of this sort is to be found in many leading theologians throughout the late patristic period and the Middle Ages. In the Renaissance it is taken up by Erasmus, for example, who in his attack on late mediaeval superstition repeats the point that 'it was the flesh of Christ which stood in the way' of an authentic spiritual faith among the apostles; that 'the physical presence of Christ is of no profit for salvation'.[21] Origen's more contentious ideas may have been anathematized by the fifth ecumenical council, and his basic framework roundly condemned.[22] But, in spiritual treatises especially, orthodox schemes of descending and ascending frequently give way to Origenist theses and assumptions, as others work the seams explored by his own *On Prayer*.

Maximus the Confessor (whose commitment to orthodoxy, like Augustine's, cannot be doubted) may be allowed to represent the East, in words that might just as easily have been spoken by the Alexandrian:

> For those who search according to the flesh after the meaning of God, the Lord does not ascend to the Father; but for those who seek him out in a spiritual way through lofty contemplations he does ascend to the Father. Let us, then, not always hold him here below though he came down here out of love to be with us. Rather, let us

---

21 *The Handbook of the Christian Soldier*, Fifth Rule (*The Erasmus Reader*, ed. E. Rummel, 146). Erasmus goes on to appeal to 2 Cor. 5:16, which he misreads in terms of a matter–spirit dualism. What he fails to see is that the superstition he is rightly criticizing is grounded not so much in an ignorance of the superiority of spirit, but in an ignorance of the way in which the physical humanity of Christ, as well as the spiritual, does indeed profit for salvation.

22 See *The Seven Ecumenical Councils* (*Nicene and Post-Nicene Fathers*, 2nd ser., vol. 14), 316ff., especially Anathemas 9–11:

'If anyone shall say that it was not the divine Logos made man . . . [who] descended into hell and ascended into heaven, but shall pretend that it is the *Nous* which has done this, that *Nous* of which they say (in an impious fashion) that he is Christ properly so-called, and that he is become so by the knowledge of the Monad: let him be anathema.

'If anyone shall say that after the resurrection the body of the Lord was ethereal, taking the form of a sphere, and that such shall be the bodies of all after the resurrection; and that after the Lord himself shall have rejected his true body and after the others who rise shall have rejected theirs, the nature of their bodies shall be annihilated: let him be anathema.

'If anyone shall say that the future judgment signifies the destruction of the body and that the end of the story will be an immaterial *physis* [nature], and that thereafter there will no longer be any matter, but only *nous* [mind], let him be anathema.'

go up to the Father along with him, leaving behind the earth and what is earthly . . .'[23]

Only beginners, says Maximus, cling to him who 'had no form or beauty', that is, to the earthly Jesus; we rightly follow Christ's new trajectory 'if we know him not in the limited condition of his descent in the incarnation but in the majestic splendour of his natural infinitude'. This prepares us for the further claim – the implications of which deserve some pondering in the light of subsequent developments – that it is no longer necessary 'that those who seek the Lord should seek him outside themselves'.[24] If such remarks are anything to go by, the internalization and marginalization of which I have spoken proceeded apace in the Middle Ages, softening the edges of the church's official commitment to the biblical story, or at least turning the thrust of its inquiries into the ascension in another direction: that of the ancient Hellenic quest for the *visio dei*.

In the West, of course, the doctrine of the ascension quickly got caught up in the eucharistic controversies, which themselves owed something to Augustine's handling of the ascension. The difficulty here was not created by the undeniably awkward notion of Christ's body rising up into the heavens, but by the necessity of bringing it down again at manifold times and places, not only for the sake of eucharistic realism but also, perhaps, as a sacramental counterweight to the subjectivist shift just described. Ironically it was in support of 'Christ's humble descent into the hands of the officiating priests'[25] that another Origenist idea, already evident in the East, gained considerable currency in the West as well. That the ascended Lord was ubiquitous or omnipresent Augustine himself had taught, but only in connection with his divine nature; in his human nature he was said to be absent.[26] But an increasingly self-conscious sacramentalism, which found it necessary to insist that in the eucharist the physical body of

---

23 *Gnostic Centuries* 2.47. We must pass over the mediating influence of Denys the Areopagite.

24 *Gnostic Centuries* 2.35; cf. 2.62: 'Those who bury the Lord with honour will also behold him gloriously risen, while to all those who do not he is unseen. For he is no longer caught by those who lay snares, having no longer the external covering by which he seemed to allow himself to be caught by those who wanted him, and by which he endured the Passion for the salvation of all.'

25 Ray Petry, *A History of Christianity*, 277.

26 *On John's Gospel* 50.13; cf. A. Heron, *Table and Tradition*, 72. See also, e.g., Question 47 in the Heidelberg Catechism.

Christ was present on the altar, obviously required some explanation. Aquinas provided what became the official answer, which sought to make sense of Jesus' physical presence without abandoning the belief that he had departed bodily into heaven. His answer was both ingenious and durable. The special pleading it involved at certain points, however, was not convincing to everyone, and the notion of the ubiquity of the incarnate Christ appealed to some as an alternative.[27] Of course this too involved special pleading, and it was recognized that a ubiquitous body was a strange sort of thing – even more difficult to reconcile with the adjective 'human' than a body said to have 'passed through the heavens'. Yet such objections were not necessarily regarded as decisive.

The intuition that deep christological and anthropological issues were at stake did as much to drive the eucharistic controversies as did the more explosive issues of church polity to which they were attached. Old questions were pressed: Where is the ascended Christ? Who and what is he? Is his humanity swallowed up in his divinity through the ascension or is it somehow preserved? What implications are there for our own humanity? If Rome did not fully resolve these questions, neither did the Reformation. Indeed, in its own concerted efforts to face them squarely, it came to grief on them. Luther claimed to have put Origen once again 'under the ban', for a defective understanding of God's grace. But Calvin and his followers saw in Luther's embrace of the ubiquity concept powerful Origenist tendencies[28] – tendencies that quite arguably helped to prepare the ground for a yet more radical turn toward 'ascension of the mind' in the modern period. To give a brief account of that turn is the next task; we will have occasion to return to the Reformation in a later chapter.

THE SIGHT THAT GREETED GALILEO when he first laid eyes (with the aid of that irreverent instrument, the telescope) on the lunar imperfections, which so plainly contradicted the notion that the heavens were a gateway to the divine, is sometimes taken as a convenient marker for the modern period. It might also be taken as the point

---

27 See *Summa Theologiae* 3.57f. and 3.75ff. on the substance–accidents distinction around which Aquinas builds his ingenious solution. We will take up this topic again in chapter five. Cf. also Farrow, 'Between the Rock and a Hard Place', *International Journal of Systematic Theology* 3.2, July 2001, 167ff.

28 See *Institute* 4.17; cf. Trigg, *Origen*, 256.

at which a reconsideration of the doctrine of the ascension became imperative. As it happened, however, it was not the progress of the Copernican revolution which brought about that reconsideration, but rather the programmatic rejection of the miraculous that prevailed among the *philosophes* during the Newtonian period, with its (now largely outmoded) mechanistic world view. This was a rejection which the new Kantian epistemology also helped to justify, and it was indeed Newton and Kant who created the conditions under which a major turning was taken regarding the doctrine in question.

When it came, it came in a rather shocking fashion. Early in the nineteenth century at the University of Berlin, for the first time in the long history of the church, an eminent Christian theologian was heard making the astonishing claim that Christ's resurrection, ascension and parousia 'cannot be laid down as properly constituent parts of the doctrine of his person'.[29] Friedrich Schleiermacher, who revised and restated the Christian faith for his own 'cultured' contemporaries with a finesse equal to that of the early Alexandrian, chose to do so in a way that neither required nor allowed the miraculous. What is more, in his struggle with the general conditions for theology imposed by Kant, he substituted Christian subjectivity (the church's shared intuition of God) for God's own being and acts as the proper object of dogmatic reflection.[30] Needless to say, all this had a profound effect on the entire dogmatic enterprise, but in particular it ruled out any kind of eschatological realism. Talk about the resurrection, ascension and parousia of Jesus could henceforth serve only to indicate something about the 'peculiar dignity' which he has in Christian eyes, and about the Christian's longing for unity with him in his own unsurpassed intuition or God-consciousness. Augustine's objective pole – perhaps even Origen's – was abandoned altogether.

Now it would not be wrong, for this and other reasons, to regard Schleiermacher's move as a decisive break with the past rather than a mere turning. But it is nonetheless true that it involves a genuine extension of the Origenist tradition as mediated by Augustine. For what emerges here is not so much a rejection of the doctrine of the ascension as the first really thoroughgoing attempt to interpret it in terms of *the effect Jesus has on us*, rather than in terms of events which

---

29   *The Christian Faith* §99; see also §157ff.
30   Cf. *Christian Faith* §3f., §15, §50, e.g., and the second of his *Speeches on Religion*.

belong to his own personal history.[31] Let me try to explain the shape that took, by way of further reference to Kant.

Immanuel Kant, who had no more time for Jesus than did Saccas or Plotinus, identified the human race 'in its complete moral perfection' as the real Son of God, arguing that it is 'our common duty as men to elevate ourselves to this ideal' through the right exercise of reason.[32] This was ascension of the mind viewed horizontally as well as vertically, corporately as well as individually; in short, the doctrine of progress which more or less defined the Enlightenment project. Schleiermacher did not reject that project, but sought to preserve for it a religious character. He too dislodged the doctrine of the ascension from christology, but made it over somewhat more narrowly into a doctrine about the church. Which is to say, he tipped it on its side in the manner already suggested by Kant and Lessing, while attempting to preserve its specifically Christian content. Through the church, he declared, the spirit of Jesus would make itself felt as 'the ultimate world-shaping power'.[33]

It was remarked at the time, and it bears repeating, that Protestantism *à la* Schleiermacher is somewhere considerably to the right of Rome. It is the *church* that now 'appears before God' in the high-priestly role that once belonged to Jesus.[34] In one sense, it might even be said that the ascension of Jesus is made to rely upon our ascension (that is, upon the success of the church) rather than

---

31 Just what to do with the resurrection accounts as such, Schleiermacher was uncertain; he even toyed with the swoon theory. As for the ascension, it had but meagre attestation, one highly doubtful view being piled upon another. Declining attention to the doctrine of the ascension – which Calvin, commenting on Acts 1, had referred to as 'one of the chiefest points of our faith' – can be traced back to this point, and to Schleiermacher's decree that it 'is not directly a doctrine of faith' at all (§158.1). Cf. W. Walker, *The Ascension of Christ in Reformed Theology*.

32 *Religion Within the Limits of Reason Alone* 2.1. Kant did allow to Jesus an epistemic value, for we can learn from him to identify this ideal as a *gift* to human reason rather than its product – as something that 'has come down to us from heaven and has assumed our humanity' – even if its actualization can only be the result of our collective self-elevation.

33 *Christian Faith* §169.3

34 Cf. *Christian Faith* §104.6. Schleiermacher recognized that Kant's nod in the direction of Jesus was merely a polite dismissal. But in proclaiming Christ's lordship afresh he himself was forced to limit that lordship to a posthumous influence over the inner life of his followers, who would in turn guide the world towards its proper destiny by pointing back to Jesus as the one who represents the perfect conjunction of the historical and the ideal. It is really the church, then, that sits in heavenly session.

our ascension on the ascension of Jesus. Schleiermacher's more renowned colleague in Berlin, G. W. F. Hegel, completed the theological inversion: since it is not as living but as dead that Jesus ascends into heaven, the church too must be prepared to die, having served its purpose of helping to open up humanity to the truth of its own intrinsically divine nature.[35] That is something we will pursue in a later chapter, when we take up again the question of the eucharist. At the moment what is of interest is the way in which both Schleiermacher and Hegel (not to mention their Enlightenment predecessors) are anticipated by Origen at certain key points.

First of all, in the Alexandrian's theology the history of Jesus is already conflated with that of the human race, introducing a doctrine of progress. Since his eschatology was circular or restorationist, operating on the principle that the end will be like the beginning, it required a horizontal as well as a vertical vector; only through the course of many aeons would the victory of the cross be won, and all restored to God.[36] Second, there as here the advance of mind or spirit is governed by a monist drive, so that it necessarily involves the sublation (to use Hegel's word) of the particular, the capitulation of 'all that is special' in favour of the universal. This is already implicit in the principle that the end will be like the beginning; arguably, it is the real reason why Origen long ago employed his 'sanctified perspicuity' to re-imagine the ascension of Jesus as an ascension of the mind only.[37] Third, and quite tellingly, Origen anticipates Schleiermacher and Hegel by letting go of Acts 1:11 along with Acts 1:9. Though Jesus himself provides the key that unlocks the door to the universal, there can be no question of a return of 'this same Jesus', for that obviously would mark a backward rather than a forward step. The ubiquitous one must remain ubiquitous; he must not compromise the advance of spirit through the re-presentation of his flesh. Such is the implication of Origen's eternal gospel, and the obvious corollary

---

35   See his *Philosophy of History*, 318ff.

36   See *First Principles* 1.6.2. For all these thinkers the original – and final – perfection of the world belongs to a timeless eternity. In Origen and Hegel it is quite clear that history, centred on the cross, is a self-cancelling affair.

37   For Schleiermacher, too, Christianity 'stands for the victory of the Infinite over finitude and sin', since 'its central interest lies in "the universal resistance of all things finite to the unity of the Whole, and the way in which the deity overcomes this resistance"' (H. R. MacKintosh, *Types of Modern Theology*, 56).

of any doctrine of ascension of the mind.[38] It has been freshly grasped in the modern era.

THE TURNING TAKEN BY SCHLEIERMACHER, I said, was something of a shock to the credal faith which he professed to interpret for a new age, and so it was, however natural and appropriate it may have seemed at the time. Perhaps the only way to account for it as a phenomenon of Christian theology, and a widely influential one at that, is to recall that it follows a long history of quietly marginalizing the human Jesus, which in my judgement is the chief characteristic of the Origenist option. But surely this is to level a strange charge at one who is widely regarded as the father of modern theology. Is not modern theology largely about a *recovery* of Jesus' humanity, which ever since the Arian crisis has been threatened by a too-fervent belief in his divinity?

Here is a little game which it is past time to expose! Modern theology in its Origenist mode has compromised all its attempts at such a recovery, however helpful some of them may be, for the simple reason that it has no room for the risen, ascended and coming one as a man; that is, as a particular human being. Eschatological docetism – looking only for the divine in Jesus by way of his effects on the church or on the race as a whole – is still docetism after all. Besides, loss of interest in Jesus' humanity did not and does not stem from belief in his unique divinity. It stems rather from setting up a false competition between his divine or 'superior' nature (as Origen used to call it) and his merely human or historical nature, a competition which one way or another will be resolved in favour of the former. That is something which a study of the history of the doctrine of the ascension makes plain; it is a further very basic way in which Origen anticipates his modern counterparts.

Now it is certainly the case that Schleiermacher himself resisted the starker forms of this competition and marginalization that are to be found among his contemporaries. Take, for example, Hegel's disciple, D. F. Strauss, author of the epoch-making *Life of Jesus*. Strauss took it as read that we must distinguish carefully between the ideal Christ and the historical Jesus, approaching the latter with caution while openly

---

38   In his outline of the apostolic teaching at the outset of *First Principles,* Origen quietly
      passes over the parousia. On his treatment of it elsewhere (e.g., in his commentary on
      Matthew) see Kelly, *Early Christian Doctrines,* 472f.

exalting the former.[39] What is more, he recognized with all candour that if the story of what happened to the rabbi from Nazareth is just a story, albeit a powerful one – if, to be more specific, bodily resurrection and ascension are events conjured up by the overwrought imagination of the disciples; if they are among those impossible things which, according to Origen's hermeneutical theory, are included in scripture only to point us to a higher, mystical meaning – then the story itself and all the dogmatic convictions that go with it can and should be transferred away from that rabbi to man as such.

> *Mankind* is the union of the two natures – god become man, the infinite manifesting itself in the finite, and the finite spirit remembering its infinitude; it is the child of the visible Mother and the invisible Father, Nature and Spirit. *Mankind* is the worker of miracles, in so far as in the course of human history the spirit more and more completely subjugates nature, both within and around man, until it lies before him as the inert matter on which he exercises his active power. *Mankind* is the sinless existence, for the course of its development is a blameless one; pollution cleaves to the individual only, and does not touch the race or its history. It is *Mankind* that dies, rises and ascends to heaven, for from the negation of its phenomenal life there ever proceeds a higher spiritual life, and through the abrogation of its finitude as personal, national and secular spirit it is exalted into unity with the Infinite Spirit of heaven.[40]

Strauss, in other words, sided with the thoroughgoing secularism of Kant; 'it is that carrying forward of the Religion of Christ to the Religion of Humanity to which all the noblest efforts of the present time are directed', he insisted. But Schleiermacher, for all his obvious efforts to reserve a permanent place for Jesus in the hearts of the faithful, was peering over the same horizon.[41]

---

39 Strauss set out to restore 'the image of the historical Jesus in its simply human features', while referring us for salvation 'to the ideal Christ' – read 'divine' for ideal – that is, to that moral pattern which Jesus helped to reveal but which belongs in fact 'to the general endowment of our kind' (*The Life of Jesus for the People* §100).

40 *The Life of Jesus Critically Examined* §151 (translation modified). The sexual – and sexist – conceptuality of this passage is unmistakable.

41 Since Kant we have returned to the task of seeking the divine in everyone. Schleiermacher accepted that quest, finding in Jesus one who is only quantitatively, not qualitatively, different from the rest of us.

The eclipse of Christ – that is, of Jesus of Nazareth – in Christian eschatology is by no means a recent phenomenon. But in its recent and more radical form lies the doubtful blueprint (note that it is the legacy of theology, not of biology) for the doctrine of the ascent of Man, that fully mythical figure who in spite of recent setbacks continues to stand in for the marginalized Jesus. Later on we will observe that the Religion of Humanity, like all myths of descending and ascending, is ultimately hollow and leads nowhere but to the triumph of forces that are destructive of human beings, even where it continues to disguise itself in Christian garb; indeed, that it is essentially gnostic, and that Origen's antidote to gnosticism has failed.[42] Meanwhile, there is another task at hand, *viz.*, to explore the other main option for the interpretation of Jesus' ascension. For this we need to pay a second visit to the Chapel of the Ascension.

---

42 It should be remembered, however, as Mark Edwards reminds us in *Origen against Plato,* that Origen was trying both to combat gnosticism and to offer a version of Christianity that would encourage those trained in Platonism to go beyond Plato and discover Christ. It should also be recalled that – even if some of his ideas were later censured by the church, astutely and for good reason – Origen suffered for the faith and retains a place of honour in it.

CHAPTER THREE

Raising the Stakes

THE CHAPEL OF THE ASCENSION was constructed for pilgrims by the crusaders nearly a thousand years ago, and still serves that purpose.[1] But what does this modest shrine, hallowing a sandal-sized depression in the Olivet rock, represent today? Nothing, perhaps, but the petty resistance of the literal-minded to Christianity's true genius, which is progressive and forward-looking, transcending all particularities of time and place and culture, even of religion itself? Or might it remain, even for the new millennium, a symbol of that steadfast devotion to particularity – the particularity of the God of Israel and of Jesus the Christ – which arguably is the *sine qua non* of Christian religion?

A theologian who lived just prior to Origen, one equally admired for a progress-oriented world view (yet how different a world view), may be called upon as a first witness in support of the latter. His name was Irenaeus, and he was bishop of Lyons late in the second century. In his celebrated struggle with gnosticism he poked fun at the delusions of grandeur already entertained by its earliest proponents; that is, at their foolish ambition to become gods before their time, as he put it.[2] Not that Irenaeus rejected the idea of human deification. Does 2 Pet. 1:4 not speak of our becoming participants in the divine nature? But he saw that it was necessary to develop an ascension theology that would articulate a decidedly more biblical version of that aspiration.

It will help here to understand just a little of the Irenaean critique of gnosticism. Asserting the native divinity of their own intellects, and ungraciously refusing the creaturely formation they had been given,[3] the gnostics were hoping to recover their lost primal unity with God. But this, according to the bishop, was something they had never had. As a matter of fact, they were only repeating – and magnifying – the mistake made long ago in the garden by Adam and Eve. By rejecting the incarnation of God's Word, and denying themselves the gift of the Holy Spirit, the gnostics were stubbornly thrusting away the very hands by which God gradually moulds human beings for union and communion with himself.[4] Spurning the church as the

---

1   See Jerome Murphy-O'Connor, *The Holy Land*, 112f.

2   *Against Heresies* 4.39.2

3   These premises explain their rather Bultmannian conclusion, pointed out by Irenaeus, that 'the resurrection from the dead is simply an acquaintance with that truth which they proclaim' (*Against Heresies* 2.31.2).

4   See *Against Heresies* 4.20ff. (esp. 4.38f.), 5.19f.

present site of that moulding, and despising the heritage of Israel with her scriptures and covenants, these pretentious individuals were not actually climbing upward to salvation but sinking downward into an abyss of futility. They had no knowledge (γνωσις) at all of the patient and wise plan of God for human deification.

So how did Irenaeus himself interpret the ascension within the context of a sound doctrine of deification? The first and most important thing to be said is that he interpreted it in trinitarian terms. Read backwards from Pentecost, the whole biblical story of man's creation and redemption seemed to demand a trinitarian interpretation, which went something like this: The uniqueness of man[5] among God's creatures consists in the fact that he is uniquely formed by the Word and the Spirit for communion with the Father. In man alone the Word himself becomes incarnate, that human beings may behold God and live; in man, as in a temple, the Spirit of God comes to dwell.[6] As already observed, man's fall consists in his rejection of the training for communion which is essential to the process of his formation. For God long ago 'bestowed the faculty of increase on his own creation, and called him upwards from lesser things to the greater ones which are in his own presence, just as he brings an infant which has been conceived in the womb into the light of the sun.'[7] But this faculty was not exercised; the Word from the beginning was mistreated, the Spirit ill received, the discipline of creation resisted. The wonderful message of the gospel, however, tells of the Word's persistence – of his pursuit of fallen man into the bondage of sin and death, in order personally to reacquaint him, at every stage of his existence and in every facet of his being, with the Holy Spirit.[8] When this pursuit has taken him as far as the cross on Golgotha, and into death itself, his resurrection completes the conditions for the realization of God's interrupted anthropological project; to wit,

---

5   It is not necessary, I hope, to remind the reader that in patristic literature 'man' is an inclusive term, as are its pronouns.
6   See *Against Heresies* 5.6ff., 5.16.2.
7   *Against Heresies* 2.28.1; cf. 5.36.3.
8   See *Against Heresies* 3.17ff., 5.19ff. 'When Adam had hid himself because of his disobedience, the Lord came to him at eventide, called him forth, and said, "Where art thou?" [Likewise] in the last times the very same Word came to call man, reminding him of his doings, living in which he had been hidden from the Lord' (5.15.4, altered).

'that man, having embraced the Spirit of God, might pass into the glory of the Father'.[9]

Ascension, in other words, is deification, and deification nothing but the fulfillment of man's creation. It is not a return to the eternal past after an unhappy episode in time. It is the setting of man, once and for all, within the open horizons of the trinitarian life and love, where he may flourish and be fruitful in perpetuity. Whether in Jesus' case or in ours – *mutatis mutandis*, of course – it is a transformative relocation by the Spirit into the inexhaustible *Lebensraum* generated for us through full communion with the Father.

> For man does not see God by his own powers; but when he pleases he is seen by men, by whom he wills, and when he wills, and as he wills. For God is powerful in all things, having been seen . . . indeed prophetically through the Spirit, and seen, too, adoptively through the Son; and he shall also be seen paternally in the kingdom of heaven, the Spirit truly preparing man in the Son of God, and the Son leading him to the Father, while the Father, too, confers incorruption for eternal life, which comes to everyone from the fact of his seeing God. For as those who see the light are within the light, and partake of its brilliancy; even so, those who see God are in God, and receive of his splendour.[10]

Mutatis mutandis, I say, because the divine light itself is mediated by the one man who belongs by nature rather than by adoption to the divine economy.[11] He it is who gives definition to this living-space, who centres its boundless horizons in a way appropriate to creatures. It is by way of his passage into the Father's glory that a passage is also opened for us.

We will return shortly to the obvious and necessary question as to what meaning can be given, especially in a scientific age, to this idea of ascension as a transformative relocation into a time and space

---

9   *Against Heresies* 4.20.3. While some contend (appealing to Irenaean speculation in 4.38f.; cf. 5.2f.) that the fall does not interrupt God's creation of man but is indeed essential to it, they must still allow that it is essential only *as* an interruption. For my part, however, I think this line of thought a dangerous one.

10 *Against Heresies* 4.20.5. Cf. 5.36, and *Proof of the Apostolic Preaching* §7: 'For those who are bearers of the Spirit of God are led to the Word, that is, to the Son; but the Son takes them and presents them to the Father, and the Father confers incorruptibility.'

11 Cf. *Against Heresies* 4.20.2.

and form of life defined by participation in the trinitarian economy. Meanwhile the second thing that must be said – lest in all of this we revert to the mythological speculation of the gnostics, which repudiates any spatio-temporal mode of existence – is that Irenaeus very definitely understood the ascension to be bodily ascension, ascension in the flesh. The flesh of man as well as the soul of man is God's handiwork; 'confirmed and incorporated with His Son', it is made recipient of the Holy Spirit, and so brought to perfection as the image and likeness of God.

Now Irenaeus had never heard of Bultmann, naturally, and certainly had never listened to a radio or seen pictures of Mars on television. Yet we would be quite wrong to imagine that in making this claim he was oblivious to the problem of mythology. Just the reverse. From his perspective, it was his gnostic opponents who were clinging stubbornly to mistaken and outmoded assumptions (learned largely from the Greek philosophers) about the universe. The trouble with 'these perverse mythologists', he complained, was precisely that they twisted the gospel narrative to fit those assumptions. Instead of allowing what happened with Jesus to correct and reform their world view, they allowed the latter to correct and reform what could be said about Jesus.[12] That is why they could not tolerate either the doctrine of the incarnation or the apostolic testimony to bodily resurrection and ascension. For his part, following the gospel narrative – especially in its Lukan form, the form later enshrined by the Chapel of the Ascension – allowed Irenaeus to challenge the reigning orthodoxy in Hellenic cosmology. Heaven was not opposed to earth, and creatures of flesh and blood had real hope of heaven. For the heavenly shepherd, who descended to seek out God's lost sheep, also ascended 'to the height above, offering and commending to his Father that human nature which had been found, making in his own person the first-fruits of the resurrection of man'. Here we observe an entirely different tack from that afterwards taken by Origen, for whom the gates of heaven remained closed to the flesh, even to the flesh of Christ. Irenaeus plainly thought it impossible to overthrow gnosticism without fully affirming the flesh, and with it all the rich diversity of the spatio-temporal world whose creator the gnostics

---

12 See *Against Heresies* 3.16ff., 4.1ff. My suspicion is that Irenaeus would have found Origen's proposal deeply flawed in pretty much the same way. (See further Farrow, 'The Doctrine of the Ascension in Irenaeus and Origen', *ARC* 26, 1998, 31ff.)

had maligned. He was not shy about saying that the whole man, not a part of man, is rendered participant in the divine nature.[13] It was his considered judgement, in fact, that those who refuse to allow that the whole man provides the substance of Christ's offering in heaven fall into the same error as those who deny that the whole man constituted his offering on earth. Likewise, those who do not allow that the bodies of the righteous are to be resurrected as beneficiaries of that offering 'entertain heretical opinions', even if they are otherwise regarded as orthodox.[14] As for the gnostics, opined the bishop, *they* entertain a kind of homicidal mania, for their thoroughgoing rejection of the body (and of its present faithful suffering) encourages and promotes the abortion of that spiritual humanity which God in Christ is bringing to birth.[15]

AMONG THE FATHERS there is no shortage of further witnesses who might be called. For in spite of the immense influence of Origen, and the gradual slackening of the cords which bound this Irenaean verdict – ascension in the flesh – to trinitarian insights, the verdict itself was regularly reaffirmed by the church and its great theologians. I have already mentioned Augustine, whose eschatology and soteriology are made to rest on orthodox cornerstones, and whose *City of God* eschews a Greek vision of human destiny for one based on the story of Jesus. Nowhere does Augustine attempt to alter or to deviate from that story in establishing the Christian hope: 'The resurrection of the Lord is our hope; the ascension of the Lord is our exaltation', he remarks, and the exaltation in view is an exaltation of the whole man.[16] 'Although he descended without a body, he ascended with a body and with us who are destined to ascend, not by reason of our own virtue,

---

13   'For by the hands of the Father, that is, by the Son and the Holy Spirit, man, and not [merely] a part of man, was made in the likeness of God' (*Against Heresies* 5.6.1). It is this line of thought that Douglas Hall appears to have overlooked in his comments on the resurrection in *Professing the Faith*, 387ff.

14   See *Against Heresies* 5.31. Tertullian (who with respect to his Montanism was numbered among the heretics) concurs: 'For as he has given us "the earnest of the Spirit", so has he received from us the earnest of the flesh, and has carried it with him into heaven as a pledge of that complete entirety which is one day to be restored to it. Be not disquieted, O flesh and blood, with any care; in Christ you have acquired both heaven and the kingdom of God' (*Resurrection* §51).

15   Their theological dismembering of Jesus Christ himself – a feature shared by modern gnosticism – is ample evidence in support of this charge. Cf. *Against Heresies* 3.16, 3.25.

16   Sermon 261 (the same in which he develops the views we earlier questioned).

but on account of our oneness with him.'[17] This thinking is followed through, as it must be, to the parousia: 'He shall come to men; he shall come as a man; but he shall come as the God-man. He shall come as true God and true man to make men like unto God.'[18] That it is really *man* who is to be made like God is plain from Augustine's defence of the body – including its sex or gender, both male and female – in his description of the life of the world to come.

Here, in answer to the criticism earlier levelled at Augustine, it is also right to notice what he says in *On the Trinity*. In showing how Christ's death and resurrection address 'this double death of ours' (that is, the death of the body and of the soul) he points out that 'both in death and in resurrection, his body served as the sacrament of our inner man and as the model of our outer man, by a kind of curative accord or symmetry.' The *noli me tangere* of John 20:17 is no warning that his body is now to be dispensed with, either literally or for faith. Rather it is a warning 'not to have materialistic thoughts about Christ'; that is, about his soul or his body. And that means, *inter alia*, to recognize that in our own resurrection 'he will transfigure the body of our lowliness to match the body of his glory'.[19] Later on in the same work Augustine writes that 'through him we go straight toward him', and there is no thought here of a mental transition that leaves anything behind other than misconceptions. Faith and truth, knowledge and wisdom, temporal things and eternal things – all these dualities, even indeed the ontological duality of man and God, are conjoined and mediated by one and the same Jesus Christ, the God-man. His single 'matches our double' in every way. It is through him 'who became what by nature we are and what by sin we are not', preserving us whole in his resurrection, 'that we are to obtain happiness with all the potentialities of human nature, that is both of body and soul'.[20]

Let us appeal to Gregory Nazianzen as well, who sets aside

---

17 Sermon 263

18 Sermon 265. If in his early work, *Faith and the Creed* §24, Augustine appears to move in Origen's direction with respect to the resurrection body, that is not true of his later work; see *Retractions* 1.17, *City of God* 22.16ff., and *Trinity* 13.3, where bodily immortality in the traditional sense is made essential to the universal goal of happiness.

19 See *Trinity* 4.4ff (preferably in Edmund Hill's fine translation).

20 *Trinity* 13.24f.; cf. 4.4. This, then, is how Augustine intends us to read other formulae, such as concerned us earlier. I was remiss not to point this out in *Ascension and Ecclesia*.

some residual Origenist instincts of his own to make the following declaration respecting the permanency of the incarnation:

> If any assert that he has now put off his holy flesh, and that his godhead is stripped of the body, and deny that he is now with his body and will come again with it, let him not see the glory of his coming. For where is his body now, if not with him who assumed it? For it is not laid by in the sun, according to the babble of the Manichaeans, that it should be honoured by a dishonour; nor was it poured forth and dissolved, as is the nature of a voice or the flow of an odour, or the course of a lightning flash that never stands. Where in that case were his being handled after the resurrection, or his being seen hereafter by them that pierced him, for godhead is in its nature invisible. Nay; he will come with his body – so I have learnt – such as he was seen by his disciples in the mount, or as he shewed himself for a moment, when his godhead overpowered the carnality.[21]

John Chrysostom may speak for a multitude of others, employing his eloquence on a motif which was a great Ascension Day favourite among the fathers, whether eastern or western. Christ, he says, 'offered the firstfruits of our nature to the Father' and 'the Father admired the gift, and on account of the worth of the offerer and the blamelessness of that which was offered, he received it with his own hands and placed the gift next to him, and said: "Sit thou on my right hand." To which nature did God say: "Sit thou on my right hand"? To that which heard: "Dust thou art, and unto dust thou shalt return."'[22]

This bold understanding of Jesus' ascension, as the exaltation of his and our humanity, body and soul, was upheld by church councils[23] and appears repeatedly in later writings. Maximus of Turin, for example, sounds in the fifth century very much like Irenaeus did in the second: 'Having fought in man and having conquered through man,

---

21 Letter to Cledonius (*Nicene and Post-Nicene Fathers*, 2nd ser., vol. 7, 440).

22 Sermon on the Ascension (*PG* 50.441–52); quoted and summarized by J. G. Davies, *He Ascended into Heaven*, 115ff. Cf., e.g., John of Damascus, *Orthodox Faith* 4.1f.

23 It is worth observing that the council in AD 381 added to the Nicene Creed much of the eschatological detail missing in the AD 325 version (with which compare *Against Heresies* 1.10.1). Particularly noteworthy is the clause, 'whose kingdom shall have no end', which for anti-Sabellian purposes highlights the permanency of the incarnation. Explicit rejection of Origenism, by the way, was not confined to the fifth ecumenical council (AD 553). It was echoed, e.g., at the eleventh council of Toledo in AD 675 (see J. F. Clarkson *et al.*, *The Church Teaches*, 346).

Christ made him the inhabitant of heaven and the Lord of earth'.[24] In the eleventh century, Bernard of Clairvaux is still thinking along the same lines when he describes the festival of the ascension as 'the consummation and fulfilment of [all] the other festivals, and a happy ending to the whole journey of the Son of God'. So too is Thomas Aquinas in the thirteenth, when he identifies Christ's ascension as the true cause of our salvation, because 'the very showing of himself in the human nature which he took with him to heaven is a pleading for us'.[25] Likewise John Calvin in the sixteenth, who famously (if controversially) extends the same tradition by turning it against contemporary forms of Pelagianism as well as Origenism.[26]

We must pause to be clear about what is being said. The fathers, on the one hand, established belief in bodily ascension as essential to Christian orthodoxy, both because that is the biblical teaching and because a holistic soteriology – forged under pressure from docetic or gnostic heresies – demanded it. On the other hand, we have already seen that they themselves were much attracted to talk of ascension of the mind as a staple of spiritual and pastoral theology. As Augustine puts it, echoing the call of St Paul in Colossians 3, 'let us ascend with him and let us have our hearts lifted up to him'.[27] Or in the later words of the collect for Ascension Day in Cranmer's *Book of Common Prayer*: 'Grant, we beseech thee, Almighty God, that like as we do believe thy only-begotten Son, our Lord Jesus Christ, to have ascended into the heavens; so we may also in heart and mind thither ascend, and with him continually dwell . . .'[28] This, then, is the place to ask: If the fathers had no difficulty holding these two ideas together, should we have any difficulty? Is the contrast between an Origenist and an Irenaean approach overdrawn, if not actually mistaken?

On the contrary, what is at stake between the two approaches is quite fundamental. For the models we are contrasting under the rubrics 'ascension of the flesh' and 'ascension of the mind' are an

---

24 Quoted by Davies, *He Ascended into Heaven*, 144 (PG 57.625–26).

25 *Summa Theologiae* 3.57.6; cf. Bernard, *Asc. Dom.* 2.1 (*Saint Bernard on the Christian Year*, Mowbray, 1954, 99f.).

26 See, e.g., *Institute* 4.17.27ff. and the commentary on Hebrews.

27 Sermon 261; cf. Col. 3:1ff.

28 Cf. the Gelasian Sacramentary: 'Grant, we beseech thee, almighty Father, that in the intention of our mind we may ever tend thither, where the glorious Author of today's festival hath entered in, and that to the place whither we reach forward by faith, we may come by our holy conversation' (see P. Toon, *The Ascension of our Lord*, 135).

inclusive and an exclusive model, respectively. To speak of the former is to speak (with Irenaeus) of the ascent of body and mind together, where 'ascent' indicates the elevation, or rather the transformation, of both: 'King of glory, in you our mortal flesh has been lifted to the heights; deliver us from the corruption of sin and restore us to immortal life.'[29] To speak of the latter is to speak (with Origen) of a spiritual ascent which entails a rejection of the body. Of course this way of putting it will be regarded by some as unfair to Origen. Like Irenaeus before him and Augustine after him, there is good reason to think that he believed rather that in Christ's ascent, as in ours, the body is newly integrated with the soul and thus transformed rather than annihilated.[30] But for Origen integration really means sublation; transformation means elimination of everything related to the body's material mode of existence. And it was against just this sort of soteriology that the conciliar anathemas were pronounced.[31]

Now on both models ascension may be regarded as a present activity and, as such, exclusive to the inner man.[32] But we should not be deceived into thinking that this activity must be the *same* activity. To inhabit in a provisional way the spiritual realities indicated by Paul in Corinthians or Colossians is one thing; to seek out those indicated by the writings of Philo or Origen, Plotinus or Proclus, is another, since a different salvation is in view. Obviously a good deal of mental energy has been expended trying to get round this point (most impressively by the pseudo-Dionysius, perhaps, but some of the West's own great mystical theologians also come to mind). The results have not been uninteresting, but too often they have marked a retreat from the task of considering the implications of bodily ascension precisely

---

29 *Daily Prayer from the Divine Office* (Collins 1974), 415: The Ascension, Evening Prayer I.

30 As Irenaeus has it: 'bodies, too, which have participated in righteousness, will attain to that place of enjoyment, along with the souls which have in like manner participated, if indeed righteousness is powerful enough to bring thither those substances which have participated in it' (*Against Heresies* 2.29; cf. 5.5f.). Or Augustine, more briefly: 'The body will easily be lifted to the heights of heaven if the weight of our sins does not press down upon our spirit' (Sermon 263). On Origen's views, see *Ascension and Ecclesia* 94f. and the literature indicated there.

31 To take the orthodox view, of course, is not immediately to decide which dimensions of human experience are essential to embodied life, and which are transitory; nor is it to know exactly what effects the fall has had on us or what will be the effects on our bodies of their redemption and re-creation. Such questions must be approached with considerable reserve (cf. Augustine, *Faith and the Creed* §13).

32 Thus, e.g., 2 Cor. 4:13-18; but cf. 12:1ff.

for ascension of the mind, another matter to which we will need to return. It is time now, however, to take up our question about the meaning and viability of the Irenaean option in the age of modern science, with its shifting cosmological paradigms.

WE DO WELL, as T. F. Torrance has advised, not to be too hasty in supposing that the significance of these shifts is as transparent as the likes of Bultmann have imagined. Arguably contemporary cosmology is more compatible with orthodox christological thought than was anything available before Faraday, Maxwell and Einstein, though certain interpretations of post-Einsteinian physics clearly are not.[33] That said, our present understanding of the cosmos may cause us to flinch a little at the sort of thing one used to find among the orthodox on the subject of the ascension. Consider what is urged upon us in that classic work on the creed by the great Cambridge scholar, John Pearson: 'I am fully persuaded, that the only-begotten and eternal Son of God, after he rose from the dead, did, with the same soul and body with which he rose, by a true and local translation convey himself from the earth on which he lived, through all the regions of the air, through all the celestial orbs, until he came unto the heaven of heavens, the most glorious presence of the majesty of God. And thus I believe in Jesus Christ who "ascended into heaven".'[34] Surely a passage such as this justifies the suspicion that a traditional understanding of the ascension, as a literal bodily event, will no longer do.

Or does it? Pearson's *Exposition of the Creed* was published just prior to Newton's *Principia* and betrays the lingering influence of old assumptions about the structure of the universe. But what of that? The church, after all, did not invent Ptolemaic cosmology. If to some degree it accommodated its beliefs *to* that cosmology, overlooking the seminal insights of its best thinkers in pursuit of what could only be – even in the deft hands of a Denys or a Dante – a wonderfully improbable redemptive hybrid, that is hardly surprising.[35] Nor indeed

---

33 Some of these, indeed, in their denial of objective reality, lapse back into a philosophically violent oscillation between monism and dualism and are incompatible with the scientific enterprise itself. See further T. F. Torrance's helpful essay, 'The Church in the New Era of Scientific and Cosmological Change', *Theology in Reconciliation*, 267ff. (cf. Jeffrey Satinover's *The Quantum Brain*).

34 *An Exposition of the Creed*, 396.

35 The doctrine of *creatio ex nihilo*, as expounded by the likes of Irenaeus, Basil and John Philoponos – and dogmatized at Lateran IV – had a great deal to do with the advances

is it surprising, though it is certainly lamentable – because unredemptive and poetically infertile – that in some sectors the church soon repeated its mistake by accommodating itself to the far more dangerous dualism of the Newtonian cosmology.[36] No, if we find Pearson's rendering of the ascension embarrassing, it should not be the demise of the Ptolemaic universe that is the source of our discomfort. For what is wrong with it is not its outworn cosmological baggage (that charge might stick better if levelled against Origenist or Enlightenment alternatives) but rather its eschatological deficiency.

That deficiency is admittedly a long-standing one, going back as far as the fanciful, extra-canonical accounts of the ascension in both gnostic and anti-gnostic literature of the second century, which first began to treat Jesus' departure in terms of a vertical progress through successively higher cosmic strata. Wherever the deficiency is overcome, however, traces of these fanciful accounts are eliminated from the orthodox view, showing it in a quite different light. And in that light it sits askew not to modern science but only to what some call scientism, which vainly proposes to bring all things into subjection to a highly reductionist form of human reason or to technological prowess.[37] To scientism the doctrine of the ascension is a firm rebuke. To science it is simply a reminder of limitations, for it speaks of a miraculous event, beginning in the resurrection and completed in the ascension, whereby creation is set upon a brand new footing. I spoke of that event earlier (in connection with Irenaeus) as a transformative relocation into a time and space and mode of life defined by full participation in the trinitarian economy. A short exegesis of this statement will help to fix the eschatological frame of reference which has always been requisite to the doctrine of the ascension, and which today is especially vital to our understanding of it.

Let us begin with the phrase 'transformative relocation', which is intended to say at least two things. First, that in the ascension God does something quite new with and for the man Jesus, as a basis for that which he intends ultimately to do with and for us. Second,

---

in experimental science that eventually overthrew the ancient cosmology. Cf. S. Jaki, *The Saviour of Science*, chap. 2.

36 Newton and his disciples led the way; cf. Michael Buckley, *At the Origins of Modern Atheism*.

37 On which see, e.g., Bryan Appleyard, *Understanding the Present*. Scientism, which is barren not only from a theological and even from a poetical point of view, is in part the result of a collapsed Newtonian dualism which has not been recognized as such.

that this something (though assuredly ecclesial and even universal in scope) is not contrary to his or our creaturely particularity. One way to put the latter point is to say that the ascension is not merely removal *from* a place but also *to* a place. If in the resurrection Jesus is already transfigured and transformed – the mortal being made immortal – in the ascension he is also translated or relocated. That is, he is taken up and placed by God where he properly belongs, just as God once took Adam and put him in Eden.[38] He is 'placed' in the sense that he is ranked first rather than last; having become the first-born from the dead, he is now made lord of all things. But he is also placed in a more literal sense. He is permitted to leave behind those things which are passing away, to which in his resurrection life he no longer belongs. He is received into heaven – that is, into the Father's house – and set at the Father's right hand.

Thus far, then, with Pearson's 'true and local translation'. In the ascension Jesus really is relocated or given a new place, for it belongs to God's creatures to have and to make and to be in a place. But how exactly shall we understand this relocation, if not as Pearson does?[39] Where is the Father's house? Where is God's right hand? These questions are not altogether easy, and immediately put in doubt our second and more literal sense of the word 'place' or 'placed'. Not because modern cosmology recognizes no such place, but because, as John of Damascus says, 'we do not hold that the right hand of the Father is an actual place'.[40] On the other hand, in going to this place which is not a place, Jesus (as the Damascene makes clear) remains who and what he is, a specific human creature to whom God affords time and space and whose bodily return we await.[41] He must, then, have a place. Indeed, any suggestion that he does *not* have a place can only be regarded as a form of Marcionism, for it posits a kingdom that has little or nothing in common with the kingdom the prophets taught Israel to long for.[42]

---

38 With Gen. 2:8, cf. Lk. 9:51 (the Latin *translatus* and the Greek αναληψις have here roughly the same sense). See also *Against Heresies* 5:31.2.

39 Pearson seems to understand it in terms of motion through space, which – even if we allow for a *variegated* space that may be open in its interpretation to some more modern form of cosmological speculation – is a non-eschatological construct.

40 'For how could he that is uncircumscribed have a right hand limited by place?' (*Orthodox Faith* 4.2).

41 'His ascent from earth to heaven, and again, his descent from heaven to earth, are manifestations of the energies of his circumscribed body' (*Orthodox Faith* 4.1).

42 N. T. Wright has been a sturdy defender of the Jewish character of the early Christian

Here is one puzzle or paradox, to which we may add another: It was of course the function of the ark of the covenant, and of the temple in Jerusalem, to give earthly definition to the right hand of God – to locate the unlocatable. According to the Gospels that function has passed to Jesus. So Jesus himself represents the right hand of God. How then can he be said to go there?

Addressing the second paradox in an eschatological way will help us with the first as well. Just as the ark, with its mercy seat, moves from Sinai to Zion, tracing out the mighty arc of God's saving hand among the people of Israel, so Jesus moves from the earthly Zion to the heavenly, tracing out God's salvation both for Israel and for all the world. That is to say, he moves, not upwards in space, but from the old creation to the new. He moves from Egypt and Golgotha and Olives – from those places in which we ourselves become old, says Irenaeus, because of transgression[43] – to the place in which righteousness reigns and all the life-giving promises of God are fulfilled. As the incarnate one, our saviour, Jesus is always at the right hand of God; indeed, he *is* the right hand of God.[44] Yet his entry into the new creation is also his entry into the full dispensation of the divine power he mediates as saviour, and from this perspective his ascension is rightly spoken of as an exaltation *to* God's right hand.

But what exactly is the relation between old and new? And is this, or is it not, ascension to a place? At this point we need to ask the adjective 'transformative' to help bear the weight of Christian eschatology. It will already be clear that the entry of which we are speaking does not entail admission to an already existing place but the creation of a new one. But we must say something more than that. We must say that it entails the creation of a new time and place and mode of life, and that not *ex nihilo* (out of nothing) but *ex vetere* (out of the old one). Is this not what the saving right hand of God is committed to doing? In Luke's Gospel Jesus begins his public ministry with a reading from Isaiah 61, and the whole course of Isaiah 60ff. is the course he undertakes to follow and fulfil. 'Behold', says Yahweh,

---

hope, which anticipated a marriage of heaven and earth (cf. Isaiah 60ff.), not the earth's abandonment. In *Surprised by Hope*, however, he draws many conclusions from this that do not necessarily follow, combining an anti-Marcionite ascension theology (109ff.) with a too-Marcionite resistance to tradition.

43   *Against Heresies* 5.36.1
44   In Jesus and the Spirit God has *two* hands, as Irenaeus liked to say.

> I will create new heavens and a new earth.
> The former things will not be remembered,
> nor will they come to mind.
> But be glad and rejoice forever in what I will create,
> for I will create Jerusalem to be a delight
> and its people a joy. [45]

The way in which he fulfills it greatly magnifies the force of Isaiah's allusion to Genesis 1, with its stress on eschatological discontinuity, yet without sacrificing the element of continuity. The place to which Jesus goes in his ascension – the 'there' and 'then' of the life which he lives at the Father's right hand as founder of the new Jerusalem – is really a place, yet it is not one to which we can refer on our own terms, cosmological or otherwise. It is not somewhere in this world, a *Lebensraum* attainable by political or technological conquest; nor yet is it somewhere in addition to this world, an 'outside' to which one escapes. Rather it exists by virtue of the transformation or reconstitution of this world in the Spirit. Hence it can be referred to only indirectly. Which is to say, from the place and time in which we ourselves remain it can be recognized only by faith, and can be touched only sacramentally. [46]

Let me expand on this point, approaching it from a slightly different angle, since there is endless confusion about it. The marriage of heaven and earth, foreseen by the prophets and witnessed to by the apostles and martyrs of Jesus Christ, entails the complete deconstruction and regeneration of earthly reality. [47] It will hardly do, then, to think about the place which Jesus occupies in his ascension as a distant

---

45 Isa. 65:17f. (*NIV*); see Lk. 4:14-20.

46 If in some sense this is already the case on the Emmaus Road, *a fortiori* it is so after the ascension. It seems to me, then, that neglect of the sacramental aspect (which I will develop in chapter five) may result in the misdirection of a project such as Sleeman's in *Geography and the Ascension Narrative in Acts*. I am sympathetic, naturally, to his attempt to read Acts with a view to mapping, as far as they appear there, the spatial continuities and discontinuities produced by the ascension; for we agree that 'place and space, although relativised by the ascension, still very much matter in this new order' (137). Yet this cannot be done without committing to some cartographical method. Sleeman wants to avoid any 'anachronistic' sacramentalism, but his still more anachronistic collectivism and verbalism (cf. 5, n. 17, and p. 20) invite him to deploy postmodern spatial dialectics such as Ed Soja's, which (however useful) can hardly be expected to take us to the root of the matter.

47 The prophets did not necessarily understand this, but with Isaiah 60ff. cf. Revelation 19ff. (esp. 21f.); see also 2 Peter 3.

place, whether spatially (above 'all the celestial orbs', in Pearson's words) or temporally (a far-off future, gradually approaching). Nor yet will it do to think of it as a place interior to this world, whether psychologically interior ('the kingdom is within you', in the famous mistranslation)[48] or even sacramentally interior (hidden somehow in the spiritual interstices of the cosmos, which may be pried open by liturgical manipulation).[49] These common misconceptions arise from a *non*-eschatological way of thinking about the ascension, and represent misguided attempts to say what cannot be said. All that can be said is that the time and place which Jesus occupies are those in which, and by way of which, God's sovereign act of recreation is extended through him to all times and places.[50] And what kind of act is that? One that we can never hope to describe, certainly, but one whose foundation – the cross and resurrection – we have already witnessed and whose consequences we will all experience at the parousia. To the church it is given here and now to point to the most important consequence: it is an act that, having overcome sin and death, results in a time and place and mode of life defined by full participation in the trinitarian economy. For in and through the ascension of Jesus the earthly things of human existence (whether physical or mental) are so united with the heavenly things of the Spirit that the former are made fully subject to the goal of communion with the Father.[51]

Here it is worth pausing a second time to make plain what is, and is not, being asserted. Many different attempts have been made to establish the coordinates of the heaven to which Jesus goes, and each of these is important theologically by reason of the nuances or alterations it introduces into the church's perception of its participation

---

48 εντος υμων (Lk. 17:20f.) indicates that the kingdom, in the person of Jesus, is among them, within their reach, not within their own persons.

49 The 'sacramental', used in this sense, amounts to little more than religious psychology. It too is connected to a misreading of biblical eschatology, in which οτι χρονος ουκετι εσται in Rev. 10:6 is taken for an announcement of the cessation of time – the removal of its obscuring veil over eternal verities – when in fact this phrase indicates rather the end of God's patience with the wicked and the beginning of his promised kingdom: 'There will be no more delay!'

50 In short, it is a time and place suited to the judging and reconfiguring of other times and places, which manifests itself provisionally in our own age by the reordering power of the gospel.

51 This is the goal towards which the incarnate one moves, and as such the goal towards which God also moves in the incarnation. It is therefore the perfection of the trinitarian economy itself.

with God in Christ. But the entire enterprise is wrong-headed. 'Where I am going you cannot come', declares Jesus, nor indeed can we comprehend or locate where he is going on any map of ours.[52] To insist on this does not mean, however, that we must revert after all to the Origenist model, or to one on which Jesus is thought to be atopic or atemporal (that is, without place or time of his own, beyond those of his life from Bethlehem to Golgotha). Nor does it mean that we wish to seek refuge in an acosmic point of view, contenting ourselves with 'the meaning of the Christ-event' or some such thing – that is, with spiritual psychology. On the contrary, an eschatological approach to the doctrine of the ascension insists with scripture and creed on bodily ascension. It is Irenaean and incarnational, hence completely committed to speaking about every dimension of creaturely existence. At the same time it remains stubbornly independent of any merely natural cosmology or anthropology, which, without knowledge of the resurrection or ascension, must lack knowledge even of its own limitations.

It may be remarked in conclusion that, while the Irenaean option may be the more traditional option, it is hardly the more conservative. Its embrace of bodily ascension obviously raises the stakes substantially in the contest between the gospel and conventional world views. That of course is its strength, and the secret of its resiliency. With a little good humour, perhaps even the 'footprint' in the Olivet rock may serve as a cipher of its commitment to taking the story of Jesus much more seriously than our own, the reality of Jesus as far weightier than our own.

---

52 Jn 13:33. Calvin remarks that where he goes 'is opposite the frame of this world' (*Commentary on Acts* 1:9); but even this is to say too much, or rather too little. A direct answer to the 'Where?' question *cannot now be given*.

CHAPTER FOUR

A Question of Identity

THAT THE CHAPEL OF THE ASCENSION is today in Muslim hands (if under Israeli control) reminds us that there is a third form of ascension theology, to Christians quite foreign, that may be mentioned in passing. According to the *Qur'an* it was said in boast by the Jews,

> 'We killed Christ Jesus
> the son of Mary, the Apostle of God' –
> but they killed him not, nor crucified him,
> but so it was made to appear to them . . .
> For of a surety they killed him not –
> nay, God raised him up unto Himself . . .
> And there is none of the People of the Book
> but must believe in him before his death;
> and on the Day of Judgment
> he will be a witness against them.[1]

This account is foreign to Christianity because it contains a doctrine of the ascension without a doctrine of the cross and resurrection. In that respect, at least, it is not unlike what one finds among the gnostics, some of whom claimed that the heavenly Christ abandoned Jesus at his death, returning to the pleroma before his awful cry of dereliction. In Islam, of course, we do not have the docetism that characterizes gnostic thought, nor any such patent separation between Jesus and the Christ; its view is closer to what we call adoptionism, and indeed it is the man Jesus who ascends to heaven. But we do have a religion that denies the cross, and that sees in his ascension not the completion of a great act of atonement, nor yet his exaltation as King of Kings, but only his preservation as a witness for the prosecution of the Jews. And that amounts to a form of docetism after all, not only because the cross is a mere seeming, but because Jesus and his story are stripped of all soteriological significance.[2]

If this foreign account bears some similarity to gnostic accounts, however, so also to that modern version of the Origenist option traced in chapter two. What we were offered there, by Christian thinkers such as Schleiermacher and Hegel, was a doctrine of the ascension that made allowances for the cross but deliberately bypassed

---

1   Surah 4, 157ff.
2   In its own curious fashion, then, Islam, in repudiating Judas, also betrays Jesus with a kiss.

the resurrection: 'Christ dies; only as dead is he exalted to heaven and sits at the right hand of God; only thus is he Spirit.'[3] The effect, as we saw, was still another form of docetism. Whether one denies with Islam that Jesus died, or (with western liberalism) that having died he rose bodily from the dead, ascended in the flesh, and will return in like manner, one comes pretty much to the same place; that is, to the conclusion that Jesus himself is no divine saviour but merely a messenger, an apostle of one sort or another.[4] What matters about Jesus is what he represents, not who or what he is. What matters is his abiding effect on us, his profound influence on our culture, not his own person or what he himself has accomplished.

THAT THE LIBERAL TRADITION adopted this conclusion as one of its premises, thus reducing the question of Jesus to the status of an historical problem,[5] needs no further elaboration here. Nor does the Islamic approach to Jesus, which sets him up as an object of pious affection but cannot even make of him a problem.[6] What does require our attention is the fact that many Christians today have become confused about the person and work of their risen Lord. His identity has been badly occluded through the distorting lens of an ascension theology too much influenced by liberalism's rehabilitation of the Origenist model. Not to put too fine a point on it, it has become difficult – even among those who recoil from the conclusion of Strauss – to distinguish him from ourselves. The 'risen' Christ is today the immanent Christ, the Christ who has ascended into the dynamic of history, into the future of the race, into the evolving cosmos. He is the Aryan Christ, the black Christ, the feminist, queer, communist, capitalist or eco-Christ – whoever we need or desire him to be. He is a

---

3   Hegel, *The Philosophy of History*, 325.

4   If Jesus rose from the dead, of course, he would be something more than an 'accidental truth of history'; hence also something more than a messenger whose articulation of divine truth is necessarily partial, inadequate and transient (cf. Rom. 1:1-7). But western liberalism cannot tolerate the claim that he rose from the dead, because its God-concept does not permit any such full articulation of the eternal God in temporal form. In this it is at one with Islam, which invents a long extension of Jesus' life rather than admit his death and resurrection, and which transfers to the *Qur'an* his absoluteness as God's Word, much as the Enlightenment transferred it to Reason ('the very voice of God'). And here it is worth noting that the *Qur'an* is rescued from transiency by a theory of divine dictation.

5   See E. Troeltsch's 'The Significance of the Historical Existence of Jesus for Faith', *Writings on Theology and Religion*, 182–207.

6   On Islamic reverence for Jesus, see Tarif Khalidi's *The Muslim Jesus*.

spirit, a force, an ideology. What he is not is what he once was: Jesus, son of Mary, the Jew from Nazareth, who departed from us into the very presence of God until he comes in judgement on the Last Day.

The dated but still popular ascension theology of Teilhard de Chardin may be offered by way of illustration. Karl Barth rightly described it as a 'giant gnostic snake', since it is immanentist through and through. For Teilhard (as for Origen and Hegel) the human Jesus is important, but only as the temporal point on which the eternal Word pivots, as it were, after its painful parsing in the realm of creaturely transience. The risen Christ is no longer a particular man. He is rather the theanthropic power that is slowly transforming our transient nature into a unified spiritual one, as once he turned water into wine. It is not 'the man who lived two thousand years ago' that interests Teilhard, but the divine Christ who now 'shines forth from within all the forces of the earth'.[7]

This occlusion of the human Jesus makes possible a synthesis of Christ, church and cosmos that many have found highly attractive. It is a Pauline doctrine, central to Catholicism as to Orthodoxy, that the church is, in some sense, Christ in action bringing the cosmos to its destiny:

> I pray also that the eyes of your heart may be enlightened in order that you may know the hope to which he has called you, the riches of his glorious inheritance in the saints, and his incomparably great power for us who believe. That power is like the working of his mighty strength, which he exerted in Christ when he raised him from the dead and seated him at his right hand in the heavenly realms, far above all rule and authority, power and dominion, and every title that can be given, not only in the present age but also in the one to come. And God placed all things under his feet and appointed him to be head over everything for the church, which is his body, the fullness of him who fills everything in every way.[8]

But whereas in Paul and in conciliar documents it is always clear that we can and must talk about Jesus Christ and the church much in the

---

7   See especially *Hymn of the Universe*. This would make a certain sense of Eph. 4:10, were it not for the fact that he who descended into the bowels of the earth 'is the very one who ascended higher than all the heavens, in order to fill the whole universe' with his gifts and glory.

8   Eph. 1:18-23 (*NIV*)

way that the psalmists, say, spoke of Israel and her king – as distinct if inseparable[9] – in Teilhard this is not at all clear. Christ and church, indeed Christ and cosmos, become confused and run together. This leads inevitably to a distortion of all three, but generates a seductive vision of the church as the cutting edge of evolution. The church, that is to say, helps to pioneer humanity's corporate advance from the material to the spiritual. As such it embraces everything that unites and transcends. It has learned that spirit advances along the same axis as history, that 'onwards' and 'upwards' are the same thing, that the divine is to be found in human progress.[10]

Standing sacramental theology on its head, Teilhard maintains that it is the function of the *church* to build up Christ. Has the church not been busy for two millennia transforming the world, through its daily masses, into an extension of his body? 'As our humanity assimilates the material world, and the Host assimilates our humanity, the eucharistic transformation goes beyond and completes the transubstantiation of the bread on the altar. Step by step it irresistibly invades the universe . . . [For] in a secondary and generalised sense, but in a true sense, the sacramental Species are formed by the totality of the world, and the duration of the creation is the time needed for its consecration.'[11] The church is thus engaged, if we have the courage to believe it, in bringing to birth for God an immense body 'worthy of resurrection'. Her arduous labour will produce a holy offspring of cosmic proportions. Through the church a universal consciousness (such *is* the Christ) will eventually emerge to overcome all the constraints of finitude, and in a final apocalypse make its escape from the prisonhouse of the temporal–material world into the divine fullness or pleroma.[12]

As an attempt to marry the Origenist tradition to today's dominant Darwinism, Teilhard's scheme has the merit of pointing up

---

9 This without prejudice to the doctrine of the *totus christus* and the new ontology – the ontology of union and communion, of deification – which informs the whole orthodox tradition.

10 For Teilhard, the true witnesses or martyrs are the heroes who embody the spirit of what he calls *homo progressivus*.

11 *Le Milieu Divin*, 125. One must not overlook the blasphemous reversal of the eucharistic dynamic; cf. *Hymn of the Universe*, 134: 'To allay your hunger and slake your thirst, to nourish your body and bring it to full stature, you need to find in us a substance which will truly be food for you. And this food . . . I will prepare for you by liberating the spirit in myself and in everything.'

12 See *Hymn of the Universe*, 149, *The Future of Man*, 122f., *The Phenomenon of Man*, 316, etc.; cf. *Ascension and Ecclesia*, 198ff., for a thorough treatment of Teilhard.

the flexibility of the former and the gnostic potential in the latter. Darwinism seeks to describe, even to explain, the emergence of mind from matter; all it lacks is the recollection that matter itself emerged from mind and that the end of the whole process will be like its beginning. The ancient suspicion of matter, indeed, the gnostic animosity towards bodily life, here reasserts itself however, revealing the vaunted Teilhardian love for nature as a consuming lust, if not a secret hatred:

> As the years go by, Lord, I come to see more and more clearly, in myself and in those around me, that the great secret preoccupation of modern man is much less to battle for possession of the world than to find a means of escaping from it. The anguish of feeling that one is not merely spatially but ontologically imprisoned in the cosmic bubble; the anxious search for an issue to, or more exactly a focal point for, the evolutionary process; these are the price we must pay for the growth of planetary consciousness; these are the dimly-recognized burdens which weigh down the souls of christian and gentile alike in the world of today.[13]

In other words, the circularity of pagan cosmology reappears even as Jesus disappears. Salvation is ultimately a denaturing deification, hard-won after many descending and ascending aeons.

The spirit of Teilhard de Chardin is at once grand and silly, triumphalist and despairing. Though the Holy Office warned against it in 1962, it continues to percolate in the liturgical experiments of post-Vatican II liberal Catholicism, breaking out especially among the impatient such as Matthew Fox. It is one of the legion of demons driving the remnants of liberal Protestantism (with which Fox now runs) down a steep slope and into the sea. Universalism and immanentism render it highly susceptible to the *Zeitgeist*, be that communist or fascist (both movements attracted Teilhard) or globalist or transhumanist or what have you. Under its influence the church, like Teilhard's Jesus, becomes infinitely malleable, able to absorb whatever is happening around it but unable to confront itself or its neighbours with the gospel once for all delivered to the saints. Which means, of course, that it soon ceases to *be* the church.

---

13 *Hymn of the Universe*, 138f.

I HAVE CHOSEN AN EXAMPLE with a Catholic (or at least a Jesuit) provenance, but there can be no doubt that Protestants have led the way in the development of this train of thought. Teilhard is as inexplicable without Schleiermacher and Hegel as he is without Darwin. Protestants have also led the way in criticism, however. It was that Danish detective,[14] Søren Kierkegaard, who first flagged the danger. Kierkegaard recognized that the doctrine of the ascension had been falsified through collusion with Enlightenment triumphalism and especially by Hegelianism. He pointed out the link between that falsification and the abatement of Christianity's devotion to particularity – the particularity of Jesus as the God-man, of God as none other than the Father of Jesus, and of the disciple as one whose allegiance to Jesus must single her out. It was Kierkegaard, in other words, who recognized that what is really at stake in modern ascension theology is a basic question of identity. Who is Jesus Christ for us today? Conversely, who are we – who am I – for Jesus Christ today? Kierkegaard pressed such questions with disturbing urgency, even as the walls went up around the cathedral of modernity in which the Religion of Humanity would be conducted.

Kierkegaard knew that there could be no retreat to a naked *theologia crucis*; that would be to concede everything. What was required was a fresh engagement with the double entendre in the dominical saying, 'And I, when I am lifted up from the earth, will draw all men to myself.' Only by maintaining the twofold reference – to the cross first of all, yes, but also to the ascension of the risen one – could the lie of immanence be exposed and both triumphalism and despair be resisted.[15] Otherwise put, Kierkegaard knew that the objective pole of ascension theology was just as crucial as the subjective pole. Indeed, he knew that the most important implications of that doctrine for human subjectivity depended upon its objective truth. Only by insisting on the fact of the resurrection and ascension of the man Jesus

---

14 'I am not what the age perhaps demands, a reformer – that by no means, nor a profound speculative spirit, a seer, a prophet; no (pardon me for saying it), I am in a rare degree an accomplished detective talent' (*Attack upon Christendom*, 33).

15 *Practice in Christianity*, of which Kierkegaard remarks that 'without a doubt it is the most perfect and truest thing I have written' (Princeton ed., xviii–xix), concludes with a series of meditations on Jn 12:32. What sets Kierkegaard apart from Hegel in his treatment of the twin *scandala* to which that passage points – the cross and the ascension – is that he does not conflate them. That is because he holds to the middle term between them; that is, to the good news of the resurrection. 'Christ is the one who died – more than that, who was raised – who also is at the right hand of God . . .' (Rom. 8:34).

could the significance of 'the moment' be re-established in the face
of the threat posed to it by 'history' and the cult of progress. Only
thus could 'the single individual' be rescued from the threat posed by
the category of 'man' or the race.

Kierkegaard saw that, by means of such categories, Christendom
was returning full circle to paganism, with its attendant perils.
According to Hegel, the individual is of 'too trifling a value' compared with the race, something to be sacrificed and abandoned on
the altar of history; but for Kierkegaard 'one is worth more than a
thousand'.[16] The difference is strictly christological. Kierkegaard's
Christ is God as a living man, whereas Hegel's is man as the living
God. Kierkegaard's individual is the person rendered infinitely important through encounter with Christ – the Christ who by his ascension
has suspended history and made his own life what no other life can
be, 'a test for all people'.[17] Kierkegaard's individual therefore stands
out from the race and from its collective history, refusing to take any
false refuge therein. She is a genuine subject in her own right, directly
related to eternity by virtue of her contemporaneity with Jesus.
Hegel's individual, on the other hand, is but a transitory instrument
for the self-realization of Spirit, of which the advancing collective
(especially 'the State' with its universal aspirations) is a higher form
of expression. To stand out is to fall; to assimilate is to rise or ascend.
Both movements may be necessary, to be sure, but the latter is more
necessary. Judaism and Christianity have hitherto emphasized the special, but now all that is special (especially dogma, which is by nature
divisive) 'retreats into the background'.[18] Or as we might say today:
exclusivism is the only sin, inclusivism the only righteousness.

Kierkegaard's critique of Hegel, though more persistent and
profound than Augustine's *en passant* critique of Origen, was also to
mixed effect, however. His emphasis on truth as subjectivity, in the
relational sense required by his christology, was torn away from that
christology and expropriated for other purposes.[19] In this way it fed

---

16   See Hong and Hong, *Søren Kierkegaard's Journals and Papers*, vol. 2, no. 2004ff.; cf. Hegel,
     Philosophy of History, 32f.

17   *Practice in Christianity*, 202

18   *Philosophy of History*, 334f. For Kierkegaard, the choice that has to be made here is not
     (as Hegel would have it) a choice between religion and philosophy or statecraft, that is,
     between a lower and a higher form of Spirit, but between Christianity and paganism.

19   Heidegger and Sartre were particularly predatory; so also Derrida, who was unable to
     adopt Kierkegaard's own christological ground *coram Deo*.

rather than arrested the growth of immanentism, which among the Romantics and those of gnostic bent was beginning to take openly pagan forms.[20] Nor was that emphasis alloyed, as it is in Orthodoxy and in Catholicism, with an ontologically robust ecclesiology, or with any obvious alternative to the compromised ecclesiology of his opponents in the Danish People's Church. Kierkegaard confessed that such an ecclesiology was required,[21] but he himself was, after all, a detective rather than a reformer. What he detected was apostasy. 'Certainly things will be reformed', he ventured in his journal,

> and it will be a frightful reformation compared with which the Lutheran reformation will be almost a joke, a frightful reformation that will have as its battle-cry, 'Whether faith will be found upon earth?' and it will be recognisable by the fact that millions will fall away from Christianity, a frightful reformation; for the thing is that Christianity really no longer exists, and it is terrible when a generation which has been molly-coddled by a childish Christianity, fooled into thinking it is Christianity, when it has to receive the death blow of learning once again what it means to be Christian . . .[22]

The hope of reintroducing Christianity into Christendom, as he put it, was about as far as his ecclesiology got.

A century later Dietrich Bonhoeffer, having witnessed a disastrous apostasy in Germany, was wrestling anew with the basic question. 'What has been bothering me incessantly', he wrote from his Nazi prison cell in 1944, 'is the question what Christianity really is, or indeed who Christ really is, for us today.'[23] The church has always to ask this question, and has done so again and again in the course of the last two thousand years. But it has never allowed what today is rampant in mainstream Protestantism and in liberal Catholicism. That is, it has never allowed that it is possible to give any answer that contradicts the witness of Hebrews to the ascended one: 'Jesus Christ is the same yesterday, today, and forever.' Such a contradiction

---

20 Wagner (whose *Parsifal* was purportedly conceived on Good Friday in 1857, two years after Kierkegaard's death) comes to mind, for example.

21 See further *Ascension and Ecclesia*, 226ff.; cf., e.g., John Zizioulas, *Being as Communion*, chap. 2.

22 A. Dru, *The Journals of Søren Kierkegaard*, no. 1407 (1854). As for Luther, he may have been a reformer but he was 'the absolute opposite of an "apostle"' (Dru, no. 1406).

23 *Letters and Papers from Prison*, 30 April 1944.

is what constitutes apostasy. When the immanentist asks this question, however, 'Who is Jesus Christ for us today?' really means 'Who do we *say* today that Jesus is?' And almost any answer at all might be forthcoming – except a truly martyrial answer such as Kierkegaard and Bonhoeffer aspired to give – because what is aimed at is not contemporaneity with Jesus but, in Ernst Troeltsch's phrase, 'to grasp the divine as it presents itself to us in our time'.[24]

For grasping the divine as it appears in our time, faith in a Jesus who lived two thousand years ago might be useful, but not faith in a Jesus who is the same yesterday, today and forever. *That* faith is no longer possible. In fact, it positively inhibits any grasp of the divine, whether as universal providence ('history', in Hegel's sense) or as self-realization (worship of the God that is in me, to paraphrase Schleiermacher).[25] It is the ahistorical faith of the fundamentalist, who is opposed to progress and to liberty.[26] Conversely, it is the faith of one who imagines that there was a real fall and a unique act of redemption, a demonic and a divine intervention in time. Such a one may believe in the ascension of Jesus to the Father and in his impending return, just as she believes in a lost righteousness and in original sin. But with such beliefs she can take no meaningful part in the modern world, which she has thereby bracketed off.

There is a choice to be made here, a choice that cannot be evaded. Who is Jesus Christ for us today? Is he someone to remember only, or someone to remember and to expect? If only to remember, whom or what shall we expect? Who is the 'Christ' who will come? The similarity of our situation to that of Israel at Sinai should strike us immediately. While Moses was absent on the mountain, hidden away in some secret recess where no one else could go, Israel deliberated. What had happened to him? Was he gone for good? Should he be awaited or left behind? If left behind, who would lead? Would Israel be identified by its newly confirmed covenant with Yahweh or would

---

24 *Writings on Theology and Religion*, 206; cf. Mt. 16:13ff., Heb. 13:8ff.

25 From the final line of the fifth speech, in which Schleiermacher argues that 'the whole of religion is nothing but the sum of all relations of man to God, apprehended in all the possible ways in which any man can be immediately conscious in his life' (*On Religion*, 253).

26 John Hick compares the christocentric faith of Heb. 13:8 to the outmoded geocentric cosmology. It must give way to a 'heliocentric' faith in God *qua* God, which can be shared by everyone. The price to be paid for this, of course, is that exacted by gnosticism: God becomes unnameable and unknowable. Cf. John Bowden, *Jesus: The Unanswered Questions*, 173.

it march under the banner of one of the fertility gods of Egypt or Canaan? Most western Christians, of course, are not thinking about such things, nor do they see the need for taking any very important decisions of this sort. Their leaders are not occupied by the question that so obsessed Dietrich Bonhoeffer. No doubt that has something to do with the fact that their circumstances are not yet quite so dire, but it also has something to do with a faulty understanding of the ascension, the same that Kierkegaard warned us against.

The doctrine of the ascension, properly understood, does not articulate an optimistic faith in an era of progress in unity or equality, and of advance towards God. It articulates the most primitive and the most costly of all Christian confessions: Κυριος Ιησους, Jesus is Lord. Pastorally speaking, its primary purpose is to put to the church the Sinaitic question, which is the question of faithfulness to its Lord in the time of his hiddenness.[27] Faithfulness can have no meaning, however, where there is no continuity of identity. 'Men of Galilee,' asked the angels, 'why stand you gazing up into heaven? This same Jesus, who is taken up from you into heaven, shall so come in like manner as you have seen him go into heaven.' Or as Hebrews puts it, 'Christ was sacrificed once to take away the sins of many people; and he will appear a second time, not to bear sin, but to bring salvation to those who are waiting for him.'[28] Only as the church persists in expectation of this same Jesus will it prove faithful. Only as it bears witness to both the ascension and the parousia will it recognize our age for what it is – a test – and reject the seductive vision of Origen's modern heirs, who would have us believe rather in an era of ever-increasing divine presence.

---

27 That was the question so forcefully put – and answered – by the famous Barmen Declaration of 1934, to which Bonhoeffer was a signatory. The whole issue between the pro-Nazi German Christian movement and what came to be called the Confessing Church was lodged (as Karl Barth, the Declaration's primary author, made clear) in the question of the authority of Jesus. See chapter six.

28 Heb. 9:28 (*NIV*); cf. Acts 1:11 (the King James Version captures well the force of ουτος ο Ιησους).

CHAPTER FIVE

Presence in Absence

ASCENSION THEOLOGY TURNS at this point to the eucharist, for in celebrating the eucharist the church professes to know how the divine presents itself in our time, and how the question of faithfulness is posed. Eucharistically, the church acknowledges that Jesus has heard and answered the upward call; that, like Moses, he has ascended into that impenetrable cloud overhanging the mountain. Down below, rumours of glory emanate from the elders, but the master himself is nowhere to be seen. He is no longer with his people in the way that he used to be. He has gone to the Father. Yet he *is* with them, in the Spirit. He has not abandoned them or left them orphaned in their time of testing. They themselves have a place with him on the mountain, which they ascend at his invitation in their holy feasts. Before 'angels and archangels and the whole company of heaven' they eat and drink together in the very presence of God. They eat and drink in his memory and in his honour and in anticipation of his return. They hear his word and receive anew their commission to be his witnesses in the world. They know him as their Moses, their Aaron, their David, and they rejoice over him:

> Alleluia! King eternal, Thee the Lord of lords we own;
> Alleluia! born of Mary, Earth Thy footstool, Heav'n Thy throne:
> Thou within the veil hast entered, robed in flesh our great
>   High Priest;
> Thou on earth both priest and victim in the Eucharistic feast.[1]

In other words, the church becomes with Jesus a community of ascension and oblation, sharing in his heavenly offering to the Father, and manifesting the Spirit who reorganizes created reality around him. It is in the eucharist that the church's identity is properly established, then, and its relation to the world decided.

The twin mysteries of the ascension and the eucharist are joined together already in a proleptic passage from John's Gospel, where objections to the latter draw the retort from Jesus: 'Does this offend you? What if you see the Son of Man ascend to where he was before!'[2] But many, unable or unwilling to receive these mysteries, turn back from following him; only Peter and his colleagues remain. An interpretive template for this passage can be found in the story of Elisha,

---

1   William C. Dix, *Altar Songs, Verses of the Holy Eucharist*, 1867 (fourth stanza).
2   See Jn 6:25ff., especially verses 53-58.

whose request for a double portion of Elijah's spirit is to be answered only if he witnesses Elijah being taken from him into heaven. Elisha therefore refuses to turn back from following his master, accompanying him over the Jordan to his rendezvous with the heavenly chariot, where he catches up the cloak that falls from Elijah's shoulders.[3] So, too, with Jesus' disciples. Only those who 'see' Jesus ascend are ready to receive the gift of the sacraments, which, as the Venerable Bede somewhere remarks, are like the miracle-working mantle that Elijah left behind.

In what sense and to what end is the eucharist, which is the condition of possibility for the other sacraments, left behind by the ascending Jesus? First, it is left behind as a lifeline to his own person and heavenly resources. The ascension, viewed from below as the incomprehensible absence of Christ – the divergence of his history from ours that leaves us gazing, dumbfounded, into the heavens – creates the eschatological tension that characterizes the present age and makes it a time of testing. The eucharist, on the other hand, through the equally incomprehensible presence *in* the absence – the conjoining of our histories to his in a communion of body and soul – provides that the present age should not be altogether without Christ and so without hope or experience of the age to come. 'Unless you eat the flesh of the Son of Man and drink his blood, you have no life in you.'

Second, the eucharist is left behind as a witness to the world of what actually happens in the ascension, namely, that the entire cosmos is fundamentally reordered to God in Christ. Ephesians tells us that the appearance of the church – that is, of the eucharistic community of Jew and Gentile, male and female, slave and free – is the sign of this eschatological fact. It is a signal even to the heavenly powers that something has changed profoundly as a result of the ascension; a signal to the earthly powers also that their situation is disconcertingly different than it was before. The appearance of the church brings to light 'the plan of the mystery hidden for ages in God who created all things', the plan to unite all things under one head.[4]

---

3   See 2 Kgs 2:1-17.

4   'Remember that at that time you were separate from Christ, excluded from citizenship in Israel and foreigners to the covenants of the promise, without hope and without God in the world. But now in Christ Jesus you who once were far away have been brought near through the blood of Christ . . . In him, and through faith in him, we may approach God with freedom and confidence' (Eph. 2:12-13, 3:12, *NIV*; cf. 4:18): In Ephesians, as in Hebrews or the Apocalypse, this language – 'brought near', 'approach', etc. – has ritual

Its local liturgy (which in the Great Thanksgiving sums up this plan) is itself a synaxis that anticipates the fact that all things are to be united in and under Christ, 'things in heaven and things on earth'.

Third, the eucharist is left behind as the means of participation in the offering that Christ in his ascension presents to the Father – himself and the people whose redemption he has won – and in all the benefits that flow from the Father's reception of that offering. It is left behind as the residue of earthly sacrifice and the hint of heavenly glory. If indeed Christ is present in his absence, present in a manner distinct from the parousia that is yet to come, he is present precisely in his freedom to include us in his offering and in his glory. He is present, that is, in his freedom to create by the Spirit a community of ascension and oblation, which is what the church is. 'Is not the cup of thanksgiving for which we give thanks a participation in the blood of Christ? And is not the bread that we break a participation in the body of Christ?'[5]

LIKE THE ASCENSION, however, the eucharist has been an occasion of offence within the church as well as outside it; large numbers of Christians no longer see in the eucharist what the Venerable Bede did. Doubts about the eucharist in John's day may have troubled potential converts from Judaism or from paganism, but today they trouble many even of the baptized. The sixteenth-century reformers are partly responsible for this, of course. Not that they themselves doubted that the eucharist somehow linked the church to its ascended Lord, or that such a link was necessary, but they did doubt, indeed they denied, the doctrines by which the church had articulated its faith in the gift of communion in Christ's body and blood. Talk of transubstantiation seemed to them to undermine the dialectic of the presence and the absence – the element of deferral in the prayer 'Thy kingdom come' – and so to impute to the church an unqualified power and authority that do not rightly belong to it in this world. Talk of oblations or sacrifices seemed to obscure the once-for-all nature of what God had accomplished in Christ, and so to challenge the *sola gratia* principle that for other reasons they had

---

connotations. It lends itself quite naturally to the notion that the church's eucharist is a participation in the self-offering of Christ to the Father, which begins on earth but is completed in heaven.

5   1 Cor. 10:16 (*NIV*)

found it necessary to defend. Turning away from such talk, they unintentionally turned many away from the eucharistic mystery itself, for which they substituted a subjectivism that Hegel approvingly labelled the principle of inwardness or 'spiritual enjoyment'.[6]

It may be helpful here to go back behind the Reformation controversy and revisit the eucharistic teaching of Irenaeus, an heir to the Johannine tradition who has proved himself a worthy guide. Irenaeus can already be found exploring the link between the ascension and the eucharist, and thinking about the role of oblations in the renewal of the world under the new covenant. Just as the Word of God gave orders to the people at Sinai to build a tabernacle and to offer up once again the firstfruits of creation, he argues, so did that same Word leave instructions with his disciples that the oblation of the church should be offered in every place throughout the world. Irenaeus, like Jesus himself, lays much stress on the inward disposition of the offerer,[7] setting great store by the liberty of spirit in which the Christian oblation is made. But he does not make the mistake of pitting the spirit against the flesh, or the New Testament against the Old, as if Christ had brought sacrifice and offering to an end in bringing the Levitical system to an end. That would introduce a gnostic discontinuity into salvation history and into the very nature of our humanity, dividing us from God rather than uniting us to him. 'The class of oblations in general has not been set aside; for there were oblations there, and there are oblations here. Sacrifices there were among the people [of Israel]; sacrifices there are, too, in the church: but the species alone has been changed, inasmuch as the offering is now made, not by slaves, but by freemen.'[8]

It is God's will, insists this early bishop, 'that we, too, should offer a gift at the altar, frequently and without intermission' (adding, lest anyone miss the eschatological nature of the act, that the altar is in heaven where our Lord is preparing a place for us).[9] Why?

---

6   See *Ascension and Ecclesia*, 188f.

7   See, e.g., Mk 12:33; cf. *Against Heresies* 4.16.5.

8   *Against Heresies* 4.18.2. Cf. 4.17.5 and Heb. 10:19-22 (*NIV*): 'Therefore, brothers, since we have confidence to enter the Most Holy Place by the blood of Jesus, by a new and living way opened for us through the curtain, that is, his body, and since we have a great high priest over the house of God, let us draw near to God with a sincere heart in full assurance of faith, having our hearts sprinkled to cleanse us from a guilty conscience and having our bodies washed with pure water.'

9   *Against Heresies* 4.18.6. Irenaeus draws here on the Apocalypse, in which light (rather than

Not because God stands in need of our offering, but because *we* stand in need of making one. Sacrifice is for our sake, not God's. It allows us to participate in the divine nature, which is one of giving and receiving. And since God has no need of sacrifice, the only really fitting sacrifice is the one that is offered in pure freedom. Such freedom can only come to us from God, of course, who therefore lends to us so that we will have something to return to him and will not appear in his presence empty-handed. It is God's intention thus to place himself in our debt, as it were, so that he may pay this debt with interest, 'counting out the increase'.[10] What he lends to us to begin with is the creation, from which we are to offer the firstfruits. But he also lends to us his own Son. Jesus Christ, who alone is capable of offering himself in perfect freedom, both as a sin-offering and as a thank-offering, *is* the firstfruits of creation. He, then, is what we offer, as well as the one through whom we offer. Creatures of bread and wine are brought forth by the church from the increase of the earth, but when 'they receive the Word of God', by the power and grace of God 'the eucharist of the blood and body of Christ is made'.[11]

This eucharist, continues Irenaeus, sets forth before all people an *indicium libertatis*, a token or proof of liberty. Not simply because it can be and is offered throughout the whole world – so that in every place, and not only in Jerusalem, people should have access to heaven – but because it is offered by those who really are free; free not only from Egypt, but from a defiled conscience and spiritual oppression, from political impotence and social fragmentation, from the fear of death and the threat of Sheol. By bringing our human nature, once for all, into the presence of the Father, Christ liberates us for life in the Spirit. The Spirit in turn causes us to cohere around Christ, so arranging his members as to create from their various relations the 'many mansions' of the Father's house that Jesus has gone to make ready. In this way, by being fitted together ecclesially, we ourselves are made ready to bear the fullness of the divine presence and so to receive the gifts of which we are otherwise incapable, including the gift of immortality. Rendering thanks for these gifts in and through Christ, we fulfill the purpose of God's people both doxologically and

---

that of a naïve cosmology) should also be understood our liturgical prayer, 'Almighty God, we pray that your angel may take this sacrifice to your altar in heaven.'

10   Cf. *Against Heresies* 3.17, 4.14, 4.18.
11   *Against Heresies* 5.2.3

ethically; that is, we fulfill the law of liberty. For God is not doing away with the law, says Irenaeus, but 'fulfilling, extending, and widening it among us', and counting out the increase of life eternal.[12]

Approached in this way, it is difficult to see how the eucharist could ever be reduced to a mere act of remembrance or of inner enjoyment, as it has in many Protestant communities and among poorly taught Catholics. Such a reduction can lead nowhere but to the very Pelagianism that the reformers feared and hoped to eradicate. Where there is a refusal to allow that a real offering of Christ is made by the church in, with and through Christ, it is allowed instead (implicitly or explicitly) that some other offering should take place. Where it is not Christ himself who both offers and is offered, something else must be offered and someone else must do the offering. Where there is not a real communion in his body and blood, where our offering is not conjoined to his and mediated by his through that communion, we ourselves become the primary offerers and offering. That is a theological, moral and material mistake of the utmost consequence, and I will say more about it later on.

But what of transubstantiation, to the denial of which the drift into subjectivism and thence into doxological Pelagianism was closely tied? Debates about transubstantiation might have come out differently if the western tradition had paid more attention to the eschatological features of the eucharist; indeed, if it had paid more attention to the eschatological nature of the church itself, the mystical body that with Christ's own body is the reality (*res*) in the eucharist.[13] A lack of eschatological analysis left room in practice for the tendency that worried the reformers – the tendency to fetishize pure presence through liturgical and devotional practices not entirely transparent to the gospel. They were right to be worried; of that there is ample evidence from the Counter-Reformation, in the rise of orders such as the Capuchins, and indeed in conciliar decisions from Trent to Vatican II, which sought in a more balanced way to correct

---

12 '[For] the more extensive operation of liberty implies that a more complete subjection and affection towards our Liberator had been implanted within us. For he did not set us free for this purpose, that we should depart from Him (no one, indeed, while placed out of reach of the Lord's benefits, has power to procure for himself the means of salvation), but that the more we receive his grace, the more we should love him. Now the more we have loved him, the more glory shall we receive from him, when we are continually in the presence of the Father.' (*Against Heresies* 4.13.3)

13 Cf. Aquinas, *Summa Theologiae* 3.73.1, obj. 2.

the abuses. Unfortunately, the reformers' own response once more exacerbated, rather than solved, the problem.

Both sides in the controversy faced the same challenge: how to maintain the eucharistic dialectic and, with it, the proper relation between church and world. If the Catholic temptation in insisting upon transubstantiation was (as the reformers thought) to underestimate the absence of Christ, it was not thereby to overestimate the presence, but rather to misconstrue the dialectic. And if the Protestant temptation in denying transubstantiation was to underestimate the presence, it was not thereby to overestimate the absence, but again to get the dialectic wrong. Without adequate recognition of the real absence of Christ, the church itself has no real absence; knowledge of his presence renders it prone to self-glorying and to illusions of worldly power, to making martyrs of others rather than walking the path of martyrdom itself. Without the real presence of Christ, on the other hand, the church has no real presence either. It is not sufficiently potent in or against the world; it falls prey first to sectarianism, then to Erastianism.

The Protestant temptation should be judged the more pernicious, however, because it is not, like its counterpart, self-correcting; indeed, it is self-defeating. Denying transubstantiation, Protestantism denies the specificity of the real presence, and in denying the specificity of the real presence it denies the specificity of the church, beginning of course (but not ending) with its Petrine discipline. In the search for new or substitute forms of specificity it divides and multiplies itself into many denominations, each claiming its own authority; that is, into orders or movements not subject to the church. These compete and decay and die just like ordinary human organizations. In so far as they are motivated by the gospel they are not ordinary, of course, but ecclesial. Sooner or later the Erastian element asserts itself, however, and political or cultural compromise ends in utter confusion between church and world. For where absence goes unchallenged by an eschatologically decisive presence – head and members in true communion of body and soul – 'presence' will eventually be found wherever we choose to find it, even in secular principalities that demand conformity from the church and the silencing of its gospel. Against such principalities Protestant martyrs continue to give witness to this day; yet it is doubtful that the Protestant temptation can be resisted without recourse to something very like the doctrine of transubstantiation, in a concession that must finally undermine Protestantism as such.

But what about the Catholic temptation? Can transubstantiation be affirmed without injury to the dialectic of the presence and the absence? If we are thinking eschatologically, the answer is yes, though it requires some patience to hear it.

Let us begin by observing that in the Catholic faith the real presence means (as the reformers also believed) our presence in Christ with God, not merely Christ's presence with us. 'Truly this is the *mysterium fidei* which is accomplished in the Eucharist: the world which came forth from the hands of God the Creator now returns to him redeemed by Christ.'[14] The eucharist, in other words, like baptism, is a participation in Christ and therefore in his whole redeeming activity from conception to cross, from descent to ascent, until at his parousia all is offered up to the Father in such a way that God is truly 'all in all'. The eucharist is also, however, in distinction from baptism, a means of sharing directly in Christ's intercession before the heavenly altar. The real presence effected in the eucharist that is celebrated on the earthly altar *is*, as the liturgy indicates, a presence in and with Christ in heaven, where he stands before God as our great high priest.[15] Acknowledging that, the reformers objected to the notion that Christ

---

14 John Paul II, *Ecclesia de Eucharistia* [On the Eucharist in its Relationship to the Church, 2003] §8.

15 Drawing on the Apocalypse as well as the Gospel of John, Dix's famous Ascensiontide hymn points us in the right direction. The first stanza conveys us immediately into the company of those who worship God and his Christ in heaven:

> Alleluia! sing to Jesus! His the scepter, His the throne.
> Alleluia! His the triumph, His the victory alone. -
> Hark! the songs of peaceful Zion thunder like a mighty flood.
> Jesus out of every nation has redeemed us by His blood.

The following stanzas insist that members of the church militant do indeed belong to that heavenly company by virtue of Christ's determination to remain in their midst:

> Alleluia! not as orphans are we left in sorrow now;
> Alleluia! He is near us, faith believes, nor questions how;
> Though the cloud from sight received Him when the forty days were o'er
> Shall our hearts forget His promise, 'I am with you evermore'?

> Alleluia! bread of angels, Thou on earth our food, our stay;
> Alleluia! here the sinful flee to Thee from day to day:
> Intercessor, Friend of sinners, Earth's Redeemer, plead for me,
> Where the songs of all the sinless sweep across the crystal sea.

The final stanza – sometimes omitted by Protestants for want of a Mass, or by Catholics in a hurry to leave Mass – leads us, as we saw earlier, to the earthly altar where the heavenly manna is received. Meanwhile, however, Dix slides over the controversy about transubstantiation with 'faith believes, nor questions how'.

should be brought down from heaven to the earthly altar; but there is in fact no bringing Christ down that is not by its very nature a lifting up of that to which he comes down. 'Christ in you, the hope of glory', in Paul's language: that is what the *hoc est corpus meum* and *hic est sanguis meus* promise and deliver.[16]

This is the place to remind ourselves, however, that the dialectic of the presence and the absence cannot be content with the symbolic language of up and down, which, though useful, is a mere concession of speech, corresponding to the concession of physical movement in the drama of Jesus' departure from his disciples.[17] The dialectic of the presence and the absence is about conversion or transformation, about the power of the Spirit to bring into being a new creation out of the old; it is not about movement from one place to another within the present creation. Because the eucharist is an advance on the new creation, its theological articulation can and must take the form of an eschatological proposition, as it does, for example, in this passage from Irenaeus:

> Just as a cutting from the vine planted in the ground fructifies in its season, or as a corn of wheat falling into the earth and becoming decomposed, rises with manifold increase by the Spirit of God, who contains all things, and then, through the wisdom of God, serves for the use of men, and having received the Word of God, becomes the Eucharist, which is the body and blood of Christ; so also our bodies, being nourished by it, and deposited in the earth, and suffering decomposition there, shall rise at their appointed time, the Word of God granting them resurrection to the glory of God, even the Father, who freely gives to this mortal immortality, and to this corruptible incorruption . . .[18]

Which is to say, the eucharistic mode of Christ's presence is itself

---

16   What does 'Christ in you' mean (Col. 1:27), if not that 'you have been raised with Christ' so that you may 'appear with him in glory' (cf. 3:1-4)? Though some doubt the Pauline authenticity of Colossians on the grounds that Paul himself had no such realized eschatology as this, that is simply to misunderstand the dialectic, in which realization and reserve – 'glory' has a future as well as a present connotation – are both operative.

17   The reformers, of course, thought that the *hoc est* itself was a mere concession of speech, which worried even Luther.

18   *Against Heresies* 5.2.3

eschatological, and transubstantiation, rightly understood, is an eschatological concept.

According to the Catholic faith, communion with the ascended Christ is accomplished through a variety of means, each dependent on an advance or down payment of the Holy Spirit, who is the Spirit of the kingdom that is coming with the new creation. But all of these means, sacramental or otherwise, are supported by way of the Spirit's conversion of bread and wine into Christ's body and blood, so that 'the mystery of unity' may be fulfilled even now as we 'receive from his (nature) what he himself received from ours'.[19] That the eucharist deepens participation in what Calvin called the *mirifica commutatio* is not in dispute:

> This is the wondrous exchange made by his boundless goodness. Having become with us the Son of Man, he has made us with himself sons of God. By his own descent to the earth he has prepared our ascent to heaven. Having received our mortality, he has bestowed on us his immortality. Having undertaken our weakness, he has made us strong in his strength. Having submitted to our poverty, he has transferred to us his riches. Having taken upon himself the burden of unrighteousness with which we were oppressed, he has clothed us with his righteousness. To all these things we have a complete attestation in this sacrament, enabling us certainly to conclude that they are as truly exhibited to us as if Christ were placed in bodily presence before our view, or handled by our hands.[20]

What is in dispute is the change that takes place in the bread and wine when, as Irenaeus puts it, they receive the Word of God. Is the change in question one of signification only – note Calvin's 'as if' – or is it one of substance? Otherwise put, do bread and wine really become Christ's body and blood, or only serve as pointers to a participation in his body and blood that is achieved through some more secret, 'spiritual' means?[21]

---

19 Lateran IV (H. Denzinger, *The Sources of Catholic Dogma* §430). 'Receive from his (nature)' should not be understood too narrowly as referring to Christ's divinity only, but to his whole character as the God-man and redeemer, our true priest.

20 *Institute* 4.17.2f.

21 An assumption whose subscribers were anathematized at Trent (Denzinger §890; cf. §878), but which lay behind Anglicanism's controversial Black Rubric, for example.

With a view to safeguarding the point that union with Christ is 'in truth of substance' and not merely 'in sign or power',[22] and that truth of substance includes the substance of flesh and blood, the church has always affirmed the former. That is why it has found it necessary to speak of transubstantiation and hence of a suspension of the ordinary relations between what a thing is (its substance) and how it presents itself (its accidents or appearances).[23] But might we not ask whether its understanding of the miracle of transubstantiation should be pressed further in an eschatological direction? Should it not be said, in fact, that this miracle does something more than make Christ present bodily in a way that overrides the bonds between substance and accidents, bodies and spaces, etc.? May it not be admitted that in converting bread and wine into body and blood the Spirit brings forward the regeneration of all things and in some fashion redeems spatio-temporal reality as such? Should we not move, that is, from arguments that still suggest something akin to an up-and-down mentality to those that reflect more fully an old creation/new creation mentality?[24] For we know that the eucharistic presence of Christ aims at, and actually effects in the faithful, a second *conversio* along with the first – a still more wonderful *conversio*. It aims, as Irenaeus indicates, at a transformation of the communicants, assimilating their fallen here and now to the glorious there and then of Jesus Christ, by means of a process that is only complete in the resurrection of the dead and the life of the world to come.[25]

---

22 Berengarius' oath (Denzinger §355).

23 See especially *Summa Theologiae* 3.75–77. 'Christ's body', says Aquinas, 'is not in this sacrament in the same way as a body is in a place, which by its dimensions is commensurate with the place; but in a special manner which is proper to this sacrament' (3.75.1, *ad* 3).

24 Only by a series of the most severe mistakes and willful misrepresentations can one regress from orthodox eucharistic theology to the sort of caricature one finds, e.g., in Joyce's *A Portrait of the Artist as a Young Man*, when Stephen is told that 'the great God of Heaven' is made to 'come down upon the altar and take the form of bread and wine' (171). Yet the tradition supplies considerable evidence (cf. Denzinger §§578–80) that mistakes abounded that might have been avoided by way of a more deliberately eschatological account of the eucharist.

25 Cf. Col. 3:1ff. with *Against Heresies* 4.18.5: 'For as the bread, which is produced from the earth, when it receives the invocation of God, is no longer common bread, but the Eucharist, consisting of two realities, earthly and heavenly; so also our bodies, when they receive the Eucharist, are no longer corruptible, having the hope of the resurrection to eternity.' Though Irenaeus wrote this long before it had become necessary for the church to resist various forms of offence at the eucharist by employing the term 'transubstantiation', we need not interpret him as do those who think that he may be taken in support

Now I do not mean to overlook the fact that there is an important difference between these two conversions. The conversion of the communicant is not, like the conversion of the bread and wine, a transubstantiation. It concerns *how* something exists, not whether it exists; the communicant is not changed into something else but receives, through the eucharistic nurture of his or her baptismal union with Christ, a new principle, power and mode of existence. That new mode of existence is based on a 'consubstantiation', if we may put that unfortunate Lutheran term to a good use. It rests, that is, on the Spirit's work of generating and perfecting a communion of being with Christ – of generating the church.[26] Against the Lutheran use it must be said that bread and wine do not enter into a communion of substance with Christ, people do.[27] Bread and wine, being useful but dispensable things, are transubstantiated; believers are consubstantiated, and in being consubstantiated they are also resurrected. For they cannot be joined to Christ without that *conversio* which is a passing with Christ from the old creation to the new. But is it not in the service of that joining, and that passing, that the eucharist is celebrated? 'The substantial conversion of bread and wine into the body and blood', writes Pope Benedict XVI, 'introduces within creation the principle of a radical change, a sort of "nuclear fission", to use an image familiar to us today, which penetrates to the heart of all being, a change meant to set off a process which transforms reality, a process leading ultimately to the transfiguration of the entire world, to the point where God will be all in all.'[28] We are right, then, to attempt to understand transubstantiation itself in a time-transforming and eschatological way.

---

of something like consubstantiation or impanation. This fails to take seriously the 'so also'.

26 It is in this sense that we should understand the claim that 'the partaking of the body and blood of Christ does nothing other than make us be transformed into that which we consume' (*Lumen gentium* §26, quoting St Leo the Martyr, Sermon 63, 7), for through the eucharist the church itself is constituted as the body of Christ and the Christian's union with Christ is nurtured (cf. Denzinger §698).

27 See in this connection Denzinger §§1843–46.

28 *Sacramentum caritatis* [Sacrament of Charity, 2007] §11. We may ask here why this should be said of the eucharist and not already of baptism. Baptism brings a person into the sphere of the Spirit (making possible what John Zizioulas calls the 'ecclesial hypostasis' and what I am calling 'consubstantiation') where this transformation is taking place through union with Christ. But in baptism water is not transubstantiated, as are bread and wine in the eucharist, for it is not the Spirit but the Son (to whom the Spirit leads us) who is incarnate for our salvation.

What difference does that make? It does not necessarily require us to let go of the substance–accidents language of Thomas Aquinas, language which the church has seen fit to employ in protection of the *de fide* assertion that in the eucharist there is a real eating and drinking of the body and blood of Christ. Neither does it require us to let go of the before-and-after distinction, marked in Catholic liturgies with the ringing of a bell, by which the church has tried to make clear that the conversion of substance is a consequence of the act of consecration. Both distinctions serve to repudiate sacramental nominalism and to defend the priority of grace. Yet there is also a danger in both – the danger that the conversion in question is subjected to a concept of substance, or of time, that is not itself subject to the truth of the resurrection and ascension. At the consecration bread is wholly converted to body in the twinkling of an eye, just as we who receive that body will, in the twinkling of an eye, be wholly conformed to Christ at his coming in glory. But since both comings (the coming of the Word in the consecration and his coming in the parousia) are comings from heaven, that is, from the new creation, neither is explicable in terms of the old creation.[29]

We must be careful, then, with our concepts and categories – with our up and down, our before and after, our substance and accidents, and so forth – and see to it that they do not control our eucharistic theology and sacramental piety in a naïve or uncritical way. The eucharistic coming, the coming that converts bread and wine to body and blood, is a coming in hiddenness, perceived only by faith. To say that the substance is converted is to say: body and blood are given to us here and now, where to the perception of the senses there is only bread and wine. To say that only the accidents remain is to say that this 'here' and this 'now' are no longer the common here and now, but the here and now of the ascended one as he embraces us, the here and now that we will discover fully when we awake on the resurrection morn.[30]

---

29   Cf. *Catechism of the Catholic Church* §§1402–05.

30   It is not the case, I hasten to add, that consecrated bread and wine remain in the old creation, having all the usual properties of bread and wine, even as we, in receiving them, remain in the old creation, having all the usual properties of mortals. They don't and we don't. Faith knows them to be the body and blood of Christ, just as faith knows the communicants really to be renewed as members of Christ and as participants in his heavenly session. And in the resurrection our new mode of being and experience will reveal both these things to have been true. Otherwise it might be objected that what

We may pause at this point to benefit from the insight of the late Fr Herbert McCabe, the Dominican theologian who has argued that many Catholics, as well as many Protestants, have misunderstood the Thomistic and Tridentine teaching about transubstantiation. That teaching, though expressed with the language of Aristotle, can no more be understood within Aristotelian categories and presuppositions than can the doctrine of *creatio ex nihilo*, say, or the doctrine of the incarnation itself. The consecration of bread and wine effects no mere chemical change, like the conversion of water into wine. It is not a change of substance in that sense. 'The bread does not turn into the body by acquiring a new form in its matter; the whole existence of the bread becomes the existence of the living body of Christ.' The consecration effects a change at the much deeper level of *esse* or existence, in other words.[31] 'Something has happened as profoundly different from chemical change as creation is. It is not that the bread has become a new kind of thing in this world: it now belongs to a new world.'[32] It belongs, that is, to the kingdom of God, to the risen Christ whose personal presence and self-giving activity constitute the kingdom. Christ 'does not literally "come down" on altar after altar', insists McCabe. The conversion of the bread and wine is not merely 'a re-adjustment of our world' but the sacramental means by which there is, proleptically, an advent of the world to come.

---

we are really saying is that bread and wine are no more the body and blood of Christ here and now than we ourselves are resurrected here and now, and, therefore, that we have fallen unintentionally into the worst sort of eucharistic nominalism. God forbid. Transubstantiation, as I am trying to understand it, effects the conversion of what we are here and now into what we shall be in the resurrection through the prior conversion of bread and wine into body and blood. All that needs to be said to defend against this objection is said with that 'prior'. The liturgy as tradition has given it to us is fully warranted by that 'prior', which itself is warranted by the distinction between the ascension and the parousia, between the heavenly session and the last judgement. Our conversion rests on the conversion of the bread and the wine, not *vice versa*, though the mystery of both conversions is an eschatological one.

31 *God Still Matters*, 119 (see 115ff.). That, he claims, is what is meant when it is said 'that there takes place a change of the whole substance of the bread into the body of Christ our Lord' (Council of Trent; *Catechism of the Catholic Church* §1376). But to shift the focus from *substantia* to *esse* may beg certain questions about the *conversio*, namely, whether it does indeed involve a conversion from one thing to another as well as this change from old creation to new.

32 'In fact, St Thomas says it is not a change (*mutatio*) at all, for such a change means a re-adjustment of our world – as when one thing is altered or changes into something else' (*God Still Matters*, 120).

McCabe's account, which is not afraid to appeal to miracle and to mystery, avoids any lapse into nominalism while also avoiding the mistakes of a crude materialism. The appearances of the consecrated elements, he says, are not to be thought of as accidents detached from their proper substance, or as the accidents of one substance arbitrarily attached to another, so as to disguise the latter and make it more palatable. Rather they cease to be appearances at all and become sacramental signs. This way of putting the matter, however, may not be entirely satisfactory. Should we say only that the appearances become sacramental signs? Or should we say rather that bread and wine themselves become sacramental signs as they are converted into what they were not, *viz.*, the body and blood of Jesus Christ? When we remember that this conversion takes place not so much *in* our time (as would a chemical change) but *to* our time, as a transformation and redemption, we may ring the bell in all devoutness without implying any actual separation of substance and accidents.[33] The whole act of blessing and sharing, consecrating and communing, becomes an eschatological participation in the body and blood of the saviour, both as he goes to the cross and as he goes to the Father, and indeed as he comes again in glory with resurrection power.

McCabe's eucharistic dynamism, so interpreted, captures the best insights of Calvin, who tried valiantly in his eucharistic theology to hold together *signum* (the promises implicit in the signs and explicit in the gospel), *res* (the reality of Jesus Christ in his death and resurrection) and *virtus* (his redemptive power). But it does so without rejecting or neglecting the real conversion of bread and wine into body and blood. This is vital, not only for the sake of faithfulness to the tradition, and for maintaining a meaningful distinction between eucharist and baptism, but because otherwise a flesh–spirit dualism may well prevail, in which the eucharistic *res* is not offered or received in the flesh and for the flesh. Was that dualism not the main concern of Irenaeus when he insisted that the gnostics should cease celebrating the eucharist?[34] It is precisely the taking up of the stuff of this old creation into the life of the new creation that the eucharist entails. The conversion of bread and wine into his own body and blood is the

---

33 And without implying that movement or mastication of the host is movement or mastication of Christ, which does not follow from the fact that reception of the host is reception of Christ, or from eucharistic adoration of Christ (cf. McCabe, *God Still Matters,* 118).

34 See *Against Heresies* 4.18.

act by which the incarnate Son of God continues to affirm his own intimate connection with this world, while proclaiming the necessity of its passage from the old creation to the new.[35] Conversely, it is the act by which he grants to us a role in making the old new, through participation in his heavenly offering, and renders concrete the reality of the church as his body in something more than a metaphorical sense. Neither Calvin nor the other reformers manage to do justice to these things.[36]

Now when we approach the problem of transubstantiation in this way, what we are calling the Catholic temptation is more readily resisted. Belief in the real presence does not diminish awareness of the absence; it does not diminish but strengthens our longing for the coming in glory. Neither then does it invite us to suppose that our union with Christ is a union in which the cross is only behind us and not also before us. (No supposition could be *less* Catholic, as the ubiquitous crucifix testifies.) The power of the real presence is the power of Christ's 'suffering and kingdom and patient endurance', as the Apocalypse puts it – the kingly being found, in this world, only on the way of the cross.[37] Belief in the real presence invites no false triumphalism and does not prescind from grace through any attempt to manipulate heavenly realities for earthly gains. It does not invite what Jean-Luc Marion calls an idolatry of presence. 'Whoever fears that an idolatry of presence according to the *here and now* might ensue from the theology of transubstantiation admits by this very fact that he does not see that only the eucharistic present touches, in the consecrated host, the "real", and that what he fears as overvalued only plays there the role of *sacramentum*.'[38]

Otherwise put, the doctrine of transubstantiation does not deny, but insists, that the church is an eschatological reality. To say that the church is an eschatological reality is not to say that it only really

---

35 If, again, the bread and wine are *replaced* by body and blood rather than converted into body and blood, then it is not at all clear that it is the stuff of this world that is taken 'up' into the world to come.

36 Though many have not done justice to Calvin, as Julie Canlis makes clear in *Calvin's Ladder: A Spiritual Theology of Ascent and Ascension*. Canlis brings Calvin into dialogue with Irenaeus, highlighting his pneumatological theology of participation and his eucharistic spirituality.

37 Cf. *Ecclesia de eucharistia* §5: 'In this gift Jesus Christ entrusted to his Church the perennial making present of the paschal mystery. With it he brought about a mysterious 'oneness in time' between that Triduum and the passage of the centuries.'

38 *God Without Being*, 181; cf. *Prolegomena to Charity*, 150ff.

exists in the eschaton, however, or that it exists here and now merely as a community of faith and obedience, or as a prophetic voice crying in the wilderness. The priestly and kingly power of the church cannot be reduced to the prophetic. The *hoc est corpus meum* cannot be reduced to a figure of speech. Bread and wine do become the body and blood of Christ in a miraculous event that, joining earth and heaven, carries the here and now into the eschaton.

The liturgy itself teaches us to understand things in just this way. Not for nothing does the eastern rite begin with the cry, 'The doors, the doors!' – as if to shut the faithful into the ark of salvation as it is carried above the waters of judgement into heaven itself. 'O God, the King of Glory, who hath exalted thine only Son Jesus Christ with great triumph unto thy kingdom in heaven: we beseech thee, leave us not comfortless; but send to us thine Holy Ghost to comfort us, and exalt us unto the same place whither our saviour Christ is gone before.'[39] 'God our Father, make us joyful in the ascension of your Son, Jesus Christ. May we follow him into the new creation, for his ascension is our glory and our hope.'[40] Such prayers are answered in the present tense as well as in the future tense. The reformers' doubts notwithstanding, they are answered through the transubstantiation of bread and wine in an event that converts the present *into* the future, the eschatological future that begins with the parousia.

IN SO FAR AS PROTESTANTISM denies transubstantiation – not the term or the particulars of mediaeval metaphysics, but the eucharistic realism of John 6 – it wittingly or unwittingly denies the reality of the church on earth. In so far as it attempts to ground that reality strictly in the preaching of the Word, or regards the sacraments as no more than 'visible words', it tends either to an idealist or to a subjectivist conception of the church.[41] This is evident in its refusal to allow the church to be subject to a particular discipline, to be Petrine; but it is evident in other ways as well. In Protestant realms there has been a pronounced tendency to turn eschatology into utopianism, ecclesiology into a branch of secular politics or of social history, sacramental theology into semiotics, and doxology into

---

39   Collect for the Sunday after Ascension Day (*Book of Common Prayer*).
40   Ascension Day prayer (Catholic missal).
41   It was Augustine who famously spoke of the sacraments as 'visible words', which of course they are, but not (as Augustine knew well enough) *merely* that.

ethics. Academia has responded to this by inventing the modern university, with its disciplinary instruments (called by the Germans *Religionsgeschichte* and *Religionswissenschaft*) for dissecting and classifying the remains: a university with no soul for a society with no church. The false inflation and subsequent devaluation of reason, to which Pope Benedict averred in his Regensburg address, is one result. The hollowing out of western culture is another.

The hollowing out of culture is not our first concern, however, but the hollowing out of the church. Rejection of transubstantiation already entails rejection of the eucharist as an offering or sacrifice, leaving us empty-handed when we appear before God. Either that, or it makes the offering something of our own, something offered alongside of Christ rather than in, with and through Christ. 'Unholy fire' upon the altar of God:[42] that is the danger that lurks in sacramental nominalism, even where the intentions are all in the other direction.

That they were in the other direction has already been granted. From Peter Martyr Vermigli, who synthesized Reformation thinking on the eucharist and the sacraments as well as anyone at the time, to Karl Barth, who turned to the task of questioning the Reformation heritage too late in life to complete it, the Protestant way is to reduce the sacraments to two of their constituent elements, *viz.*, revelation and response. The whole point is to preserve the priority of revelation, and so of grace, but the unfortunate effect is to render response an independent work.[43] Whether we regard the response with Vermigli as a movement of the mind based on a spiritual mutation of the bread and wine (akin to what some today call 'transignification'), or with Hegel as an act of inner enjoyment based on a mutation of the collective mind, or with Barth as an act of faith elicited by the living Word, the result is pretty much the same: the true matter of this sacrament is 'to be eaten with the mind, not the bodily mouth'.[44] What is left to the mouth is simply (to borrow the language of Hebrews) 'the fruit of lips that acknowledge his name'.

---

42 Cf. Lev. 10:1-3; also Numbers 16.

43 Barth, indeed, describes the sacraments 'as nothing but response' (John Webster, *Barth's Ethics of Reconciliation*, 116; quoting *Fragments Grave and Gay*, 88).

44 Vermigli, *Treatise on the Sacrament of the Eucharist* (1549), 124: 'For we declare and insist that these symbols signify, offer, and most truly exhibit the body of Christ, although spiritually, that is, to be eaten with the mind, not the bodily mouth . . .; what is done in this sacrament is not corporeal but spiritual.'

Sacrament as revelation and response, revelation being God's offering and the response being ours – such is the conclusion to which Protestantism has come. Barth, the greatest of the modern Protestant anti-Pelagians, made what he called his 'poor exit' by moving somewhere to the left even of Zwingli, dropping the very term sacrament. He decided that both baptism and the eucharist must be treated under the rubric of ethics.[45] Two of his prominent anglophone disciples, Thomas Torrance and his brother James, recognized the Pelagian trend in Protestant doxology and lamented the liturgical emphasis on what *we* do rather than on what Christ does, but even they did not grasp fully the nub of the problem. Thomas put it down to a form of Apollinarianism, which he traced back into the patristic period and believed (rightly, I think) to have had a Pelagianizing effect on Catholic, Orthodox and Protestant doxology alike.[46] He also appealed to Barth not to publish the final fragment of his famous *Church Dogmatics*, arguing that its rejection of sacramental theology undermined Barth's own christological insight. But Barth cannot be corrected by Barth on this score, nor Vermigli by Vermigli. Only by refusing the dualism between 'carnal' and 'spiritual' eating is it possible to prevent the hollowing out of eucharistic theology and with it the hollowing out of the sacramental basis of ecclesial life. The reduction of offering to ethics is otherwise inevitable.

We will follow up this reduction in the next chapter, touching on certain substitutes for eucharistic humanity that are popular today. But there is something more that must be said in the present chapter; namely, that the reduction in question is bound up with a rejection of Mary as mother of the church.[47] Eucharistic communion is understood in the tradition (one thinks, for example, of Rublev's icon) as access to the tree of life, which is what the dead wood of the cross becomes, thanks to him who hangs upon it. Mary stands at the foot of that tree and is given by Jesus, in place of Eve, as 'mother of all the living'.[48] While it is possible, as Protestants aver, to allow Mary to

---

45   *Church Dogmatics* IV.4, preface.
46   See *Theology in Reconciliation*, chap. 4.
47   Barth (*Church Dogmatics* 2.1, 138–146; cf. 4.2, ix), himself observing the intimate connection to the sacraments, was as decisive – and decisively wrong – in this as it is possible to be. In my critique of Barth ('Karl Barth on the Ascension', *International Journal of Systematic Theology* 2.2, (2000), 127–50), I failed to see this, or to show how the whole nexus of concerns I was trying to articulate is linked to the problem of Mary.
48   Indeed, in the scheme of recapitulation she becomes 'the patroness (*advocata*) of the

displace the human Jesus soteriologically, thus making her do service to Pelagianism, it remains the case (a case Catholics have not argued as well as they might) that only a proper Mariology can decisively overcome the Pelagian tendency. For a proper Mariology, coupled with a sound ascension theology, prevents confusion between Christ and the church, while grounding the latter's perfection entirely in the prevenient grace of the former.

Jesus is the head of the church, Mary the pre-eminent member of the church and the prototype of its free participation in the pure offering of Jesus.[49] The 1854 dogma of the immaculate conception testifies to the fact that this participation is wholly the product of grace, a grace prior to any works. The grace in question is not another grace than the grace that is vouchsafed in Christ. It is the same grace retroactively operative, for Mary's immaculate conception is based, as *Ineffabilis Deus* says, on 'the merits of Jesus Christ'. (There is no reason to object to the notion of retroactivity; that would be to object also to baptism and the eucharist and the resurrection of the dead.) Yet it is operative in Mary in a special way, overcoming even in this life the effects of original sin, so that in this life she may be the help to Jesus that God requires and desires.[50]

Her assumption into heaven should be seen in the same light. If in Mary Eve is recapitulated, as Irenaeus contends, then Mary's ending,

---

virgin Eve', says Irenaeus (*Against Heresies* 5.19).

49   Vatican II's decision to treat Mariology within ecclesiology already settles the question about her relation to Christ. Mary is the first convert as Eve is the first sinner; she is joined to Christ, not as an equal, but nonetheless as a helper fit for the new Adam whose work is not the proliferation of the race but its redemption. She assists in the work of redemption (as do the martyrs, for example) but is herself among the redeemed; she belongs to the church, which is the company of the redeemed, for which reason the council did not call her 'co-redemptrix'. It will not do, however, merely to say (as some Protestants do) that she is first among the redeemed, and as such the pre-eminent member of the bride of Christ. For she is something else as well. She is his mother, not only in the metaphorical and spiritual sense of Mt. 12:49f. but also in the literal sense with which the theology of the incarnation (and of the resurrection) cannot dispense. Mary's pre-eminence among the redeemed is therefore based on something more than being first. We may (as Edward Sri has argued) think of her as Queen mother; we must think of her as *Theotokos*.

50   The retroactive power of what God accomplishes in Christ is implied in the definition of *Ineffabilis deus* (1854) that 'from the first moment of her conception the Blessed Virgin Mary was, by the singular grace and privilege of almighty God, and in view of the merits of Jesus Christ, Saviour of mankind, kept free from all stain of original sin.' A robust notion of retroactivity is something I have pursued in *Ascension and Ecclesia* (see especially chap. 3 and appendix B); this notion, I think, has application to Mariology both in the matter of the immaculate conception and of the assumption into heaven.

like her beginning, is rightly understood to be uniquely open to perfecting grace.[51] Moreover, if Mary is the prototype of the church's free
participation in Christ, then her end will be of the kind that is proper
to the church, displaying in advance the principle that the members
follow where the head leads. To speak of her assumption into heaven
is to speak of the fact (revealed to the church theologically rather than
by the testimony of witnesses) that Mary enters the new creation first
of all those who are lifted up out of death by him who holds the keys
to Death and Hades. Her death (if death it was) was indirectly rather
than directly connected to human sin, and her assumption entails a
direct rather than an indirect transition to the new creation.[52]

---

51 It may be helpful here to make a distinction between the grace of creation (creation, if
    ex nihilo, is already an act of grace), the grace of perfection (which lifts rational creatures
    beyond themselves into communion with God), and the grace of redemption (which
    overcomes sin and its consequences, the essence of sin being to attempt to raise oneself
    to God by purely natural means; that is, to refuse perfecting grace). Mary's relation to
    the latter two forms of grace is *sui generis*. She is neither in Eve's prelapsarian position
    (still open to perfecting grace without yet needing redemptive grace) nor in her post-
    lapsarian position (a sinner in need of forgiveness). As a daughter of the lapsed Eve, she
    does indeed need and receive redemptive grace, but this grace, 'by the merits of Jesus
    Christ', functions in her case to keep her open to perfecting grace even in a race that has
    closed itself to that grace. She is kept open for his sake, that she might receive and bear
    the saviour – who comes not to Eve as she was, but in her need as she now is, yet who
    comes precisely to undo that which has been done, to open to God that which has been
    closed to God. The dogma of the immaculate conception declares no more and no less
    than that Mary is opened *ab initio*. The dogma of the assumption declares that one who
    is so opened is opened also to an end partly analogous (but only partly) to the end that
    Eve might have had, but did not.

52 Should we object that this dogma enshrines 'a post-biblical legend' (Tim Perry, *Mary for
    Evangelicals*, 307)? It may or may not be the worse for that. Gen. 5:24 no doubt enshrines a
    pre-biblical legend. Yet what is said of Enoch and what is said of Mary is that they attained
    to the very goal of creation, *viz.*, that man should live with God. Because that is possible
    only by way of what Christ has done, it is *his* history (including his departure) that must
    be documented by witnesses, not theirs. There is no dogma about Enoch, of course, but
    then Enoch only arrived at the goal with Mary's help; she who is the mother of our Lord
    goes first – before Enoch – into the new creation. As for the question, 'Did Mary die?',
    it was wisely left open by Pius XII in *Munificentissimus Deus* [Defining the Dogma of the
    Assumption, 1950] just as Genesis left open the same question about Enoch. Mary 'was
    assumed body and soul into heaven' after 'having completed the course of her earthly
    life'. The question must be left open because the inner logic of the dogma does not
    determine the answer; nor is there any other resource for determining it. In Enoch's
    case we might appeal to 1 Cor. 15:51ff. for a partial analogy, since Heb. 11:5 answers the
    question left open by Genesis. But the case is not exactly the same with Mary, she who
    not only 'pleased God' but was 'full of grace'. Perhaps it belonged to that grace to 'see
    death' and to laugh at its new impotence, thus conforming more closely to her son, so
    as to pray more effectively for us sinners 'now and at the hour of our death'. (*Pace* Perry

It should not be supposed that this way of putting the matter makes Mary's assumption into heaven of theological relevance only, or that the theology in question (*per impossibile*) has no bearing on prayer and piety. Mary goes first, after Christ, into the kingdom. And because she goes first, she assists those who come after, and does so in a manner distinct from that of the martyrs and saints, whose own deaths are not direct transitions to glory in the same sense that hers is.[53] Mary is 'mediatrix of all graces'. She is that on earth in as much as she is mother of the mediator and so also *Theotokos*, mother of God. She is that in heaven in as much as she enters first after the mediator and is made queen of heaven, participating in its ordering and flourishing, and indeed in its eucharistic appropriation of the old creation. But neither on earth nor in heaven is she herself the mediator. On earth her *fiat mihi* is already enabled, retroactively, by his 'not my will but Thine be done'. And she does not enter heaven before her Son or with her Son, but after him and because of him. He is her mediator; she the first of those for whom he mediates. He is her redeemer; she the first of the redeemed. Because she is the first, however, and more especially because she is the one through whom he comes to save, she is for the rest mediatrix.

That there should be a first – not Peter or John but Mary – witnesses to the relation of redemptive grace to nature, which it recapitulates and perfects. Adam is made from the dust of the ground; Jesus is born of a virgin. Adam is betrothed to Eve; Jesus is betrothed to the church. Adam and Eve are fecund through procreation; Jesus is fecund through the power of the Spirit to unite their progeny in one body. Mary is fecund through the same Spirit to be the mother of Jesus physically, and the mother of the church spiritually. Perhaps it is not surprising that Protestantism, in attempting to reaffirm the graciousness of grace through its *solus Christus* slogan, should have lost sight of the connection of grace to nature in this particular respect.

244f., all of this can be thought without falling into the error of Roschini and the 'co-redemptrix' maximalists.)

53 We should not make here the mistake we refused to make in regard to the eucharist – *viz.*, of thinking that heaven, in the sense germane, is 'up there' or 'out there' or 'in there' somewhere, rather than a new creation. That would permit us to make no more sense of Mary's departure or destination than of Christ's. Mary's assumption into heaven, if it is an assumption 'body and soul', is to be distinguished from the blessedness of the other saints, who await (whatever that means for the dead, whose temporality is certainly not our temporality) the resurrection of the body before entering the new creation.

There is certainly a recovery underway, as a spate of recent books from Protestants on the subject of Mary attests, but it has not been widely noticed that the tendency to reduce offering to ethics, and to see the church's mission in moralistic terms, is connected to the prior marginalization of Mary.[54]

The two recent Marian dogmas specify Mary as the sign of grace and hope in Christ.[55] They identify her as the sign proper to the church, just as the church itself is the sign proper to the ascension of Jesus and its eucharists the sign of his parousia. Confidence in Mary, in whom the perfection of the end is already anticipated in the present, is confidence in the church and in its freedom to make offering through Christ to God. Lack of confidence in Mary translates into doubts about that freedom, and about the sacramental vocation of the church. Moreover, it translates into doubts about the unity of the church as a single or common household. In the absence of due acknowledgement of 'our Lady', the unity of the church becomes a theoretical proposition the realization of which may with impunity be postponed to the eschaton.[56]

Legitimate concerns must be raised, of course, respecting Mariological distortions and the not infrequent excesses of Marian devotion. But it would be a curious undertaking to identify counterfeits without recognizing the true currency. The true currency of the kingdom shows Jesus' face on the obverse and Mary's face on the reverse. Anyone who doubts that would do well to reflect on the fact that the form taken by current revivals of the old christological heresies is often Marian. The advance of Marian dogma has been slow and difficult. It took time to declare Mary, for christological reasons, *Theotokos*. Given that declaration, it would seem strange not to say such things about her as the church has subsequently said, but it

---

54 Perry, for one, glances in this direction in his admirable volume (p. 237, n. 138; cf. 213f.). He does not pursue it, however, perhaps because it is counterintuitive for one who adheres to the Protestant principle, which grossly inflates Augustine's insight that 'to tread where scripture is utterly silent' is unnecessary and inadvisable (295). Certainly Perry wants to bring Mary back from the margins, theologically speaking. But this can hardly be done if no amount of reflection by the church, and no teaching office of the church, can add authoritatively to what scripture has actually said. Which begs the question, of course: What authority has the Protestant principle?

55 See the Anglican–Roman Catholic International Commission document by that title (2005), which puts the Marian dogmas in a helpful eschatological light.

56 Cf. Farrow, 'Church, Ecumenism and Eschatology', in Jerry Walls, ed., *The Oxford Handbook of Eschatology*, 356ff.

has not said them lightly or quickly. The rise of Marian pretenders, however, since the settling of the Catholic mind less than a century ago, has been remarkable for its rapidity and its theological rapacity. Feminist alternatives to traditional Marian devotion – alternatives that are openly gnostic and starkly Pelagian in their devotion to the Sophia-Christ or some similar invention – are no accident. Nor is the strategy of pretending that Marian devotion is rooted, not in christology, but in pagan mythology. Nor indeed is the hallowing of ideas and actions inimical to everything the scriptures and the church say about Mary, in contemporary discussions of marriage and family, sex and gender, contraception and abortion, etc.[57] No doubt the anticipation of the end in Mary has implications and effects that must be tested by the appearance of distortions and alternatives. But when we turn to Mary for help in saying the *fiat mihi* of the receptive heart and mind, when we come to recognize in her 'the woman whom the prayer of the Church invokes as *Seat of Wisdom*', when we honour her as the one in whom the ecclesial disposition is an actuality and not merely an ideal, these distortions and alternatives melt away.[58] Faithful adherence with Mary to the word of the Lord, and faithful adherence with the church to the mother of our Lord, enable us to see just how impoverished and grotesque they are.

---

57 We will return to this topic briefly in the next chapter, when we confront more directly the challenge faced by the eucharistic community in light of the hollowing out of western culture.

58 John Paul II, *Fides et Ratio* [Faith and Reason, 1998] §108. I am indebted to R. R. Reno, as to Anna Farrow, for warning me away here from a merely abstract Mariology.

CHAPTER SIX

The Politics of the Eucharist

THE CELEBRATION OF THE EUCHARIST is always a political act. In fact it is *the* political act. Not only because it is, to use the Orthodox terminology, a synaxis or coming together επι το αυτο, 'in one place' – the assembling of a people and the constitution of a body with its own unique authority – but because it is a coming together to participate in the ultimate political reality, the kingdom of God, from which its authority derives. It is an intervention of that kingdom within the kingdoms of this world. Through it the reconciliations are effected, and the virtues nourished, that authentic political community requires. Through it a truly public space is opened up, one that includes even the 'angels and archangels and the whole company of heaven'. Through it the King of Kings himself is present and adored. His sacramental coming, as the Apocalypse declares, is a sure and certain portent of his coming with glory to judge the quick and the dead. The celebration of the eucharist is the means by which his heavenly session, otherwise invisible and impenetrable to man, is made visible; by which his intercessions before the Father are echoed and adumbrated on earth; by which the subjects of the rulers of this world are liberated and made subjects of the kingdom not of this world.

It is tempting to appeal, by way of illustration, to the Mass celebrated by John Paul II in honour of St Stanislaus at Blonie Park, Kraków, on 10 June 1979, in the presence of some two to three million people. He began his homily with the famous Great Commission text, Mt. 28:18–20, and with Jesus' words, 'All authority in heaven and on earth has been given to me.' Celebrating Mass in the same place on Ascension Day, 2006, Benedict XVI recalled the words with which John Paul II urged the Christian people of Poland to stand firm in that knowledge:

> You must be strong, dear brothers and sisters. You must be strong with the strength that comes from faith. You must be strong with the strength of faith. You must be faithful. Today, more than in any other age, you need this strength. You must be strong with the strength of hope, the hope that brings perfect joy in life and which prevents us from ever grieving the Holy Spirit! You must be strong with love, the love which is stronger than death ... You must be strong with the strength of faith, hope and charity, a charity that is conscious, mature and responsible, and which can help us at this moment of our history to carry on the great dialogue with man and the world, a dialogue

rooted in dialogue with God himself, with the Father, through the Son in the Holy Spirit, the dialogue of salvation.[1]

They were, in other words, to understand the dignity and power of their position; to understand that the dialogue of salvation, grounded as it is in the authority of Christ and in the very life of the Holy Trinity, is a dialogue between ambassadors of a kingdom that is coming and ambassadors of a kingdom that is passing away, a kingdom whose authority is merely secular.[2] And they did understand, and were strengthened, and prevailed. The Mass at which these words were spoken marked the end of John Paul II's first visit to Poland as pope; it also marked the beginning of the end of the Soviet empire.

I say 'tempting', however, because this example can obscure as well as reveal the political nature of every eucharistic celebration, which may just as easily end (as did the last Mass said by St Stanislaus) in bloodshed or even, by worldly standards, in impotence and apparent defeat. If, as Ambrose once said, the church is 'the form that justice takes, the common right (*ius*) of all',[3] we may be quite certain – as certain as we are of Jesus' ascension and heavenly session – that the church will prevail. Its prayer, 'Thy kingdom come, thy will be done on earth as it is in heaven', will be answered. But if the Apocalypse is anything to go by, to see the kingdom coming is the privilege not of historians, or even of popes, but of martyrs:

> And I saw what appeared to be a sea of glass mingled with fire, and those who had conquered the beast and its image and the number of its name, standing beside the sea of glass with harps of God in their hands. And they sing the song of Moses, the servant of God, and the song of the Lamb, saying,
>
> 'Great and wonderful are thy deeds, O Lord God the Almighty!
> Just and true are thy ways, O King of the ages!
> Who shall not fear and glorify thy name, O Lord?

---

1   John Paul II, 10 June 1979, quoted by Benedict XVI ('Homily by the Holy Father: Mass in Krakow–Blonie', 28 May 2006).

2   'What has greater might than a confession of faith in the Trinity?' asks Ambrose in his *Sermon against Auxentius* (§34).

3   *The Duties of the Clergy* 1.29. 'Justice, then, enables the association of the human race and its community' (1.28). It is not merely critical or retaliatory, in other words, but is allied with goodwill (i.e., with generosity; cf. Mt. 10:1-16) in its service to the social principle.

> For thou alone art holy.
> All nations shall come and worship thee,
> for thy judgments have been revealed.'[4]

'Not my will but thine be done', was Jesus' prayer in Gethsemane, and Gethsemane provides the hermeneutic for the prayer he taught his disciples. The kingdom is coming, it is in our midst, but it is not coming openly 'with signs to be observed'. It is coming sacramentally, which means that it remains hidden even in its revelation. It remains embedded between the 'tribulation' and 'patient endurance' in which we are called to be the companions of Jesus.[5] The politics of the eucharist, as the author of the Apocalypse is at pains to point out, is the politics of the cross, the politics of the martyrs.

In order to grasp this properly and to read aright the Apocalypse (which with Acts and Hebrews forms a biblical trio of books on the heavenly session, blending in the eucharistic note) we must grasp two somewhat paradoxical principles. The first is that justice precedes power. The second and more difficult principle is that the good things that come into the world through the church provoke and permit the maturation of evil. Without these principles ascension theology cannot provide an adequate account of Christ's heavenly session or of the church's eucharistic witness to it. It cannot put the church's martyrial politics in a clear light. It is still in danger of succumbing to a false and illusory triumphalism.

The first principle is well articulated by Augustine. Christ, he says, sets before us both power – 'what could be a greater show of power than to rise from the dead and ascend into heaven with the very flesh in which he had been killed?'[6] – and justice. But not in that order. The economy of salvation puts justice first, the justice of the cross. This shows us that 'power should follow justice and not precede it'. Power is not the prerequisite of justice, in other words, as men are inclined to think. Rather justice is the prerequisite of power: 'The essential flaw of the devil's perversion made him a lover of power and a deserter and assailant of justice, which means that men imitate him all the more thoroughly the more they neglect or even detest justice and studiously devote themselves to power, rejoicing at the

4   Rev. 15:2-4 (*RSV*)
5   See Rev. 1:9 (4ff.); cf. Lk. 17:20f.
6   *Trinity* 13.18

possession of it or inflamed by the desire for it. So it pleased God to deliver man from the devil's authority by beating him at the justice game, not the power game, so that men too might imitate Christ by seeking to beat the devil at the justice game, not the power game. Not that power is to be shunned as something bad, but that the right order must be preserved which puts justice first.'[7]

This order is not superseded by the heavenly session, but reinforced. It is true that all authority in heaven and on earth has been given to Jesus Christ in his ascension. He sits at the right hand of God. He moves with invincible majesty among the golden lampstands. He holds the keys of Death and Hades. He has seven 'eyes' and seven 'horns', a plenitude of knowledge and power. He shares the glory that the four living creatures and the elders give to the Almighty. But when he takes in his hand the scroll of history, to break open its seals, he does so as 'the Lamb that was Slain' – as the one who chose the way of the cross as the path to glory, who chose the cross itself as the earthly seat of his power. It is equally true that the church participates in his heavenly authority and is upheld by it. Like Christ, it binds and it looses. But it does this on earth only by continuing Christ's own strategy of 'beating the devil at the justice game, not the power game'. For the church militant, as for Christ in the days of his mortality, that means submitting willingly to injustice, albeit with the sure knowledge that injustice is self-defeating, and that the unjust death of the Just One has broken forever the power of death and of injustice.[8]

What then of the judgements that are meted out as the seals on

---

7   *Trinity* 13.17. Augustine's argument could be strengthened by an appeal to Anselm's treatment of the atonement in *Cur Deus Homo?* Anselm, while downplaying the contest with the devil (who, according to Augustine, makes a fatal error by marshalling the forces of death against one on whom death could make no proper claim), better highlights the justice-loving freedom of Christ in his self-offering to the Father.

8   Revelation 11 provides the picture when it says of Christ's two witnesses, who epitomize the prophetic ministry of the church: 'They have power to shut the sky, that no rain may fall during the days of their prophesying, and they have power over the waters to turn them into blood, and to smite the earth with every plague, as often as they desire. And when they have finished their testimony, the beast that ascends from the bottomless pit will make war upon them and conquer them and kill them, and their dead bodies will lie in the street of the great city which is allegorically called Sodom and Egypt, where their Lord was crucified . . .; and those who dwell on the earth will rejoice over them and make merry . . . But after the three and a half days a breath of life from God entered them, and they stood up on their feet, and great fear fell on those who saw them. Then they heard a loud voice from heaven saying to them, "Come up hither!" And in the sight of their foes they went up to heaven in a cloud.' (vv. 6-12, *RSV*)

the scroll of history are broken – the judgements of which Christ warns us from heaven, as indeed he warned us on earth? Do they speak only of power or do they also reinforce the priority of justice, the justice of the cross? These are the judgements of natural law, in the going forth of the four horsemen, and of divine law, which rewards fidelity and punishes infidelity.[9] The Apocalypse indicates that all these judgements (even those of natural law) have been restrained by God until the arrival of one who is fit to oversee them: one who has stooped to rescue us, as Irenaeus puts it, 'by means of persuasion' not violence; one who has defeated the devil 'in a manner consonant with reason'; one who is just, as God himself is just.[10] The appearance of such a one in heaven is the signal that heaven's judgement may now proceed. But how does it proceed? With power, yes, but not without continued persuasion. And the persuasion is of two kinds: the increase in discipline that results from the gradual removal of restraint on the agents of judgement ('Release the four angels who are bound at the great river Euphrates!') and the increase in love that flows from the martyrs who 'fill up the sufferings of Christ'.[11] The first form, taken by itself, is ineffective; the second form has power to bear fruit in repentance and salvation. Those who learn from Christ to conquer as he conquered – who ascend to him in heaven with heart and mind, that on earth they may also descend with him on the *via crucis* – put justice before power and, just so, are given the power to win for God their fellow man.

Semen est sanguis Christianorum. The blood of Christian martyrs is the seed of the church. That Christ's appearance in heaven is not simultaneously his appearance on earth, that his coming on the clouds to the Ancient of Days is not already his full parousia to the world, that there is a sacramental pause or delay in which the kingdom remains hidden, is for the sake of their victory and their harvest.

---

9   See Revelation 6 and 16:5-7; cf. Gen. 4:1-16. We ought, I think, to see the present financial crisis in that light, and to recognize that the preconditions for the fulfillment of Revelation 18 (cf. Isaiah 24) have already been met on the economic level.

10 Irenaeus (*Against Heresies* 5.1.1) anticipates Augustine. The Word of God, he says, 'powerful in all things, and not defective with regard to his own justice, did righteously turn against that apostasy, and redeem from it his own property, not by violent means, as the [apostasy] had obtained dominion over us at the beginning, when it insatiably snatched away what was not its own, but by means of persuasion, as became a God of counsel, who does not use violent means to obtain what he desires; so that neither should justice be infringed upon, nor the ancient handiwork of God go to destruction.'

11 Col. 1:24; Rev. 9:14.

The generosity that is the gospel, the justice that is the church, the judgements that discipline rather than destroy, the conquering love of the martyrs – these are the true reasons for the saeculum, in which there is no more politically profound act than the eucharist, and no truer expression of the politics of the eucharist than martyrdom. We have not yet understood this, however, until we have turned it over and looked at it from the other side; that is, until we have recognized our second principle also.

The second principle – that the goodness embodied in the church enables evil to grow and mature – introduces us to the shadow side of the saeculum and to what Paul calls 'the mystery of lawlessness' and John 'the spirit of antichrist'. Aware that the devil does not quickly concede at either game, they portray the present age as one in which a counterfeit justice appears, and a contest of wills and of nerves ensues, in a great struggle between Christ and Antichrist. They do not present this struggle as a gradual triumph for the former, at least not in worldly terms. The history of the saeculum is for the church (as J. R. R. Tolkien put it) a 'long defeat'.[12] The present age cannot be brought to a close, Paul tells us, 'unless the rebellion comes first, and the man of lawlessness is revealed, the son of perdition, who opposes and exalts himself against every so-called god or object of worship, so that he takes his seat in the temple of God, proclaiming himself to be God.'[13] The Apocalypse, for its part, warns us of 'a beast rising out of the sea, with ten horns and seven heads, with ten diadems upon its horns and a blasphemous name upon its heads', and of a beast rising out of the earth that looks like a lamb but speaks like a dragon, deceiving those who dwell on the earth – that is, of another 'dialogue of salvation' rooted in an unholy trinity.[14] These are different ways of saying the same thing, namely, that in the saeculum the whole world is the stage for a war identical to the one that was waged, in and around Jerusalem, between Jesus and the 'strong man'. The outlines of this war are known in advance, for they are already revealed to the

---

12 Though it contains 'glimpses of final victory' (*The Letters of J. R. R. Tolkien*, no. 195, ed. H. Carpenter and C. Tolkien).

13 2 Thess. 2:3f.; cf. Mark 13 and 1 Jn 4:3.

14 The beast that rises out of the earth, meaning out of Israel and especially out of the church, is the beast that is permitted to work signs and wonders (Rev. 13:11ff.); but neither it nor its followers can understand the song of the eucharistic faithful, which is the song of the true Lamb (14:1-5).

faithful in the story of Jesus; only the details remain to be discovered through the living-out of the eucharistic vocation.[15]

Our second principle can help us with all of this, which at first seems counterintuitive. How does it come about that a battle decisively won by Jesus should nevertheless spill over into the remainder of history and become its defining feature? How is it that the present age should be for the church, in spite of Christ's heavenly reign, a long defeat? How is it that the preaching of Christ should produce Antichrist, the law of liberty the mystery of lawlessness? Must we really expect all of this? May we not rather think of advances – with occasional setbacks, of course – towards a kingdom where God's will is truly done on earth? Or at least of an ebb and flow, of successes and failures in equal measure, while the age lasts? To suppose so is to ignore the fact that the present age will end, as Jesus and the prophets and the apostles all insist, in a definitive act of judgement, analogous to the destruction that fell on Jerusalem in AD 70, and indeed to the Great Flood that is the biblical prototype of such judgements.[16] This definitive act of judgement awaits its time, for God judges nothing before its time. And that is where our second principle comes into play. The definitive judgement awaits the time when evil is historically embodied in its most definitive form – a form it cannot achieve or assume apart from the working of the gospel and the eucharistic presence of the church throughout the world.

Now to follow this line of thought it is necessary to observe that evil does not coexist with God and is not created by God; hence it has neither form of its own nor any natural stability. Grounded only in rebellion, it is strictly parasitic on created goods, and without these goods it has no potential at all, for it is chaotic and self-destructive. The greater the good, then, the greater the potential for evil. Evil flourishes only *with* the good, indwelling what is otherwise healthy and true, building within it and upon it, until it reaches the boundary beyond which God will not permit it to go. That boundary is the day

---

15   Cf. Mt. 12:22ff., Jn 14:30–16:33, and Revelation 2–3, 13–14. Such passages require us to face the fact that, if Christ has ascended into heaven, then there is no such thing as a post-Christian culture in which Christians may settle down more or less comfortably as a minority subculture. The putatively post-Christian culture is only an anti-Christian culture in disguise, and the disguise must sooner or later be dropped.

16   See 2 Pet. 3:1-13. Is it fair to say that continued use of 'AD' (*versus* the now-preferred vagary, 'CE') is mandated by this passage? If not by this passage, then certainly by the doctrine of the ascension.

of the Lord, the day of the Rider on the White Horse, the day when divine judgement is executed in all the earth. On the far side of that boundary lies the kingdom in which every participating citizen will be made free with the freedom of God himself, who cannot do evil or be approached by evil.[17] It is also necessary to observe that in the saeculum the healthiest thing of all, the bearer of the greatest good, is the church, with its saving gospel and life-giving sacraments. The mystery of lawlessness, then, is the form that lawlessness takes when it seeks to occupy the gospel, the law of liberty.[18] It is the form that injustice takes when it tries to take shape within justice itself, which is the church. It is the form that sin takes when its self-destructive tendencies are restrained and tempered by the influence of righteousness. It is the form that evil takes when its chaotic forces converge in opposition to the unifying sacraments and so, *mirabile dictu*, achieve a harmony not their own. It is the maturation of evil that is otherwise unthinkable and impossible. Because the church does not collapse under this assault, but is held firm by 'him who holds the seven stars in his right hand',[19] the mystery of lawlessness does not collapse either, but continues to grow while the saeculum lasts. It is not through the crucifixion of the Son – who prays, 'Father, forgive them, for they know not what they do' – that it reaches its climax or takes its definitive form, but through long experience with the gospel and sacraments. When it does reach its climax, however, it will be exposed in all its emptiness and futility, and its agents will be justly judged, once and for all.[20]

The politics of the eucharist is not the politics of success by any worldly measure. We do not live in an age from which evil is gradually disappearing, as many prefer to think. On the contrary, evil continues to grow by perverting the goods that belong to the church,

---

17 *City of God* 22:30; cf. *Enchiridion* §118, and *On Rebuke and Grace* §33. That the highest form of human freedom is the God-like *non posse peccare* clearly demarcates the Christian concept of freedom from pagan alternatives and simulacra (including Nietzsche's notion of the man beyond good and evil).

18 If sin finds 'opportunity in the commandment' (Rom. 7:11), how much more in the gospel?

19 Rev. 1:16, 20; 2:1.

20 'The coming of the lawless one by the activity of Satan will be with all power and pretended signs and wonders, and with all wicked deception for those who are to perish, because they refused to love the truth and so be saved', says Paul (2 Thess. 2:9-10, *RSV*). But the lawless one will be 'destroyed by *his* appearing and *his* coming' – that is, Christ's (2:8).

often through clever and subtle parodies, like a virus mimicking the structure of healthy cells. The parousia of the lawless one – that is, of the man who has decisively repudiated the dialogue of salvation but perfected its simulation – is the precondition for the parousia of Jesus Christ. It is the precondition for the end of the dialectic of presence and absence.

This calls for wisdom. The human race has a vocation to unity, as Pope Benedict has reminded us. Knowledge of the paradoxical nature of the saeculum, of its progress in evil as in good, does not relieve us of our responsibility to seek that unity – participation in the eucharist encourages, not discourages, participation in the common life of man.[21] But it also calls for caution. 'Guard thyself then, O man,' advises Cyril of Jerusalem, for 'thou hast the signs of Antichrist . . . If thou hast a child according to the flesh, admonish him now; if thou hast begotten one through catechizing, put him also on his guard, lest he receive the false one as the true. For "the mystery of iniquity doth already work".'[22] Indeed, it calls for courage: 'If I had not done among them the works which no one else did,' says Jesus, 'they would not have sin; but now they have seen and hated both me and the Father.' And he adds, lest we miss the point, 'the hour is coming when whoever kills you will think he is offering service to God.'[23]

SPECULATION ABOUT THE MAN OF LAWLESSNESS or about the society of which he is the epitome, is speculation about the beginning of the end of history as we know it. Of that we must fight shy – it is less hazardous, says Irenaeus, 'to await the fulfillment of the prophecy than to be making surmises' – without shying at all from the

---

21 See Benedict XVI, *Caritas in veritate* [Charity in Truth, 2009] and my commentary in *Nova et Vetera* 8.4 (2010), 'Baking Bricks for Babel?' See also *Sollicitudo rei socialis* [On the Social Teaching of the Church, 1987] §30f., in which John Paul II insists that 'today's "development" is to be seen as a moment in the story of creation, a story which is constantly endangered by reason of infidelity to the Creator's will, and especially by the temptation to idolatry.' Here, then, is the vital dialectic: Since development belongs to the nature and vocation of man, we may not renounce, even in the face of such danger as the present chapter describes, 'the difficult yet noble task of improving the lot of man'. Yet we must approach that task in full awareness of the danger, and of the fact that only the resurrection can bring unadulterated development or 'unlimited progress'.

22 *Catechetical Lectures* 15.18

23 See again Jn 15:18–16:4; cf. Rev. 13:18 and 14:12. 'He who conquers, I will grant him to sit with me on my throne, as I myself conquered and sat down with my Father on his throne' (3:21, *RSV*).

task of discerning the workings of the mystery of lawlessness in our own place and time.[24] Ascension theology has its own responsibility here, which it cannot rightly avoid, since the mystery in question (as chapter seven will clarify) is a consequence of the ascension.

When taking up that task, I think that we may leave aside the rise of Islam as a rival religion. It is true that from its inception Islam has been locked in combat with Christianity over the question of the unity of God, the reign of Christ, the form of justice and the fundamentals of human nature. It is true that Islam rests on an appropriation of Abrahamic monotheism and covenantalism, and that this appropriation is without benefit of the restraint either of Jewish ethnocentrism or of Christian trinitarianism (which latter, in connection with the cross of Christ that Islam denies, qualifies both the justice and the power of God by the still more fundamental notion that God is love). It is true that its own curious mixture of particularism and universalism has helped to make Islam a fierce global competitor to Christianity and an ever-present threat to Jews,[25] a fact that accounts for many wars and rumours of war. It is true that Islam – which purports to show all mankind how to submit to the law of God – has lent itself, in its suppression of the gospel, to cultures of secret or open lawlessness. It is true that Islam, the glories of its golden age notwithstanding, has not proved itself a friend of freedom, and that Muslims (though many other Muslims abhor this) have brutally martyred numberless Christians and other infidels.[26] And it is further true that Islam has perverted the very concept of martyrdom, by placing a bloody sword in the hands of its own 'martyrs'.[27] Islam, in other words, is no accident of history about which

---

24 *Against Heresies* 5.29.2. The writer of the Apocalypse certainly was not shy – though he was shrewd, deploying a genre not easy for the imperial authorities to understand – about doing that, and he plainly intended to cultivate wisdom and courage, both of which are necessary if one is to conquer.

25 Gabriel, it is said, 'from amongst all the angels, is the enemy of the Jews' (*Sahih Bukhari* vol. 4, book 55, no. 546).

26 Not to speak of apostates – the illicit killing of which Christians have also been guilty of, as even those who defend the teaching of Aquinas or Bellarmine (cf. *Summa Theologiae* 2-2.11.3 and Bellarmine's *De Laicis* §21), or who give a sympathetic account of the late-mediaeval view that obstinate heresy is a form of treason, must admit. But is Islam guilty of a greater and more persistent disparity between its ideals and the religiously motivated outrages of its adherents? That may be a more difficult question to answer than the question as to the coherence of its ideals, which was the question put to it again by Benedict at Regensburg.

27 For which, presumably, it was admired by Nietzsche: see *Antichrist* §53, §59f.

one may be merely curious, or perhaps afraid. Islam depends upon Christian precedent, yet it confesses a counter-revelation to that on which the church relies.[28] For that reason it is theologically important to Christians in a way that other world religions are not; and, on the ground, as it is actually practised in some cultures, it is a daily threat to the peace and safety of those who will not submit. But it does not fit the bill posted by Paul or by John. The mystery of lawlessness, as I said, is the form that lawlessness takes when it seeks to occupy the Christian reality from inside.[29]

On the other hand, we may not leave aside the rise of communism and fascism, which, though also historic rivals to Christianity, are among its own bastard children. Communism and fascism bring us closer to home. They have taken Christian ecclesiology and eschatology and (without any intervening revelation but the Enlightenment) twisted them into a totalitarian vision at once more attractive to western minds than Islam and more deadly than any invading Saracen army. Binding human beings together by force of arms and of propaganda in the long march towards an ever-receding Utopia, they have dismantled, quite methodically, Christian concepts of the dignity of man and of the rule of law.[30] Their leaders have indeed exalted themselves 'above every so-called god or object of worship' in a fashion that pagan kings of old – even the late-Roman emperors in their last desperate bid to crush Christianity – never attempted. In a single century they have shed the blood of more martyrs, and of more ordinary citizens, than any other foe of humankind. Even so, they continue to have their admirers and sympathizers among lapsed Jews and Christians, who have borrowed from their techniques to advance their own lawless causes.

We may take a lesson or two in discernment from the form that fascism took in Germany. Having learned from the followers of Hegel how to transform his trinitarian dialectics into a warrant for revolution

---

28 As on a prophet whom Christians can only regard, with John of Damascus (*Heresies* §101), as false.

29 The Damascene did speak of Islam, in its capacity as *populorum seductrix*, as 'a forerunner of the Antichrist' (cf. D. Sahas, *John of Damascus on Islam*, 69, n. 7), and seemed to view it as a virulent form of the Arian heresy; but even then it is external to Christianity in a way that most Arianism, ancient and modern, is not.

30 'A society without any objective legal scale is a terrible one indeed', said Solzhenitsyn in his famous Harvard address – while adding to this critique of communism a rebuke to the West for having 'no other scale but the legal one' (*The Solzhenitsyn Reader*, ed. E, Ericson and D. Mahoney, 566).

and conquest, German fascism (its Darwinist, Nietzschean and neopagan influences notwithstanding) made much of the messianic possibilities in its own nationalized version of salvation history. This made it a direct threat to Jews, who along with faithful Christians represented the resistance of authentic salvation history to the Teutonic alternative. Against the Jews German fascism quickly turned its face in hatred, pouring water from its mouth like a river to sweep them away;[31] whereas it tried, for obvious reasons, to co-opt the Christian majority into its vile program. That it was quite successful in its efforts testifies first of all to the link between the mystery of lawlessness and anti-Semitism, which (already exacerbated by Luther) had taken a new course in the Enlightenment. Lessing, Kant and Hegel had taught Europeans to embrace the proto-heresy of Marcion as if it were the faith itself. By their lights, the essence of Christianity was to be found in its turn from the particular to the universal, and Jewish particularism was precisely what had to be left behind if Christianity were to succeed. As it had with Marcion himself, this evisceration of the gospel in search of a hidden kernel within its disposable Jewish husk produced a curious mixture of libertinism and moralism; but it also generated a certain tolerance for the obsessional hatred of Jews and Jewishness that was now surfacing.[32]

A revealing struggle ensued between the German Christian movement and what was later called the Confessing Church. The German Christians wanted, as Hitler did, a church rooted 'in the national character' and in the 'religious powers of the *Volk*' – the Aryan *Volk* who belonged to a higher stage of history. The moment was right to grasp the divine as it presented itself in their time. It was their desire, as their platform of 1932 stated, to 'develop the powers of our Reformation faith into the finest of the German nation' and to put them to work on behalf of the revival of German fortunes promised by National Socialism. Over against the German Christians stood the signatories of the Barmen Declaration of 1934, which was drafted by Karl Barth. Representing about one-third of the country's

---

31 Cf. Rev. 12:13-17.

32 What, I wonder, would be the effect of hanging Chagall's powerful rendition of Isa. 6:5-7 (Musée National Message Biblique, Nice) alongside his disturbing 1938 White Crucifixion (Art Institute of Chicago)? The hot coal that touches Isaiah's lips burns deeply into the Jewish people itself, flaring up into flames that destroy their homes and synagogues and their bodies too, when fanned by those who resent their role in the mediation of the Word of God and in the salvation of the world.

Protestant churches, they resisted this rising tide of neo-messianism by way of a potent reaffirmation of Christianity's true messiah, who in his very particularity (as a first-century Jew) was universally the Lord. Barmen's first article was bold and clear: 'Jesus Christ is the one Word of God whom we are to hear, whom we are to trust and obey in life and in death. We repudiate the false teaching that the church can and must recognize yet other happenings and powers, images and truths, as divine revelation alongside this one Word of God.' Barmen mounted no defence of the Jews, however. Though Barth and others, to their credit, would not remain silent on that score – Barth eventually going so far as to say that any attack on the Jews was an attack on Jesus[33] – they confessed that they did not connect 'the Jewish problem' as closely as they should have to the whole phenomenon they were trying to identify and overcome.

One lesson to be drawn here is that the mystery of lawlessness cannot abide the particularity of the gospel. Its anti-Semitism (which infects even some Jews) bears witness to that aversion, as does its propensity to generate grand abstractions calculated to overwhelm everything concrete and particular. Similarly, the mystery of lawlessness cannot abide the concreteness and particularity of the eucharist, which it labours to undermine. Barth understood the former point better than the latter. He rightly argued that confusion between the vocation of the Christian *Kirche* and the aspirations of the German *Volk* was in part a consequence of bad eucharistic theology; that is, of the confusion between the divine and the human generated by Lutheran notions about the real presence. There was, he believed, a dangerous immanentism at work in Lutheranism that tended to domesticate the gospel to the interests of the *Volk*, and to cultivate the illusion that the crest of the wave on which Germany was now poised could collapse on no other shore than that of God's kingdom.[34] His insight did not carry him much further than that, however. It was through Hegel that he traced a line from Luther to the German Christians, but his own later attempt at a 'respectful demythologising'

---

33   His response to the infamous *Kristall Nacht* episode was: 'Anyone who is in principle hostile to the Jews must also be seen as in principle an enemy of Jesus Christ. Anti-Semitism is a sin against the Holy Spirit' (E. Busch, *Karl Barth*, 290).

34   B. McCormack puts it this way: 'In Luther's insistence on the literal force of the "is" in the words of institution . . . Barth saw the opening of a door to every *direct* identification of revelation and history' (*Karl Barth's Critically Realistic Dialectical Theology*, 392; cf. Barth, *Theology and Church*, 230).

of the sacraments shows, as we have seen, certain affinities with that of Hegel. Hegel (who read Luther rather differently) praised the reformer as one whose entire emphasis in the eucharist fell on 'faith and spiritual enjoyment', on the internal not the external, on the subjective not the objective. That is precisely what Hegel wanted to bring to the fore, in order to liberate the church from the remnants of its Jewish particularism, from the remains of its Catholic dogmatism and, indeed, from the burden of being the church.[35] Barth would have none of that, of course. Yet he eventually embraced a view of the eucharist that was equally subjective, demonstrating that he had failed to see, not *where* the mystery of lawlessness had quietly concentrated its work, but how that work had been carried out.

There are many other chapters to this story, of course, and many other lessons that we cannot rehearse here. But it should not go unnoticed that German fascism was something more than an attempt to raise up Thor's Oak from where it had fallen under Boniface's axe. Ground that for more than a millennium had been occupied by Christianity could not be taken by returning to a pagan or pre-evangelical past; nor could it be taken through a seamless transition to some religionless future such as Ernst Haeckel envisioned.[36] It could not be taken at all without the mediation of those who were still in their fashion Christian thinkers, and the blessing and cooperation, however briefly, of a Christian public and its churches. Its conquest therefore required the colonization of the gospel, through the poisoning of its Jewish roots, and of the eucharist. It required the hubris to elevate the old gods, through philosophies of Nature and History and Spirit that were only possible as perversions of Christian truth, to something like the heights of the Christian God; and so to elevate Man (as in Strauss's panegyric) to something like the heights of the ascended Christ.

Heinrich Heine foresaw the explosive results of this deadly admixture, when he prophesied a century before Hitler: 'It is to the great merit of Christianity that it has somewhat attenuated the brutal German lust for battle. But it could not destroy it entirely. And should ever that taming talisman break – the Cross – then will come roaring back the wild madness of the ancient warriors of whom our Nordic poets speak and sing, with all their insane Berserker rage.

---

35 See further *Ascension and Ecclesia*, 188f.
36 See R. Weikart, *From Darwin to Hitler*, 11f.

That talisman is now already crumbling, and the day is not far off when it shall break apart entirely. On that day the old stone gods will rise from long-forgotten wreckage and rub from their eyes the dust of a thousand-year sleep. At long last leaping to life, Thor, with his giant hammer, will crush the Gothic cathedrals! . . . And laugh not at my forebodings, the advice of a dreamer who warns you away from the Kants and Fichtes of the world, and from our philosophers of Nature. No, laugh not at a visionary who knows that in the realm of phenomena comes soon the revolution that has already taken place in the realm of spirit. For thought goes before deed as lightning before thunder . . . There will be played in Germany a play compared to which the French revolution was but an innocent idyll.'[37]

I USE THE WORD 'explosive' advisedly. When Paul begs the Thessalonians not to be too easily alarmed about the coming of the day of the Lord, he knows that the lawless form of humanity that will come first must somehow reach maturity. That cannot be said of western communism, which imploded, or of fascism, which exploded. To witness the continued progress of the mystery of lawlessness we must look closer to home, turning our attention to the story being written today in our own democracies. For these democracies now abound in the hubris necessary to expropriate and pervert Christian goods, and to turn law itself into lawlessness. For all their fine talk about universal principles and their efforts to expand the realm of international law, they seem to believe in nothing but positive law and to acknowledge no higher jurisdiction than their own collective will.[38] Whether they know it or not, this puts them on the same page with communism and fascism. In the end they themselves may prove the more profound embodiment of the mystery of lawlessness, for it is the law of liberty itself that they are subverting, through the construction of an antinomian gospel that only pretends to put justice before power.

The leading evangelist of this gospel has been John Stuart Mill, whose *On Liberty* still has canonical status among western legal and

---

37   As rendered by J. Satinover, *Homosexuality and the Politics of Truth*, 236; see Heine's *Religion and Philosophy in Germany*, 159f.

38   J. Waldron, 'Foreign Law and the Modern *Ius Gentium*' (*Harvard Law Review* 119, (2005), 129–45, at p. 142) observes that 'positivism's general credentials are suspect today', while acknowledging that the Benthamite tradition of seeing law 'as purely a matter of will' (145) is far from moribund. Would that it were moribund, but the line of thought we are following suggests otherwise.

political theorists and the cultural elite. Mill's most seductive idea is one that turns the Christian gospel inside out. As William Gairdner observes, for Mill 'liberty is prior to truth'.[39] In Jn 8:32, Jesus promises that the truth will set us free, but Mill promises instead that freedom will lead us into the truth – the truth that is already in us, the truth of our own highest potential. This inversion, in which we may certainly discern an echo of the Serpent's flattery, should be seen for what it is: a bid to get round Jesus' claim to *be* the truth. It undoes the dialogue of sonship, and the dialectic of binding and loosing, that informs Christian thinking about liberty. It requires a different concept of liberty altogether. Liberty no longer means freedom for God or for the realization of created ends, but rather freedom from others (God especially) and the autonomy to attain one's own ends. 'The only freedom which deserves the name,' insists Mill, 'is that of pursuing our own good in our own way, so long as we do not attempt to deprive others of theirs, or impede their efforts to obtain it.'[40]

Building on these premises is like building on sand. The inverted relation between truth and freedom causes other core concepts of Christian civilization to shift and waver as well. Dignity, for example, no longer implies objective participation in the image of God, but subjective satisfaction with one's self-image. Compassion no longer means suffering together but refusing to countenance suffering at all.[41] Tolerance no longer means charity towards those who do not measure up, but refusing to measure at all. Secularity no longer implies service to God within the provisional structures of the present age (as in 'secular priest', say), but service to the powers of the present age with a studied ignorance of God and of the age to come (as in 'secular argument', for example). Revisions such as these are having a profound impact on our society, not least on its jurisprudence, which today takes Mill's harm principle – itself an inversion of dominical teaching – as its one fixed idea. 'The only purpose for

---

39 Indeed, for Mill liberty is truth's 'efficient and final cause' (W. Gairdner, *The Trouble with Democracy*, 302).

40 *On Liberty*, chap. 1, 75. Because Mill has no meaningful doctrine of the fall, we do not find here any soteriological dialectic such as that of Irenaeus, for whom (as we saw) 'the more extensive operation of liberty implies that a more complete subjection and affection towards our Liberator has been implanted within us'. Mill's dialectic is strictly that of the one and the many – society *versus* the individual.

41 In the absence of Christian liberty the dialectic of charity and tranquility, on which authentic compassion depends, also disappears; cf. Augustine, *On the Morals of the Catholic Church* §53.

which power can be rightfully exercised over any member of a civilised community, against his will,' declares Mill, 'is to prevent harm to others.'[42] And by that light, with rationalizations constructed out of the faulty conceptual materials just mentioned, one taboo after another has been abandoned or struck down, with another rising immediately to take its place.

The ripples of the gospel according to John Mill have spread far and wide, with many damaging effects.[43] Perhaps the most damaging, however, is the least looked-for and the least noticed – though some have argued that Mill intended it – namely, the growing concentration of power in the hands of the state. It could hardly have been otherwise, of course. For the alternative to the truth and freedom offered by God to man in Jesus Christ is not, as in Mill's fancy, the freedom of every man to discover that he is his own truth; it is whatever arises and asserts itself as the arbiter of all those competing truths.[44] In other words, it is not the individual who triumphs in the appeal to a freedom that is prior to truth, but that which can command the allegiance of the individual without reference to truth. Under these circumstances the state, rather than the church – the secularist state with its purely utilitarian moral and political logic – becomes our common good and the form that justice takes. It also becomes Leviathan, the beast that the dragon calls forth from the sea.[45]

---

42 *On Liberty*, chap. 1, 73. Mill abandons 'Do unto others as you would have them do unto you' in favour of 'Do what you please so long as you do not harm others.' Of course, Mill is not the inventor of the harm principle; see Farrow, *Recognizing Religion in a Secular Society*, 148, 182.

43 For the individual as well as for society: 'Not philosophers, but merely people who love an argument, say that everyone is happy who lives as he likes. But this is false. To want what is not right is itself a very unhappy situation; in fact not to get what you want is not so unhappy a state of affairs as to want to get what you have no business to' (Cicero, as given by Augustine in *Trinity* 13.8).

44 Take education, for example. 'We must be guided', says one UNESCO (United Nations Educational, Scientific and Cultural Organization) document, 'by the Utopian aim of steering the world toward greater mutual understanding, a greater sense of responsibility and greater solidarity, through acceptance of our spiritual and cultural differences' (J. Delors, *Learning: The Treasure Within*, p. 32). But what this means in practice is that the state will shoulder aside, in the name of the individual, the communities to which individuals belong and from which they derive their different identities, thus assuming direct control of the spiritual formation of the young. Which is just what Mill hoped for; cf. L. Raeder, *John Stuart Mill and the Religion of Humanity*, 12–14 and 234–67.

45 Those who appreciate the libertarian side of Mill have been slow to recognize this, though some have at least begun to observe that within the state the judicial arm, which both weighs 'harm' and eliminates unacceptable restraints on the individual, has been

A simple illustration may be offered of the way in which Leviathan grows. In recent years, a number of western democracies have reinvented the institution of marriage in a form lacking gender complementarity or procreative purposes. This was done in the name of equal dignity and equal rights.[46] What was not noticed at the time, however, and is still widely overlooked, is that the institution of marriage has in consequence very little capacity to restrain sexual behaviour (something it did for the sake of children) and no capacity at all to limit or restrain the power of the state. Being now a creature of the state, rather than of natural law and divine mandate, marriage no longer carries with it *a priori* rights and responsibilities that the state is obligated to recognize and respect. The family no longer appears before the state as the bearer of goods all its own, and with an authority of its own. In positive law one now speaks, not of 'natural parent–child relationships', but only of 'legal parent–child relationships'.[47] This gives the state a custodial authority over every child, and every parent, that it formerly did not have. It strengthens the hand of the state in determining how children shall be educated, even morally and religiously. It is no accident that we have witnessed, both in Europe and in North America, laws and punishments (including incarceration) designed to coerce parents into cooperation with the state in its determination to see that all children are trained according to state-sanctioned ideologies, which include many other parodies of Christian beliefs and practices than same-sex marriage.[48]

Leviathan grows with the mystery of lawlessness, and the mystery of lawlessness grows by grazing on the goods of the church. Without the high concept of human dignity proposed by the church, the subjectivized version that prevails today (and prevailed in the marriage dispute) would not be possible. Without the high stakes created by the church's sacramentalizing of marriage, the alternative of same-sex marriage would never have been conceived. Without the church's subordination of sexual pleasure to sexual virtue, of the sacrament of

---

replacing the legislative arm as the real source of power, thus undermining the democratic ideal.

46   See Farrow, 'Rights and Recognition', in Cere and Farrow, eds, *Divorcing Marriage*, 97–119.

47   Cf. Farrow, *Nation of Bastards*, 13f.

48   See further *Nation of Bastards*, 63ff., and R. Reno's response in 'Personal Freedom without Political Liberty' (*First Things*: On the Square, 5 June 2008).

marriage to holy orders, of fecundity in the flesh to fecundity in the Spirit, the contrast between eucharistic man (who is fruitful in both senses) and contraceptive man (who is fruitful in neither) would not stand out with so much clarity, making marriage an object of envy and a prize to be fought for. But of course the mystery of lawlessness has worked within the church itself to obscure this contrast as far as possible even from Christians, sowing the tares of sexual confusion that have choked off their spiritual and political resistance to the contraceptive mentality.[49] That was necessary in order to deprive them and their neighbours of their political birthright; that is, of the power of the natural family and of natural law to place internal limits on the state.[50]

This example, however, points us to a much wider conquest: the general refashioning of human rights discourse, which was bequeathed to the West by the church, into a formidable weapon with which to assault not only the family but the church, and those whom the church attempts to defend. It is a small thing, perhaps, that the church is pilloried for its 'homophobia' because it discriminates in favour of children. Or for repressing women, when it has produced a civilization in which women have enjoyed unprecedented freedom. Or for supporting slavery, when for two millennia it has been the primary force of resistance to slavery. It is a small thing, in other words, that little gratitude is shown to the church for developing concepts of dignity and equality that make it possible to think about human rights, and for contributing so much in the way of the faith and courage that is necessary to act against violations of human rights.[51] But it is not a small thing when the very idea of human rights becomes a cover for violence against the church and against the weak and vulnerable, the enemy rather than the friend of justice – an enemy that does not stop short, in the name of so-called 'reproductive rights', of murder on a massive scale. How did this happen? How did rights discourse, which is so deeply rooted in the Christian doctrines of creation and fall and redemption, come to be used in what John Paul II in

---

49 Perhaps the decisive moment historically was the summer of 1968, when *Humanae vitae* [On the Regulation of Human Births] was promulgated by Paul VI, and widely rejected by 'the faithful'. The following summer saw the famous Stonewall riots, which marked, externally to the church, the transition to a new moral and political order.

50 Cf. Leo XIII, *Rerum novarum* [On Capital and Labour, 1891] §6 and §§9–11.

51 See Leo XIII, *Tametsi futura prospicientibus* [Jesus Christ the Redeemer, 1900] §3; cf. *In plurimis* [On the Abolition of Slavery, 1888] §7ff.

Evangelium vitae called 'an objective "conspiracy against life", involving even international Institutions'?[52]

An adequate answer is much too long to give here, for it would have to trace over the course of centuries a gradual shift of emphasis from objective rights to subjective rights; which is also to say, from divinely imparted rights, rooted in the *ius naturae*, to self-possessed rights, which are taken to be original and underived, and as such competing and chaotic; and from thence to rights that derive strictly from human law, understood in positivist fashion as the expression not of some underlying morality but merely of the collective will to impose order.[53] It would also have to take account of the development of 'a perverse idea of freedom' that long antedates Mill or even the Enlightenment – an idea already operative among the gnostics, whom Irenaeus criticized for wanting to be gods before their time, and for thus refusing to be human. It would have to take stock of the fact that when people reject the gospel of Jesus Christ, the incarnate and resurrected one, they turn their backs not only on their God and their brother but also on their own created nature.[54] They become debased, as Paul said, inventing 'sexual rights' to cover their debasement and to demand that others be encouraged to join in their debasement.[55] They turn their backs on those who remind them of their own weakness and vulnerability, or who get in the way of their self-deification. John Paul II got it right: 'It is clear that on the basis of [their] presuppositions there is no place in the world for anyone who, like the unborn or the dying, is a weak element in the social structure, or for anyone who appears completely at the mercy of others and radically dependent on them, and can only communicate through the silent language of a profound sharing of affection. In this case it is *force* which becomes the criterion for choice and action in

---

52 *Evangelium vitae* [The Gospel of Life, 1995] §17. The path to the immodest tyranny of the man of lawlessness, it seems, does not skirt, but runs directly through, the ever more powerful judicial fiefdoms of modern western democracies and their analogues in world bodies such as the United Nations; for 'he pretends that he vindicates the oppressed' (*Against Heresies* 5.30.3).

53 It is complicated work to trace the line that leads from Marsilius of Padua, at the turn of the fourteenth century, to today's positivists, but that line is, so to say, the lifeline of the beast.

54 See *Evangelium vitae* §18ff. on this perverse idea of freedom and the correlative 'eclipse of the sense of God and of man'; cf. *Against Heresies* 4.38f.

55 See, for example, *Sexual Rights: An IPPF declaration* (International Planned Parenthood Federation, London, 2008).

interpersonal relations and in social life. But this is the exact opposite of what a State ruled by law, as a community in which the "reasons of force" are replaced by the "force of reason," historically intended to affirm.'[56]

But even were we to say all of that, we would not have said quite enough to account for the self-devouring faith that today covers itself with the mantle of human rights, a faith that sets up its own twisted versions of the Holy Office (in the form of human rights commissions) in order to hunt down dissenters and heretics, while winking even at forced abortions and coerced sterilization and implanted identification devices and other freedom-hating practices.[57] We would have to add that all of this is not possible without the gospel itself. The proponents of this new faith go beyond the sins of pagans who have not learned the gospel, simply because they are not and cannot be pagans any more. Their thinking is informed by the gospel they have rejected; therefore both their ambition and their despair are necessarily greater. Cain-like, they rise up in irrational hatred of their eucharistic brethren, on whom they project their own evil. But they do not withdraw and build fortresses to protect themselves. Instead they seize the levers of unions, presses, governments, courts, universities, and even Christian organizations, in order to grind their enemies into the ground. They have no hidden Birkenaus or Buchenwalds, but they have – and are proud to have – their embryology labs and their abortion mills. They have as yet no uniformed Brownshirts or Hitler Youth movement, but they know quite well how to intimidate dissenters and how to propagandize children against their parents' will. They practise eugenics and euthanasia, conduct outrageous experiments moral and medical, and purport to alter natural law with the stroke of a pen. They are against fascism, they say, but they promulgate deceptive abstractions like 'pluralism' and 'tolerance', which serve only to push those of particular faith and conviction – those who might otherwise resist their designs – into the maw of their homogenizing machine. They attack the rights of conscience,[58] and even deploy nanotechnology to produce, quite invisibly yet all too literally, the mark of the beast. For the city to which they intend to be

56 *Evangelium vitae* §19.
57 See M. Connelly, *Fatal Misconception*. On the human rights commissions, cf. Farrow, 'Kangaroo Canada', *First Things*, August/September 2008, 17–19.
58 See further Farrow, 'The New Moral Order', *Catholic Insight*, November 2008, 22–4.

fathers and mothers is founded, not on communion, but on conformity. Their ambitions are not catholic, but merely universal.[59]

Today's immodest democracy, in other words, is more than an accidental by-product of a Millsian self-contradiction. It is an expression of the mystery of lawlessness that shadows on earth the heavenly session of Jesus, opposing the justice of persuasion with the justice of coercion, the culture of life with the culture of death, and the love of God with hatred of God's beloved. And we may be certain that we are not yet at the end of history, that there is still more to come. Heine's prophecy may have been fulfilled, at least in part, but Huxley's brave new world is only now becoming possible. Democracy is not destiny. It is only a moment of indecision on the path to human self-deification, which (were God to permit it) would also mean human self-destruction. At the end of that path, warns Irenaeus, when the mystery has run its course and the man of lawlessness is fully formed, 'there shall be a recapitulation made of all sorts of iniquity and of every kind of deceit'.[60] In order to reach that point, however, in order to attain that leveraged maturity, lawlessness must go on borrowing from law, disobedience from obedience, wickedness from righteousness, 'rights' from right, lies from truth, Antichrist from Christ. Only the gospel can lead humans to freedom; likewise, only the gospel can equip humanity for the final apostasy.

About the end we have already declined to speculate, but it is no speculation to say that the man of lawlessness begins as the man who turns freedom into unfreedom by rejecting the truth:

> Freedom negates and destroys itself, and becomes a factor leading to

---

59 When the Reformation brought to an end the tenuous peace that Christendom had introduced into an always-fractious Europe, the attempt to restore peace consisted, philosophically speaking, in a search for universal principles that would transcend all confessionalism. While this was said to be necessary in view of the violence that had erupted along the confessional fault-lines in Europe, it was not quite what it seemed to be. Arguably it was the boldest assertion yet of the Protestant principle. Catholic Christianity (and Christendom with it) had already been broken up by the reformers' attempt to reduce it to its biblical elements; it would now be milled by still more penetrating critics for its strictly rational elements, on which a universal commonwealth might be built ('the Catholic idea with the supernatural left out', as Benson put it in *Lord of the World*, 34). These elements proving elusive, however, postmodernists, to save face, took to treating interim measures, such as 'pluralism' and 'tolerance', as if they were ends rather than means. But to make ends of means is perforce to make a mean end; to retain the notion of a universal commonwealth while doing so is perverse.

60 *Against Heresies* 5.29.2

the destruction of others, when it no longer recognizes and respects its essential link with the truth. When freedom, out of a desire to emancipate itself from all forms of tradition and authority, shuts out even the most obvious evidence of an objective and universal truth, which is the foundation of personal and social life, then the person ends up by no longer taking as the sole and indisputable point of reference for his own choices the truth about good and evil, but only his subjective and changeable opinion or, indeed, his selfish interest and whim . . . Thus society becomes a mass of individuals placed side by side, but without any mutual bonds . . .; any reference to common values and to a truth absolutely binding on everyone is lost, and social life ventures onto the shifting sands of complete relativism.[61]

Under these conditions the man of lawlessness arises (to use Nietzsche's term) as the *Übermensch*, the man beyond good and evil, the man who is bold enough to set himself above every so-called god and to forge his own truth. To this man, and to his day – the day, so to say, of his incarnation – Hitler himself looked forward in *Mein Kampf*. 'From millions of men', he wrote, 'one man must step forward, who with apodictic force will form granite principles from the wavering idea-world of the broad masses and take up the struggle for their sole correctness, until from the shifting waves of a free thought-world there will arise a brazen cliff of solid unity in faith and will.'[62] Or as John put it long beforehand: 'To the beast the dragon gave his power and his throne and great authority. One of its heads seemed to have a mortal wound, but its mortal wound was healed, and the whole world followed the beast with wonder. Men worshiped the dragon because he had given his authority to the beast, and they

---

61   *Evangelium vitae* §19

62   *Mein Kampf*, 346. Apparently he held this vision to the end. *The World at War*, vol. 9 (Jeremy Isaacs, producer; Thames Television, 1973) includes an interview in which his personal secretary, Traudl Junge, reports that in his last days Hitler replied to a question about the future of Germany as follows: 'I believe there will not any more be a National Socialist Party; the idea will die with me. But maybe in a hundred years or so there will arise another National Socialist idea, like a religion.' Hitler's valet, Heinz Linge, offered this account of Hitler's final leave-taking in the Führerbunker on 30 April 1945 (*World at War*, vol. 8): 'He said farewell to everybody, and I was the last one he came to. Hitler said to me, "I have given the order to break out. You should break out in groups. Join one of these groups and try to get through to the West." Then I asked Hitler, "For whom should we fight on for now?" And to that Hitler said, in a monotone, "For the coming Man."'

also worshiped the beast and asked, "Who is like the beast? Who can make war against it?"'[63]

GOD, OF COURSE, has already his answer to this man, who rules not over the beasts but for and as the beast. His answer is the parousia of Jesus Christ. But the church has also to give answer in the meanwhile, by way of its eucharistic witness to Jesus. And here we must recognize the part that schism plays in the rise of the beast, for the mystery of lawlessness must perforce work within the church if it is to work outside it. Of Antichrist there is, as Cyril says, 'a sign proper to the church'.[64]

An illuminating illustration, both of schism's contribution and of authentic eucharistic witness, may be drawn from an earlier struggle over marriage that took place in the sixteenth century.[65] King Henry VIII, though vigorously opposed to most Protestant ideas, broke with the church nonetheless for reasons connected to his great cause; that is, the annulment of his marriage to Catherine of Aragon so that he might produce a son and heir by another woman. In his vain attempt to control the future of the English throne, Henry followed the Erastian trajectory of the Reformation and exalted himself, Uzziah-like, before God – not by offering eucharistic sacrifice, much less by declaring himself to be God, but by taking a seat in the temple that did not belong to him, declaring himself to be the Supreme Head of the Church of England. In this way he conflated his own session and kingdom with the session and kingdom of Jesus, and presumed to govern the church, especially its sacraments of marriage and holy orders.

Henry's struggle to secure his position led in 1534 to the Act of Succession, and to a mandatory oath of allegiance that included assent to everything declared by parliament about marriage in general and about the king's personal situation in particular. Later

---

63   Rev. 13:3-4 (*RSV*, altered). We need not play off the individuality that is evident in the above quotation from *Mein Kampf* against the corporate nature of the beast or of the society it rules. We should think at one and the same time of both.

64   *Catechetical Lectures* 15.7

65   Marriage being the point at which the sacraments of the church directly overlap the ceremonies of the state, and the Christian family being both a domestic church and, with other families, the 'natural and fundamental group unit of society' (*Universal Declaration of Human Rights*, art. 16), it need not surprise us that the contest has been playing itself out around marriage. The attack on that sacrament is a thinly veiled attack on the eucharist itself, as the sacrament of sacraments.

that year the Act of Supremacy also established his ecclesiastical jurisdiction, making no mention of the proviso formerly attached to it by the bishops: 'as far as the law of Christ allows'.[66] With that Act a crack appeared in the very fabric of Christendom. The forces that produced it were many and various; some had been building for a very long time. But it is safe to say that the Protestant schism created the conditions under which Henry's course (conceivable from the days of Philip the Fair and, in England, of the Praemunire) became navigable.[67] Those who aided him in it, at home or abroad, had their own reasons, religious or otherwise. Some were more successful than Henry, some less, at achieving their goals. Those of purest motives no doubt thought that they were striking a blow for Christ and for freedom, but whatever success they enjoyed was fleeting and came with a high price tag. I am not referring to the battle of the bonfires that was just round the corner, but to something still more serious than that. For the head of the beast that had been mortally wounded by the witness of the early Christian martyrs, who by the fourth century had wrung from the Roman emperors admissions of defeat and edicts of religious toleration, was slowly being healed by the Erastian draughts proffered it by politically entangled reformers.[68]

From Ambrose to Anselm, watch had been kept over this twitching beast by courageous bishops who refused to be moved on the matter of the church's independence from the state. But the overweening ambitions of Boniface VIII, followed by the Avignon débâcle and by the rampant simony and nepotism of the Renaissance popes – whose sins were punished not only by the Sack of Rome but also by the Reformation itself – relaxed that watch. Princes arose in Christendom who knew no authority greater than their own and who were determined to make the church do their bidding.[69] Henry VIII,

---

66   See R. Marius, *Thomas More*, 480f. (I am indebted to this fine biography, though the reader must not attribute my views to Marius.) Cf. O. Chadwick, *The Reformation*, 99ff., which points out that the 1531 formulation did not yet amount to a repudiation of papal authority.

67   For the background, see O. O'Donovan and J. O'Donovan, *From Irenaeus to Grotius*, 389ff. For the foreground, see D. MacCulloch, *Thomas Cranmer*, 41ff.

68   Obviously I use the word 'Erastian' anachronistically; perhaps 'Marsillian' would be more accurate. Nor do I mean to imply that reformations can, historically or even in principle, be disentangled from politics. What is always at stake is the authenticity of the politics of the eucharist, and of the relation between church and state.

69   This was by no means a Protestant phenomenon, and it will be recalled that Rome was

by writing his right to do so into the laws of the realm, stepped forth (without really meaning to) as the representative of a new era. In that era the ongoing process of subordinating religion to the demands of the state would outrun the monarchy as such – helping indeed to bring about its degradation and defeat – and adapt itself successfully to polities supposedly more tolerant of religious liberty.[70] These new polities, of course, do not claim authority over the spiritual or ecclesial realm in the manner of Henry, or even of Hobbes. They do not ask or demand religion's cooperation, in the fashion of Hitler and the Nazis. With alarming frequency they simply deny, as a matter of principle, that religion touches in any legitimate way upon political life, and so expropriate for themselves not merely some but *all* of the church's authority over things public. Therefore they also do not shy from commanding public servants (not only clergy acting in that capacity but also doctors and druggists and the like, not to speak of soldiers and police) to contravene the laws of their religion and the dictates of their consciences in order to demonstrate their allegiance to the state.

Henry's famous chancellor, Sir Thomas More, knew all about that. More, not spared by his resignation from high office, was quickly called upon to sign the oath of allegiance. He knew that he could not do so, for his conscience forbade him to put the law of England above the law of Christ. (The oath was nothing if not a conscience-suppressing measure.) Embodying the eucharistic spirit, the spirit of the thank-offering and the sacrifice for sins, he did not flee for refuge to the Continent or to Rome. He went rather to Mass and then to face his fate. Like St Ambrose, his way was 'to defer, but not yield,

---

sacked by Protestant mercenaries under the banner of the Holy Roman Emperor, who in his perpetual war with Philip the Fair's distant successor, Francis I, had decided to put the squeeze on the dithering Pope Clement V (see R. A. Scotti, *Basilica*, 159ff., for a brief account). But Charles V and Francis I were both still subject to the Vicar of Christ in matters eternal, and hence also – albeit in an increasingly confused sense – in at least some matters temporal. In the Protestant lands, and in England first of all, this was about to change.

70 Henry set in motion, and his successors (including his Catholic and Protestant daughters) perpetuated, a process of religious repression in England that was entirely understandable in its context, though not for that reason morally justified. That here, as elsewhere, this did much to deplete the moral capital of the monarchy cannot be denied, even if there are good reasons to challenge the larger myth that surrounds the so-called Wars of Religion, which were largely the product of secular ambitions – that is, of familial and national rivals exploiting the conditions created in part by the schism.

to emperors'.[71] He was ready both to honour the king in every way possible and also to refuse the king what was demanded. Ambrose had resisted the young Valentinian II with the following words: 'I fear the Lord of this world more than the emperor of this age . . . The church is God's, and so it ought not to be given over to Caesar, because Caesar's sway cannot extend over the temple of God.'[72] That was More's conviction also, though his public statements were at first more guarded.

More, let it be said, was no crusader against the king or his new marriage, though he declined to attend Anne's coronation. He did not see himself as the judge of Henry's conscience, nor did he take lightly the duty of self-preservation. He was quite willing to sign to the succession, which after all was a temporal matter, within the powers of parliament. But he was not prepared to give assent to the supremacy. Having already cost him the chancellorship, this now cost him his freedom. While he was imprisoned in the Tower on suspicion of high treason, he had time to reflect on the public as well as the private combat in which he was engaged. His immense knowledge of the law only underlined the perversion in Henry's obsessive concern (for such is the way of lawlessness in our time) to cross every legal 't' and dot every legal 'i' in defence of his innovations.[73] The case More generated for his own defence, though itself a sophisticated appeal to the laws of the realm, cut through all of that like a knife. It came down in the end to a simple rhetorical question, containing a *reductio ad absurdum* that he had already put to the Solicitor General during the interview that precipitated his trial. Can parliament, he asked, declare God not to be God?

His point was that of Paul and the early martyrs, as of Ambrose. It was a point about jurisdiction, and it exposed the mystery of lawlessness now coiled at the core of the law itself. Both for Henry's sake and for his own – the punishment for treason being what it was – More was not eager to make this point, but in the end it had to be made. And, in making it, he left the king little choice but to repent of the

---

71   *Sermon against Auxentius* §2

72   *Sermon against Auxentius* §1, §35.

73   As Marius puts it, 'Cromwell's government had an almost compulsive desire in those years to preserve the form of the legalities, as if meticulous attention to the processes of law would cast a spell over the actuality that might ensure the survival of the new order' (*Thomas More*, 482). This tendency was evident from the beginning of the King's Great Matter.

supremacy or to make of him an example, lest his godly reasoning spread like an infection throughout the realm. Henry took the latter course. An invalid Act – invalid because *ultra vires* for a secular power and, as such, blasphemous – was enforced by the law's ultimate sanction. Against that sanction More offered no resistance. Having been liberated by Christ from his fear of what Henry could do to him, he calmly laid himself on the block at Tower Hill on 6 July 1935, and died, as he said, 'the king's good servant but God's first'.[74]

Unlike the Carthusian monks who also refused to sign the oath of allegiance, More was spared the ritual horrors of Tyburn, which were as inhuman as any devised by pagan tyrants. But he too was made like 'the wheat of Christ', ground by the teeth of the beast because he refused to live by the lie.[75] Richard Marius writes in his biography that More 'intended to set a precedent that men would remember when Henry was gone and passions had cooled and the law had resumed its grand march across English life toward ideal justice'.[76] But that takes us nowhere near the heart of it. The author of *Utopia* (as Marius knows) was no utopian. He did not presuppose any such grand march. Rome had been brutally sacked, and the Red Horse would soon be running willy-nilly throughout the fields of England as well as Europe. Whether More saw far enough to glimpse the head of Charles dropping from the block at Whitehall, or figures such as Hobbes and Du Moulin teaching men to regard the commonwealth itself as a kind of church, is hard to say.[77] Whether he

---

74 Marius, *Thomas More*, 514; see further 504ff. On the invalid nature of legislation that, though valid procedurally, contravenes the law of God, see, for example, Aquinas, *Summa Theologiae* 1-2.93.3: 'Human law is law only by virtue of its accordance with right reason; and thus it is manifest that it flows from the eternal law. And in so far as it deviates from right reason it is called an unjust law; in such case it is no law at all, but rather a species of violence' (cf. 1-2.96.4). Any state that acknowledges the link between the rule of law and the supremacy of God implicitly allows Aquinas' and More's point; failure to acknowledge it, or to attempt to act in a manner consistent with it, is to transform law into lawlessness.

75 The same may be said of St John Fisher, of course. The quotation is from *Against Heresies* 5.28.4: 'And for this cause tribulation is necessary for those who are saved, that having been after a manner broken up, and rendered fine, and sprinkled over by the patience of the Word of God, and set on fire [for purification], they may be fitted for the royal banquet. As a certain man of ours said, when he was condemned to the wild beasts because of his testimony with respect to God: "I am the wheat of Christ, and am ground by the teeth of the wild beasts, that I may be found the pure bread of God."'

76 *Thomas More*, 505.

77 See J. Collins, *The Allegiance of Thomas Hobbes*, 226; cf. *Leviathan* §§39, 42.

foresaw the foolishness of the Fifth Monarchists (who quite forgot Irenaeus' warning against speculation) or the wickedness of the Powder Treason (which betrayed everything More stood for) may be doubted. Yet he did see the sign of Antichrist growing ever blacker in the schismatic European skies. Perhaps he even saw the little red cross he carried to the scaffold as a symbol of waves of martyrdom to come. Certainly he could see that the devil himself was learning to play the justice game, and what he wanted to leave for posterity was an example of how to win when playing against the devil; that is, how to put justice before power and how to live and die by the politics of the eucharist.[78]

More was not to know, of course, that Henry's church, modified by the march of democracy, would last nearly half a millennium before coming to grief on the very same stumbling blocks – marriage and holy orders – that occasioned its appearance. But that knowledge would have changed nothing, for the politics of the eucharist that More practised is opposed to what men falsely call *Realpolitik*. More understood very well the art of the expedient; he was a consummate lawyer, diplomat and politician. But he also understood that it is never expedient to oppose oneself to God, as Henry and his collaborators were doing. The politics of the eucharist takes one into the heart of things as they actually are, and teaches a prudence that only the love of God knows. It makes of its students the best of citizens (because they are the only ones who are truly free) and hence also the most subversive of citizens. It teaches them to be governed by the sign of the cross – apart from which there is no *Realpolitik* – and to deny any claim by the state to rule over their souls or, indeed, over their bodies in abstraction from their souls. For in the eucharist both body and soul are claimed by the one who ascended in the flesh to the right hand of God.

More knew that, and needed to know no more than that. We who live in modern democracies need, however, to recognize the distinction between his day and ours. Henry attempted to rule over

---

78 To remark on the fact that More defended the practice of burning heretics, and was himself involved in it, would require comment not only on the *Dialogue Concerning Heresies* but on the death penalty as such and on the whole late-mediaeval construct of Christian society. That I cannot attempt, but neither can I avoid acknowledging the question as to whether lawlessness was not already present, in this form, in the heart of law. Certainly, though this is quite another matter, it was present in the rituals of Tyburn.

the souls of his subjects that he might rule also over their bodies. Modern democracies pretend rather to leave the soul free by dividing the realm of the body from that of the soul, a trick they learned from Rousseau and, more subtly, from Locke and Rawls.[79] That it is a trick should be obvious enough, for if Christ cannot be divided neither can we. 'We adjure all Christians throughout the world to strive all they can to know their Redeemer as he really is', wrote Leo XIII; to know him, that is, as 'the fountain-head of all good' both in this life and the next, and as the one on whom 'the salvation of all men, both severally and collectively, depends'.[80] Was that not Barth's point also in the second article of the Barmen Declaration? 'Just as Jesus Christ is the pledge of the forgiveness of all our sins, just so – and with the same earnestness – is he also God's mighty claim on our whole life; in him we encounter a joyous liberation from the godless claims of this world to free and thankful service to his creatures. We repudiate the false teaching that there are areas of our life in which we belong not to Jesus Christ but to another lord, areas in which we do not need justification and sanctification through him.'[81]

With this statement – a statement that has no force if Christ did not ascend in the flesh – Barth captured the essence of an authentically Christian political theory, the deficiencies of his dualistic sacramental theology notwithstanding. And, putting theory to practice, he fell afoul of temporal authorities who wished to extend their sway over the household of God. In the four hundredth anniversary of More's martyrdom, Barth too was asked to sign an oath of allegiance that many around him had signed. He refused. He also declined to open his classes at the University of Bonn with a salute to Hitler, offering instead, as was his custom, a prayer in the name of Christ. For these offences he was dismissed from his post and deported to his native

---

79 On Rousseau, see further my essay, 'Of Secularity and Civil Religion' (*Recognizing Religion in a Secular Society*, ed. Farrow). It is from Rousseau, more than Mill, that modern democracies have learned to regard religion as a private matter, and so to conflate freedom of religion with freedom of conscience – a very strategic weapon against the church in their present legal arsenal.

80 *Tametsi futura prospicientibus* §3f. and §13. Cf. §5: 'What hope of salvation can they have who abandon the very principle and fountain of life? Christ alone is the Way, the Truth and the Life.'

81 The operative principle here is that no one can claim our unconditional allegiance but the one who can offer us a complete salvation, which is exactly what Christ has done in consequence of his ascension in the flesh. That was Leo's view too, and no one did more than Leo to spell out its political implications.

Switzerland.[82] Those who put theory into practice a few years later, when the Nazi beast had fully bared its teeth, would not be so fortunate. A great many around the world have not been, nor will be, so fortunate, as the power of the beast grows. But I misspeak. For the high politics of martyrdom is neither misfortune nor mere fortune. It is vocation and exaltation. In its own way, it is redemption and atonement. Moreover, in the mingled blood of Catholic and Orthodox and Protestant martyrs there is hope, as John Paul II insisted in *Ut unum sint*, for the overcoming of schism itself, and hence for a united witness to the one who says, 'Behold, I am coming soon, bringing my recompense.'[83]

---

82   For details see E. Busch, *Karl Barth*, 235ff.

83   Rev. 22:12 (cf. v. 20 and chaps 14–19). On *Ut unum sint* see further my 'Church, Ecumenism, and Eschatology', *The Oxford Handbook of Eschatology*, 347ff.

CHAPTER SEVEN

Ascension and Atonement

THE ASCENSION OF JESUS CHRIST is an act of saving grace accomplished by the triune God. It is an act of Jesus the Son, who for our sake presents himself before the Father in his priestly role as offerer and offering. It is an act of the Father who receives Jesus, and so also receives us; and who also says to him, 'Sit here, while I make your enemies a footstool for your feet.' It is an act of the Holy Spirit, who places Jesus at God's right hand by reorganizing all things around him, beginning with the church, which is the first and greatest consequence of his ascension and heavenly session. In all these ways the ascension is salvific, finishing what was begun in the baptism of Jesus, namely, the defeat of sin, the overthrow of Satan, the reconciliation of Israel to God, and the founding of a royal priesthood that is catholic in scope.[1]

But the ascension of Jesus Christ is also an act of perfecting grace, completing what was begun when the Spirit, who long ago brooded over the waters and brought forth life on earth, hovered over Mary, who brought forth a son. Not only does it fully erase the alienation between God and man introduced by the fall, it fully establishes the communion between God and man at which God was already aiming in the creation itself. Indeed, it is the act by which God in principle – or rather in person – completes the formation of man and perfects his image in man. In bearing our humanity home to the Father, Jesus brings human nature as such to its true end and to its fullest potential in the Holy Spirit. He causes it to be entirely at one with God, and so to become the object (and, for other creatures, the mediator) of God's eternal blessing.

The ascension, in both senses, is atonement: the 'one-ing' of God and man that is the goal of the incarnation. For 'in him and by him we are mightily taken out of hell and out of the misery on earth, and honorably brought up into heaven and full blessedly one-ed to our essence, increased in riches and nobility, by all the virtue of Christ

---

1 The church, says Cyril, 'is called "catholic" because it extends over all the world, from one end of the earth to the other; and because it teaches universally and completely one and all the doctrines which ought to come to men's knowledge, concerning things both visible and invisible, heavenly and earthly; and because it brings into subjection to godliness the whole race of mankind, governors and governed, learned and unlearned; and because it universally treats and heals the whole class of sins, which are committed by soul or body, and possesses in itself every form of virtue which is named, both in deeds and words, and in every kind of spiritual gifts' (*Catechetical Lectures* 18.23). Its catholicity is a function of the universal reach of the atonement accomplished by the ascended Christ.

and by the grace and action of the Holy Spirit.'[2] Distinguishing the two senses is necessary, however, and useful just here. For the first or soteriological sense invites us to think in terms of the purgation of heavenly things, while the second or teleological sense invites us to think in terms of the glorification of earthly things; and these tasks we have yet to assay. Let us begin with the former, matching brevity to ignorance, lest we find ourselves trying to speak of things 'which man may not utter' – even that rare man who reportedly has seen and heard.[3]

When he was lifted up on the cross, Jesus defeated the last temptation and answered for human rebellion with an unreserved self-offering. Thus did he complete his earthly mission of purification, 'not despising or evading any condition of man' but sanctifying our humanity entirely in his own person.[4] That he did complete his mission goes without saying, since he himself cried out, 'It is finished!' But Hebrews, working by analogy with Yom Kippur, tells us that when Christ ascended he purified heavenly things as well. 'It was necessary', says the author, referring to the work of the Aaronic priesthood, 'for the copies of the heavenly things to be purified with these rites, but the heavenly things themselves with better sacrifices than these. For Christ has entered, not into a sanctuary made with hands, but into heaven itself, now to appear in the presence of God on our behalf.'[5]

---

2   Julian of Norwich, *A Lesson of Love* §58 (trans. Fr John-Julian). In Julian, the at-one-ing accomplished by the triune God is quite rightly a seamless garment of creation and salvation. We may, however, take the Scotist line that the incarnation would have taken place even had the fall not taken place, on the understanding that this counterfactual speculation serves two purposes: to insist, as I think we must, that the fall is not predetermined; and to assert that the true nature of man is realized only through *henosis* or union with God, a *henosis* that requires the mediation of the God-man.

3   See 2 Cor. 12:1-4.

4   'For he came to save all through means of himself – all, I say, who through him are born again to God – infants, and children, and boys, and youths, and old men. He therefore passed through every age, becoming an infant for infants, thus sanctifying infants; a child for children . . . a youth for youths . . . an old man for old men, that he might be a perfect master for all . . . Then, at last, he came on to death itself, that he might be "the first-born from the dead, that in all things he might have the pre-eminence," the Prince of life, existing before all, and going before all.' (*Against Heresies* 2.2.4)

5   Heb. 9:23f. We may not, with Hegel, take these words as a reference to the cross itself, for it is not the dead Christ who appears in God's presence. Nothing dead can appear before the living God! Hebrews echoes the theology of Paul: 'Christ Jesus who died – more than that, who was raised to life – is at the right hand of God and is also interceding for us' (Rom. 8:34, *NIV*).

What is meant here by heaven? Not God's own place, which is simply God himself, who dwells in unapproachable light. That is a place the Son, even as incarnate and crucified, has never left, for it is God himself who has undertaken the work of atonement.[6] The heaven to which Hebrews refers, though not a part of this world, is the place from which God's rule over the world is effected through the angels, the place where God's presence to and for creation is manifested and known, the place around which all creation is therefore ordered and arranged. This is the heaven that Jesus enters through his ascension, and he enters to prepare the new creation by conforming that place to himself; we might almost say, by revealing himself – his incarnate self – *as* that place.[7]

Why is heaven's purification necessary? It is necessary because he enters as man and on behalf of men, yet even the likes of Enoch and Elijah are sinners.[8] It is necessary because heaven has been the place of traffic between God and a fallen world. For the same reason that Isaiah's lips require to be purged with a coal taken from the heavenly altar, in that they are an avenue of the word of God to sinners, the altar itself and the sanctuary and the ministers in the sanctuary require purification. And there is nothing that can purify them except Jesus Christ, 'the Man who sinless came and sinless went'.[9] He is the only one who can traffic with sinners and effect atonement, rather than need atonement. It is necessary, moreover, because of the rebellion in heaven itself, which antecedes rebellion on earth. The angelic

---

6   Cf. 1 Tim. 6:13-16.

7   To speak, as we did earlier, about Jesus ascending into the new creation requires us to speak about him entering heaven in this sense; that is, to acknowledge, not only the preparation which is his own resurrection, but the preparation which is his self-unveiling to the angels as the King of Glory. 'From of old, yea, from the beginning, [the Son] always reveals the Father to angels, archangels, powers, virtues, and all to whom he wills that God should be revealed' (*Against Heresies* 2.30.9). But now there is a new form of revelation, a revelation in person, a revelation that will be the basis for a new order in heaven and on earth. 'Truly, truly, I say to you, you will see heaven opened, and the angels of God ascending and descending on the Son of Man' (Jn 1:51; cf. Phil. 2:5-11).

8   'Enoch was translated, Elijah was borne, but Jesu Christ by his own might is ascended into heaven' (Jacobus de Voragine, *Aurea Legenda*, vol. 1; trans. W. Caxton, 1483). But behind 'might' we must understand, first, justice; that is, 'by what merit he ascended'. Here we are referred to Jerome, though Augustine would have done as well: '"But by justice and not only by puissance, but by justice and by right thou hast delivered man, and I have withholden of thy puissance, and thy virtue shall bring thee to heaven." This said God the Father to the Son.'

9   Dante, *Inferno*, xxxiv, 15 (trans. L. Binyon).

host is divided and heaven is at war. It is not just hell (Hades) that Jesus harrows, then, when he descends and ascends, but heaven too. It is not just the world that he sets to rights, but also the place from which the world is ruled. Though he does not forgive the sins of angels, he does purge the heavens of every trace of angelic sin and shred every banner of the one whom Dante astutely called 'the Emperor of the kingdom of despair'.[10] That is what the mediaeval Rogation processions used to celebrate, is it not?[11]

The impact of his self-offering as he enters heaven is different from its impact on earth, because heaven is not earth. On earth his offering incites sin, as well as forgiving sin; that is why it comes to a cross, and why it provokes the mystery of lawlessness and the man of lawlessness. In heaven, however, it puts an end to sin and comes to glory, unadulterated glory. It restores paradise as the seed of the new creation. It generates the church and sends it forth into the world[12] – along with the devil, whom it confines to the world. The devil, scripture tells us, ceases to appear before God when the blood that speaks a better word than the blood of Abel is presented there. 'I saw Satan fall like lightning from heaven', says Jesus prophetically, and the Apocalypse takes up the theme, illuminating the great drama of salvation history:

> Then God's temple in heaven was opened, and the ark of his cove-
> nant was seen within his temple; and there were flashes of lightning,
> loud noises, peals of thunder, an earthquake, and heavy hail. And
> a great sign appeared in heaven, a woman clothed with the sun,
> with the moon under her feet, and on her head a crown of twelve

---

10 *Inferno* xxxiv, 28. Cynewulf's *Christ* does not go wrong when it associates the ascension of 'the King of glory, the Helm of heaven' with a great battle in the angelic realms, albeit a battle certain in outcome, which completes the harrowing of hell: 'Now hath the Holy One harrowed hell of tribute . . .' (trans. C. Kennedy).

11 'In this procession,' says the Golden Legend, 'the Cross is borne, the clocks and the bells be sounded and rung, the banners be borne, and in some churches a dragon with a great tail is borne. And aid and help is demanded of all Saints . . . The Cross is borne for to represent the victory of the Resurrection, and of the Ascension of Jesu Christ. For He ascended into Heaven with all a great prey. And thus this banner that flyeth in the air signifieth Jesu Christ ascending into Heaven. And as the people follow the Cross, the banners, and the procession, right so when Jesu Christ styed up into Heaven a great multitude of Saints followed Him. And the song that is sung in the procession signifieth the song of angels and the praisings that came against Jesu Christ and conducted and conveyed Him to Heaven where is great joy and melody.'

12 Irenaeus (*Against Heresies* 5.20.2) thus refers to the church as *paradisus in hoc mundo*.

stars; she was with child and she cried out in her pangs of birth, in anguish for delivery. And another sign appeared in heaven; behold, a great red dragon, with seven heads and ten horns, and seven diadems upon his heads. His tail swept down a third of the stars of heaven, and cast them to the earth. And the dragon stood before the woman who was about to bear a child, that he might devour her child when she brought it forth; she brought forth a male child, one who is to rule all the nations with a rod of iron, but her child was caught up to God and to his throne . . . Now war arose in heaven . . . and the dragon and his angels fought, but they were defeated and there was no longer any place for them in heaven. And the great dragon was thrown down, that ancient serpent, who is called the Devil and Satan, the deceiver of the whole world – he was thrown down to the earth, and his angels were thrown down with him. And I heard a loud voice in heaven, saying, 'Now the salvation and the power and the kingdom of our God and the authority of his Christ have come, for the accuser of our brethren has been thrown down . . .'[13]

We cannot neglect the fact that the cleansing of heaven produces trauma on earth. We have already observed that Christ's presence there breaks the seals of history and brings judgements upon men, even as it heaps glory upon the church through its martyrs. But what we must emphasize here is that Christ's appearance in heaven also makes heavenly things accessible on earth, even to the least of his brethren, while denying them to principalities and powers that once had access but are now reduced to borrowing what they can from the church, to producing mere simulacra of heavenly things. This produces an observable revitalization of the human race, to which the communal life of Christians in Jerusalem, and the table fellowship of Jew and Gentile in Antioch, and other signs of hope and promise (religious and cultural and scientific) bear witness. 'The salvation and the power and the kingdom of our God and the authority of his Christ have come', not only because the devil has been thrown down but because the stars of the seven churches have been lifted up,

<hr />

13   Rev. 11:19–12:10. 12:12 adds: 'Rejoice then, O heaven and you that dwell therein! But woe to you, O earth and sea, for the devil has come down to you in great wrath, because he knows that his time is short!' Cf. Lk. 10:17-20.

revealing to men and angels alike 'the plan of the mystery hidden for ages in God who created all things'.[14]

'When he ascended on high he led a host of captives, and he gave gifts to men.'[15] The churches have access to the righteous angels. They have access to the things of the Spirit. They have access to the heaven that orders earthly affairs. They have access to the world to come. They have access for the intellect, through participation in the mind of Christ; access for the soul, through the exercise of the keys; access for the body, through receipt of the bread of immortality and the cup of salvation and the oil of anointing.[16] All this may be partial and provisional, transitional and hence sacramental, mysterious and not altogether utterable, but it is nonetheless real and effective. 'Whatever you bind on earth shall be bound in heaven,' Jesus tells Peter, 'and whatever you loose on earth shall be loosed in heaven.'[17] So if there remains until the parousia a profound distinction between earth and heaven, a distinction more wonderful and more terrifying than ever it was, that distinction is already redundant now that Christ has entered heaven. To say 'heaven is not earth' is to speak of what was, not of what will be. Which leads us to think also of the glorification of earthly things, and of man himself.

IF ONE WERE TO EXCHANGE the reading of this little book (and a happy exchange it would be) for the privilege of participating in an eastern celebration of the eucharist, one would find that the latter takes its departure from this very point: 'Blessed be the kingdom of the Father and of the Son and of the Holy Spirit, now and for ever and ever.'[18] One would also discover that the eastern liturgy

---

14 See Ephesians 3. The heavens are cleansed, then, also in this sense: that the purposes of God are vindicated before the whole creation.

15 Eph. 4:8, quoting Ps. 68:18. Or as Hebrews puts it: 'Christ appeared as a high priest of the good things that have come' (9:11, *RSV*).

16 Cf. 1 Corinthians 10–15 and 2 Corinthians 3–6.

17 Mt. 16:19. Were the task of outlining an ecclesiology – a task already declined in the preface – to be taken up after all, would it not have to begin here? In the first chapter of *Ascension and Ecclesia* I agreed with those who think that ecclesiology must be rooted in the revelation of the church that is made available in the eucharist. It is no retraction or retreat to say that Matthew 16 provides a set of keys, hermeneutically speaking, without which we cannot hope to unlock the secrets of the eucharist itself. The church is a consequence of the ascension. It is eucharistically constituted, shaped, and nourished. But it is built on Peter's confession and on the power of the keys given to him, as our Lord expressly declares.

18 P. Galadza, ed., *The Divine Liturgy*, 391.

unfolds, much more dramatically that that of the Latin church, as a re-enactment of the atoning work of Christ, which it considers not only from the standpoint of soteriology but also of teleology. By stages it moves – through the little entrance, the great entrance, the epiclesis and the communion – towards the promise of eternal glory, towards the union of earth with heaven, towards the final destiny of man. The work of atonement it depicts as an anaphoric work, a work that is at once redemptive and perfective, an uplifting and transfiguring work.[19]

It is in that light that we want now to see the ascension. But since we are reading (or, in my case, writing) rather than singing and praying the divine liturgy, we may resort instead to the artificial device of a diagram, containing both eastern and western elements, the function of which is to display the ascension's place in the wider anaphoric work of Christ, which is conducted in three distinct movements, each with a descending and an ascending phase:

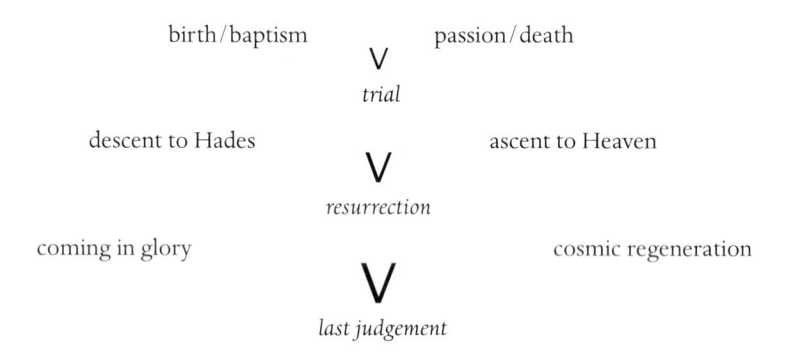

In the first movement Christ undertakes the mission that provokes his trial and conviction, and is lifted up on the cross in demonic

---

19 If the West has tended to focus on the soteriological at the expense of the teleological, that is not true of the East. One reason westerners are struck by the eastern liturgy is that it functions as a narrative of eschatological transformation and transfiguration, telling a story that includes the ascension, the parousia, and the perfection of creation. (Maximus, e.g., as H. Wybrew notes in The *Orthodox Liturgy*, p. 183, sees in the anaphora a symbol of 'our future union with the spiritual powers of heaven' and in the elevation 'the union of all the faithful with God in the age to come'.) And here I want to observe that, were I writing *Ascension and Ecclesia* today, I would write the fourth chapter differently, retreating from certain of the criticisms it offers of East and West, yet without retracting altogether.

mockery of the destiny of man. Thence, in a second movement that recapitulates the first, he descends to the dead but is restored to life – God's verdict reversing man's – and is raised up to heaven.[20] In the final movement, having set heaven in order, he descends to judge the earth, and causes us to ascend with him into the Father's presence. That is, he glorifies God by leading many sons to glory and by offering up the whole creation to be the kingdom of God. It is to that consummate offering that his ascension already tends, for by it he presents himself to God as the firstfruits, that God may in turn present him to us as the guarantee of a full harvest.[21]

Now in each of these movements we should see an accompanying work of the Spirit, without whom there can be no harvest. That is something we learned from Irenaeus, whose account of Christ's anaphoric work is aptly described as epicletic. What man lacks, in his disobedience to the Word of God, is precisely that which perfects in him the divine image and likeness, namely, the Spirit. What he receives in the gift of the incarnation is the resolute companionship of that Word, who is capable both of recalling man to the Spirit's ministrations and of recalling the Spirit to the aid of man; that is, of habituating man to the Spirit and the Spirit to man.[22] And when this is achieved, when through the invocation of the Spirit man has ascended with Christ into the presence of the Father, it is not only the condition of man that will be changed but the condition of all

---

20 'S. Ambrose saith: Jesu Christ came into this world to make a leap; he was with God the Father, he came into the Virgin Mary, and from the Virgin Mary into the crib or rack. He descended into flom Jordan, he ascended upon the cross, he descended into his tomb. From the tomb he arose, and after ascended up into heaven, and sitteth on the right hand of the Father.' (*Aurea Legenda*, vol. 1)

21 1 Cor. 15:20-28 is not contradicted by the credal statement that Christ's kingdom 'shall have no end'. Both invite us to see the achievement of Christ in anaphoric terms, and to interpret the kingdom itself that way. The credal statement was deemed necessary to resist any Marcellian reading of this passage that threatened to assign to the church, and even to the Trinity, a merely instrumental status.

22 See *Against Heresies* 3.17, a text perhaps too rich in biblical allusions: 'This Spirit he did confer upon the Church, sending throughout all the world the Comforter from heaven, from whence also the Lord tells us that the devil, like lightning, was cast down. Wherefore we have need of the dew of God, that we be not consumed by fire, nor be rendered unfruitful, and that where we have an accuser there we may have also an advocate, the Lord commending to the Holy Spirit his own man, who had fallen among thieves, whom he Himself compassionated, and bound up his wounds, giving two royal *denaria*; so that we, receiving by the Spirit the image and superscription of the Father and the Son, might cause the *denarium* entrusted to us to be fruitful, counting out the increase to the Lord.'

creation. For the Spirit takes man for a divine inheritance in Christ, and in doing so takes also the body of man. He enables the body to be rightly related to the rational soul, just as he enables the rational soul to be rightly related to the Son and hence to the Father. This makes possible the new man, and with the new man a new world order in which the glory of God shines forth in all things. As Paul says, 'the creation waits with eager longing for the revealing of the sons of God . . . because the creation itself will be set free from its bondage to decay and obtain the glorious liberty of the children of God.'[23]

The glorification of earthly things, which extends to all God's handiwork, is for the sake of man. 'For if there are to be real men', insists Irenaeus, 'there must also be a real plantation, that they vanish not away among non-existent things, but progress among those which have an actual existence.'[24] The verb *proficere* (to advance or make progress) is worth pausing over. It signifies both continuity and discontinuity between the old creation and the new. Of course it has always belonged to man to advance and progress, because it has always belonged to him as creature to go on receiving from his Creator.[25] Yet man declines and expires. The reason for that is a prideful grasping after perfection that amounts to a refusal of creaturehood, and has the ironic effect of preventing any actual growth – hence of trapping man in his own time as a mere 'creature of today'. Not only man himself, but all 'those things among which transgression has occurred', must therefore pass away, 'since man has grown old in them'.[26] But when man has been refashioned in Christ, when the very possibility of sin is behind him, when body and soul have been fully invested with the life-giving Spirit, then he shall go forth and flourish in a world that is also incorruptible. This he will do, not laying aside

---

23 Rom. 8:19-21

24 *Against Heresies* 5.36.1

25 See *Against Heresies* 4.11.2. Irenaeus is the first to develop a doctrine of *epektasis* or perpetual progress. Unlike the later version of Gregory, which also envisions 'a tending forward, an endlessly greater apprehension of divine glory by creatures who "kinetically" experience the peace of God, and finitely live in his infinity' (D. Hart, *The Beauty of the Infinite*, 110; cf. 402ff.), that of Irenaeus is untainted by the remnants of Origenism which Gregory, his opposition to Origen's notion of 'satiety' notwithstanding, retains. Cf. 'Resurrection and Immortality', *The Oxford Handbook of Systematic Theology*, 215f.; on *epektasis* see, e.g., Paul Blowers, 'Maximus the Confessor, Gregory of Nyssa, and the Concept of "Perpetual Progress"', *Vigiliae Christianae* 46 (1992), 151–71.

26 'And therefore this present fashion has been formed temporary, God foreknowing all things' (*Against Heresies* 5.36.1; cf. Augustine, *City of God* 20.14).

his creaturely nature, but ever reinvigorating it by means of 'fresh converse with God'. Under these conditions, says Irenaeus, even the least spiritual man will 'forget to die'.[27] He will be as one for whom, in his delight, time stands still, or rather runs always towards God, and therefore never runs out. 'For communion with God is life and light, and the enjoyment of all the benefits which he has in store.'[28]

The glorification of earthly things is more than their restoration, then. Restoration is certain, both for man and for his environment, for heaven has not been fully cleansed until the promises of God have been fully vindicated:

> On this mountain the LORD Almighty will prepare
> a feast of rich food for all peoples,
> a banquet of aged wine –
> the best of meats and the finest of wines.
> On this mountain he will destroy
> the shroud that enfolds all peoples,
> the sheet that covers all nations;
> he will swallow up death forever.
> The Sovereign LORD will wipe away the tears from all faces;
> he will remove the disgrace of his people from all the earth.
> The LORD has spoken.
> In that day they will say:
> 'Surely this is our God; we trusted in him, and he saved us.
> This is the LORD, we trusted in him;
> let us rejoice and be glad in his salvation.'[29]

In what Jesus referred to as 'the regeneration of all things', however, there will be a new revelation of the ends towards which creatures tend, and of the grace that perfects their nature. That is why, from the little apocalypse of Isaiah to the great apocalypse of John, witness is given that the whole creation is to be made new.[30]

Many difficult questions may be asked, of course, about the nature of this new creation, from which the shroud of sin and death has been lifted. The evocative images presented in the final chapters

---

27 *Against Heresies* 5.36.2
28 *Against Heresies* 5.27.2
29 Isa. 25:6-9 (*NIV*)
30 Mt. 19:28; with Rev. 21:1-5, 22:1-5, cf. Isa. 65:1ff., 66:10ff.

of Isaiah, and reworked on a still grander scale at the conclusion of the Apocalypse, leave a powerful impression. They stir up in us that *Sehnsucht* of which C. S. Lewis used to speak.[31] But *Sehnsucht* is indeed a longing for 'we know not what', as is that 'inborn thirst, which never is allayed, for the God-moulded realm'.[32] It is far from easy to imagine pleasure without pain, joy without sorrow, life without death. Or to envision a life in which reason and will, desire and satisfaction, are perfectly knit together – the truly happy life in which, as Augustine says, one receives everything one wants while wanting nothing wrongly.[33] Nor is it obvious, as Schleiermacher pointed out, how to posit perfection while leaving room for change, and for differing degrees or capacities of wanting rightly.[34] Or how to reconceive freedom, not as mere freedom of choice but as a truly divine freedom, the freedom of the *non posse peccare*.[35] Moreover, it lies beyond any science of ours to describe the dynamism of a body totally responsive to the soul, or to grasp the spiritually modulated physics of a world in which the saviour shall be seen everywhere, 'according as they who see him shall be worthy'.[36] It lies beyond any art of ours to tell a story with a beginning but no end, a story that is all ascent and no descent.[37]

Yet we concede nothing of significance when we concede that all of this is beyond us; that the promises of God, when we try to think them out, prove unthinkable. For there is no direct line of development between the old order, falsely parsed by sin, and the new order yielded by the anaphoric work of Christ. If that is the case for Christ himself in his resurrection, *a fortiori* it is the case with the new creation.

---

31 Or at least what Carnell refers to as *Sehnsucht* 'in its happiest form' (*Bright Shadow of Reality*, 20).

32 Dante, *Paradiso* ii, 19-20 (Binyon).

33 *Trinity* 13.8ff.

34 See further 'Resurrection and Immortality', 221ff.

35 'For it shall be more free being freed from the delight of sinning to an undeclinable and steadfast delight of not sinning. For the first free will . . . had power not to sin, but it had also power to sin; but this last free will shall be more powerful than that, because it shall not be able to sin' (Augustine, *City of God* 22.30). What is more, it 'shall both be one in all, and also inseparable in every one' – which is equally difficult to conceive but likewise implied by man's participation in the perichoretic life of the Holy Trinity.

36 Irenaeus, *Against Heresies* 5.36.1; cf. Augustine (*City of God* 20.18 and 22.29), who wrestles more directly with the novel powers of sight and movement in the new creation, though book five of *Against Heresies* better develops the underlying pneumatology.

37 Except in the memory, yet even there only in such a way as to inspire gratitude and joy (*City of God* 22.29).

The *ordo Spiritus* that will emerge on the far side of the last judgement may be an *ordo progressivus*, but it does not itself evolve from the provisional arrangements that preceded it. It is virgin born. Its appearance depends upon 'the sudden and strange power of God'[38] foreshadowed at Pentecost and in every eucharist. That does not mean, however, that we cannot speak of it at all, that we must lapse with Schleiermacher into eschatological agnosticism. It means only that, when we do speak, we must be faithful to the dialectic of continuity and discontinuity that we find in the apostles and fathers. The great and terrible day of God shall come, says Peter, 'because of which the heavens will kindle and be dissolved and the elements melt with fire'; yet 'we wait for new heavens and a new earth in which righteousness dwells'.[39] The fashion of the world passes away, explains Irenaeus, but not the essence.[40] 'Changes, conversions, and reformations will necessarily take place to bring about the resurrection,' echoes Tertullian, 'but the substance will still be preserved safe.'[41]

Jesus taught, as Tertullian reminds us, that the sabbath was made for man, not man for the sabbath. If it is agnosticism at all, it is a holy (and wholly justified) agnosticism to admit that, until the real man appears in his glory, the sabbath itself remains shrouded in mystery.[42] But a real man there will be, and a real garden too, though we know not what will grow there, or how. 'Beloved,' writes John, 'we are God's children now; it does not yet appear what we shall be, but we know that when he appears we shall be like him, for we shall see him as he is.'[43]

Equally difficult questions may be asked about the transition to glory, whether for the individual or for creation as a whole, and a similar concession must be made. Since there is presumably a different time-form (corresponding to a distinct *Existenzform*) for each of the movements of descent and ascent identified on the second and third levels of our diagram, and indeed for the transitions between them,

---

38 A power capable of perfecting even infants in the resurrection (*City of God* 22.14).

39 2 Pet. 3:12f.; cf. Phil. 1:21-24.

40 *Against Heresies* 5.36.1 (cf. 1 Cor. 7:31). The theological justification for statements such as these is the faithfulness of God to what he has made: *non enim substantia neque materia conditionis exterminatur, verus enim et firmum qui constituit illam.*

41 *Resurrection* §55. These conversions and reformations are so decisive, he allows, as to seem almost 'a complete destruction of the former self'; yet they introduce a radical discontinuity into 'our condition, not our nature'.

42 *Resurrection* §59; see again 'Resurrection and Immortality', 229ff.

43 1 Jn 3:2 (*RSV*).

much care must be taken not to speak on the basis of false assumptions generated by imposing a common time upon them.[44] This bears, obviously, on our own participation in these transitions; on what is said about purgatory, for example, or the millennium, or the events connected to the last judgement. Ascension theology reminds us to speak only from the eucharistic vantage point.

From that vantage point we may learn not to run to Origen's excess by making of purgatory another world or succession of worlds, thus dissolving both continuity and discontinuity in the acid of incrementalism.[45] Nor yet to think of it in terms of the time of the saeculum, but to see it instead as a personalized time determined strictly by the fitting of the soul to its true centre in the ascended Christ.[46] As Benedict remarks in *Spe salvi*, purgatory 'is simply purification through fire in the encounter with the Lord' – something not measurable either in time as we know it or in some putative progression of *aeons*.[47] Indeed,

---

44 That is, the time of the first level. Dante, though exercising the licence of the poetical theologian and working of course with an older understanding of the cosmos and of the formation of the human being (see *Purgatorio* xxv), structures the *Commedia* and the experiences it recounts of hell, purgatory and heaven with sensitivity to the differing characteristics of time, space, matter and spirit in those places. We ought to follow his example, seeking in ascension theology to sharpen that sensitivity and indeed to reconfigure it through closer attention to christology.

45 That is the tendency of Origen's thought, which by bridging with many intervening aeons the present world and the world to come, removes both the continuity and the discontinuity between the two. See further Farrow, 'The Doctrine of the Ascension in Irenaeus and Origen' (*ARC*, vol. 26, 1998), 40ff.

46 'Christ's resurrection has instrumentally an effective power not only with regard to the resurrection of bodies, but also with respect to the resurrection of souls . . .; *even in our souls we must be conformed with the rising Christ*'(Aquinas, *Summa Theologiae* 3.56.2, emphasis mine; cf. Dietrich Bonhoeffer's remarks on the mediating centre of human personhood in *Christ the Centre*, 59–61). This is the underlying principle that ought to guide our consideration of purgatory. Unfortunately much that is said about purgatory in the supplement to the *Summa* is unsound, by reason of introducing principles (respecting justice and punishment and satisfaction) at odds with this one and with Aquinas' mature thought. Dogmatic claims about purgatory are much more modest and do not make of the grace of purgation something that might diminish or obscure the gospel. (An instructive overview is offered by P. Griffiths in *The Oxford Handbook of Eschatology*; note his summary on p. 437 of J. Ratzinger's treatment of purgatory in *Eschatology*, 218–33, which is guided by the gospel principle above and also by a more flexible view of time.)

47 When Augustine comments in *The City of God* on 1 Cor. 3:15 – that is, on those 'saved, but only as through fire' – he declines to pass judgement as to whether, in the time between death and the parousia, 'the spirits of the dead are all that while tried in such fire as never moves those that have not built wood, straw, or stubble' (21.26). But it does not occur to him to ask how the question might look if the 'time' or 'while' under consideration

for those who are still alive at the coming of the Lord, purgation is something that will happen (viewed from the standpoint of the saeculum) 'in a moment, in the twinkling of an eye'. For, as Paul says, 'we shall not all sleep, but we shall all be changed'.[48]

Likewise, we may learn not to make of heaven a place where the soul does not rule over the body but without it, a mistake that must push us either in Origen's direction or towards some other form of dualism. The sudden and strange power of God is a power applied to soul and body alike in the definitive encounter with the risen Christ that takes place for each person after death, for Christ himself is the one through whom and with whom we make our transition to glory – the one who determines both continuity and discontinuity for us. When we consider that transition from the point of view of the soul, we call it 'purgatory' and its outcome we call 'heaven'. When we consider it under the aspect of the body's transformation, we call it 'resurrection' and 'the new creation'. How exactly these meet and correspond, since death separates body from soul, is difficult, perhaps impossible, to say. For the body, 'after death' means at the parousia. For the soul, it means something else, something related to the soul's own nature and to Christ's proclamation of the gospel in Hades.[49] Certainly it is safe to say that there is a distinct process of liberation and conformation appropriate to the soul, on which the body awaits, just as it is safe to say that in glory the body shall always wait on the soul.[50] That is why we cannot speak of our personal transition to glory without speaking of purgatory as well as of resurrection. But it must be added that the only common measure

---

were not assumed to be identical with our time. Cf. Benedict in *Spe salvi* [On Christian Hope, 2007] §47: 'At the moment of judgement we experience and we absorb the overwhelming power of [Christ's] love over all the evil in the world and in ourselves. The pain of love becomes our salvation and our joy. It is clear that we cannot calculate the "duration" of this transforming burning in terms of the chronological measurements of this world. The transforming "moment" of this encounter eludes earthly time-reckoning – it is the heart's time, it is the time of "passage" to communion with God in the Body of Christ.' Or more briefly in §48: 'in the communion of souls simple terrestrial time is superseded'.

48 1 Cor. 15:35ff.; cf. *City of God* 20.20.

49 Cf. *Against Heresies* 5.31.2.

50 The notion of a body generated by the soul in and for the intermediate state is one that ought to be treated with great wariness. It begs the question about the relation of the soul to its earthly body, as well as the question about the nature of 'intermediate' time. (What Irenaeus says in *Against Heresies* 5.31.2 about 'the law of the dead' speaks to the first question, but not so well to the second.)

of these personal times and spaces is the Spirit of God, and the only mediator capable of guaranteeing the convergence of every transitional process is Jesus Christ – hidden as he is with God yet manifest in the eucharist, in which body and soul and the whole *communio sanctorum* hold together.

We are not surprised to hear, then, by the word of the Lord, that at the parousia 'the dead in Christ will rise first', and that afterwards 'we who are alive and remain shall be caught up together with them'.[51] In the precedence of the dead lies the mystery of the millennium and of the intercession of the saints, who participate with Christ – whether in the body or out of it, or both, who can say? – in the purification of the heavens and in the corresponding judgements of God on earthly affairs. This too calls for a measure of holy agnosticism. The eucharist situates those who live in Christ, whether they be dead or alive from an earthly point of view, in the otherwise inaccessible breach between his ascension and his parousia. It situates them together in that breach, for all the real difference between them.[52] Those, then, who stand still at the foot of the holy mountain, who wait outside the entrance to the holy tabernacle, have recourse not only to the aid of angels and archangels but also to 'the spirits of just men made perfect', who rule with Christ in the secret sovereignty of his heavenly session. They have access together with them to the future of man that is already the present of the risen and ascended Christ. But how those 'just men made perfect' rule is not known to us, nor can be known, though from earliest times Christians have claimed to see some of the effects of their rule.[53] Nor is it altogether clear whether the reign of those who take part in what John calls 'the first resurrection' should be understood with Irenaeus as a time of preparation for the eternal kingdom, hence as analogous to the forty days that followed Jesus' resurrection – that is, as a time between the parousia and the last judgement – or whether with Augustine as the hidden

---

51   1 Thess. 4:13-18. Would that Paul's tantalizing remarks here and in 2 Cor. 4:13–5:10 did not leave us feeling as if we were still missing a few pieces of the puzzle!

52   'For the souls of the godly are not excluded from the Church, which [even] as it is now is the kingdom of God. Otherwise she would not mention them, nor celebrate their memories at our communions of the body and blood of Christ' (*City of God*, 20.9).

53   Heb. 12:23. The glorification of earthly things is both a miraculous work of the Spirit, whose brooding over Mary brings forth the Christ and eventually the new creation, and a function of Spirit-filled people, who inhabit earth in a heavenly way. On what grounds shall we deny that the latter receive help, *via* the former, from those who have gone before them? Cf. *City of God* 20.7-13 and 22.8.

side of the eucharistic existence of the church in the saeculum. Either way it is transitional time, and its nature not fully disclosed to us, for if it is transitional time it is the time of God's sovereign reversal of the work of sin and death and the devil.[54]

Ascension theology insists, without hint of embarrassment, that only at the parousia will such things become plain, when Christ confronts what he has left behind with what he has gone towards, passing judgement on the former in the light of the latter. They are not plain now, nor have we any model capable of making them plain. We have no model for the parousia itself – though we have heard in advance of its effects too – or of the cosmic transition that turns on the last judgement. In the memorable words of Augustine, 'The last judgment is that which he shall settle on earth, coming to effect it out of heaven. And this judgment shall consist of these circumstances, partly precedent and partly adjacent: Elijah shall come, the Jews shall believe, Antichrist shall persecute, Christ shall judge, the dead shall arise, the good and bad shall sever, the world shall burn and be renewed.' 'All this', he adds instructively, 'we must believe shall be, but how and in what order, our human understanding cannot perfectly teach us, but only experience of the events themselves.'[55]

---

54 Cf. *Against Heresies* 5.32-35 with *City of God* 20.7-13. Several things deserve notice here: First, Irenaeus' view 'might be allowed' even by Augustine (20.7) as an alternative reading of Revelation 20, for it is not subject to the main objection he offers to the chiliasm of his day. Second, both views represent plausible readings of the passage, a matter much too complicated to discuss here. Third, the two constructs are not entirely incompatible, even if the two readings are; one can reasonably posit both a spiritual and a bodily resurrection, with corresponding millennial referents, without asking John to be teaching both in the same passage. Fourth, Irenaeus seems, from our more traditional Augustinian point of view, to conflate purgatory and the millennium by envisioning the latter as a further period of preparation for glory; but we have already allowed, without endorsing it, that the idea is neither theologically naïve nor uninteresting. Fifth, Paul, John, Irenaeus and Augustine are united in giving precedence to the martyrs and the righteous dead, whatever their differences. And if the dead are to rise first, perhaps we could even posit a hybrid construct, partly Irenaean and partly Augustinian, so long as we allow that the time of the saeculum is not the measure of these transitional times.

55 *City of God* 20.30 (adapted from the 1610 translation by J. H. and that of M. Dods). If we cannot say in what manner (*quibus modis*) or even, with certainty, in what order (*quo ordine*) these things shall happen, it is because they are the consequences of changes in heaven itself. They are the consequences of the reordering that takes place through the ascension of Jesus – a reordering, as we already know from the eucharist, that does not stop short of the reordering of the space–time world itself. The world shall burn and be renewed when confronted with the ascended one in his glory.

WE DO WELL TO EMULATE Augustine's example of confidence and caution when trying to follow out the trajectories of Christian eschatology. *Caution*, because the great transition to glory is an event, or set of events, that cannot be understood save by experience. We believe in the parousia of Jesus Christ, the resurrection of the dead, the just recompense of every man's deeds and the reconstitution of the world.[56] We believe that the dew of the Spirit is able to preserve us, not only from the ravages of time but also from the raging fires that will dissolve the world and forge it anew. But when and how these things shall happen, and what their outcome shall be, we do not know exactly. We await what eye has not seen, nor ear heard, nor the mind of man conceived. *Confidence*, however, because our hope for what we do not yet see is not a blind hope, nor are we deaf to the promises of God. We await what is already being given sacramentally. We expect to inhabit a city whose foundations are already laid. We look for the renewal of the world we have already been given, not for some wholly other world, because it is this world of ours that has been redeemed, this bread and wine that is consecrated, this very creation of which Jesus – 'this same Jesus', as the angels said – has become high priest and high king.

But is all this really true of us? Is our faith still of this sort? The feebleness of so much Christianity in the West suggests otherwise. It is not that we are cautious to a fault, but that the very ground of our confidence eludes us. We no longer see ourselves as those who wait at the entrance to the tabernacle for our high priest to reappear, driving sin from the camp so that the blessings of atonement may rain down. We no longer identify ourselves as those who stand at the foot of Sinai, waiting for our prophet and guide to reappear and take us to the promised land. We no longer regard ourselves as those who man the walls of Zion, waiting for our king to reappear, bringing his rewards and recompense with him after conquering his enemies from

---

56   'For we must all appear before the judgment seat of Christ, so that each one may receive good or evil, according to what he has done in the body'(2 Cor. 5:10). To echo Pearson, to whom the reader may turn for an exposition (art. 6, chap. 6, §9): 'The testimonies . . . in the law and the prophets, the predictions of Christ and the apostles, are so many and so known, that both the number and the plainness will excuse the prosecution. The throne hath been already seen, the judge hath appeared sitting on it, the books have been already opened, the dead small and great have been standing before him; there is nothing more certain in the Word of God, no doctrine more clear and fundamental, than that of eternal judgment.'

the river to the sea. Even Catholics, whose sacraments and lections are a daily reminder of these things, seem often to have forgotten – our bishops and priests and theologians allowing word and sacrament alike to disappear into an impenetrable cloud of psychology or social commentary. Sirach remains among our scriptures, but the vision of its fiftieth chapter has become opaque:

> How glorious he was, surrounded by the people,
> when he came from behind the temple curtain.
> He was like the morning star appearing through the clouds
> or the moon at the full;
> like the sun shining on the temple of the Most High
> or the light of the rainbow on the gleaming clouds . . .
>
> Then the choir broke into praise,
> in the full sweet strains of resounding song,
> while the people of the Most High
> were making their petitions to the merciful Lord,
> until the liturgy of the Lord was finished
> and the ritual complete.[57]

We have forgotten, in other words, how the Mass actually goes, and where its time and its story lead.[58] No image of the parousia of the priest-king comes readily to us. We have lost the political sensibilities and the liturgical experience that present it to the imagination, and the theological concepts that present it to the reason. We have lost even the narrative thread, and the point at which we have lost it is the ascension of Jesus into heaven – the feast of which scarcely registers any more on our collective consciousness.[59]

---

57   See Sir. 50:1-21 (*New English Bible*); cf. Zech. 6:9ff., Heb. 10:27f.
58   Which perhaps explains in part why we have forgotten also how to sing, or how to sing in the Spirit. 'If, therefore, at the present time, having the earnest, we do cry, Abba, Father, what shall it be when, on rising again, we behold him face to face – when all the members shall burst out into a continuous hymn of triumph, glorifying him who raised them from the dead, and gave the gift of eternal life?' (*Against Heresies* 5.8.1)
59   B. Lindars remarks that 'few people can share the presupposition that the Day of Atonement defines what is required for reconciliation with God' (*The Theology of Hebrews*, 136). We may go further than that and say that we live in an era shaped quite decisively by its rejection of priests and priesthood, not to speak of kings and kingship. But is it possible to let that era shape us completely and still to be Christian? Is it not the case that the whole age in which we live, from a Christian point of view, amounts to an arraying of

The church that celebrates the ascension in spirit and in truth, or the Christian who ascends with the mind seeking him who has preceded us into heaven, can no more fail to watch for his parousia than the congregation of Israel could fail to watch for Aaron (or, in Sirach, Simon) to return from the inner sanctuary on Yom Kippur. They can no more fail to watch for his second coming than Simeon and Anna for his first. What was said or sung of the first coming, they will sing, *fortissimo*, of the second also:

> Saints before the altar bending,
> Watching long in hope and fear –
> Suddenly the Lord, descending,
> In his temple shall appear![60]

But the question that was asked with respect to the first – 'Who shall withstand his coming?' – will likewise be asked with respect to the second. For it is believed by them that 'the Son of man is to come with his angels', as he promised, 'in the glory of his Father, and then he will repay every man for what he has done'.[61]

A note of warning, therefore, will sound from them along with the note of joy. For they know and are well aware that judgement begins with the household of God.[62] They know that the cleansing of the heavens must have its consequences on earth, and that they themselves must anticipate the threshing that will occur when the feet of the Son of Man touch again the holy mountain, just as Jesus anticipated on the cross the judgement that was to fall on Jerusalem. They know that, while the ascended one does not literally sprinkle blood in the heavenlies, as the high priests once did in the holy place, he continues through his martyrs to sprinkle his blood liberally upon the earth, to make it holy. The signs of the times do not escape them. They recognize 'the urgency of penance, of conversion, of faith'.[63]

---

peoples and nations, whether friendly or hostile, outside the tabernacle where humanity's true priest is hidden with God, round the mountain on which he converses with God?

60   J. Montgomery, 'Angels from the Realms of Glory' (1816, 1825), fourth stanza.

61   Mt. 16:27

62   1 Pet. 4:17

63   As Cardinal Ratzinger wrote in his Theological Commentary on the third secret of Fatima (available on the Vatican website): 'To understand the signs of the times means to accept the urgency of penance – of conversion – of faith. This is the correct response to this moment of history, characterized by the grave perils outlined in the images that follow.'

They are prepared, if need be, to be numbered among the martyrs and to wash their own robes in the blood of the Lamb. They are prepared to lay down their lives for the sake of the great transition, for they themselves have been warned:

> Then they will deliver you up to tribulation, and put you to death; and you will be hated by all nations for my name's sake. And then many will fall away, and betray one another, and hate one another. And many false prophets will arise and lead many astray. And because wickedness is multiplied, most men's love will grow cold. But he who endures to the end will be saved . . . Immediately after the tribulation of those days the sun will be darkened, and the moon will not give its light, and the stars will fall from heaven, and the powers of the heavens will be shaken; then will appear the sign of the Son of man in heaven, and then all the tribes of the earth will mourn, and they will see the Son of man coming on the clouds of heaven with power and great glory . . .[64]

The schoolmen used to say of the parousia that it would be *personalis, visibilis, beatificus, terribilis, et gloriosus*; and so it shall be, if Christ has indeed ascended in the flesh. Which means: it is precisely this world and our humanity that will be confronted once again by God in *his* humanity. There are those who know this, of course, but do not long for it, who fear it and hate the very thought of it and are determined to eradicate every testimony to it. Their time is short, but the fervour of their attacks, on the church and on humanity as such, is intense. It is directed, as we observed, at the poor and the weak, since the poor and the weak remind them of their own poverty and weakness. It is directed at the helpless, the mute, the innocent in the womb, the aged, the infirm, the depressed – not merely for eugenic or 'environmental' or frankly pecuniary purposes, but out of loathing for life itself and for the One who gives life. It is directed especially at those who speak the truth in love, who give thanks to their Creator and Redeemer, who live in expectation of the end that is also the beginning, who stand as a sign of contradiction to the culture of death.[65]

---

64 Mt. 24:9-13, 29–31.
65 Linda Gibbons comes to mind here: a grandmother who has spent much of the last decade in Toronto prisons for her silent but visible witness against that city's culture of death.

If we ask what drives these attacks, we must answer: the spirit of gnosticism. For gnosticism is the only real alternative to the ascension theology of the church, and the secret source of lawlessness. Gnosticism rejects even the flesh itself rather than accept God's redemptive rule in and over the flesh. It refuses to hear of any atonement that restores the integrity of the human person. It prefers to speak instead of an atonement in God himself, of a loss and recovery of divine integrity. Such talk is only for the aficionado, naturally – for the 'knowing' one. To the ordinary person gnosticism has learned to present itself rather as the champion of the human. Still, it cannot accommodate man as a unity of body and soul. It therefore cannot accommodate the Jew, and particularly the Zionist. It cannot tolerate the Christian either, and particularly the papist. It cannot even tolerate man as male and female, much less man as mere fœtus. Its true secret is not refinement of spirit but refusal of flesh; that is, of the whole sphere of atonement as it actually takes place. Of gnostics Prisca rightly said, 'They are carnal, yet they hate the flesh.'[66]

The spirit of gnosticism is the spirit of Antichrist, and the spirit of Antichrist is the spirit that will not confess Christ come in the flesh or ascended in the flesh. It is foolish to think that gnosticism has disappeared, or to object that we are defining it too broadly. Gnosticism is eminently adaptable and resilient; its tares grow together with the wheat of the gospel until the end of the age. Today's gnosticism can sound down to earth, pragmatic, historically and politically astute, even scientific; it can sound quite unlike the gnosticism of the patristic period. But it is recognizable not only by its man-hating tendencies and consequences, but by its hatred above all of Jesus Christ. Just as it developed from the outset myriad mythologies with which to overwhelm the gospel, so today it develops entire epistemologies, philosophies of history, and political theories designed for the same purpose. It failed then, but it hopes to succeed now, by treating the Christian doctrine of atonement – the doctrine that addresses the fall of Adam with the lifting up of Jesus – as an outdated fiction 'that contradicts the idea which lies at the foundation of all modern thinking'. And what idea is that? Ascension of the mind only. Or, as we now say, 'a development of spirit which emerges at different nodal

---

66   Tertullian, *Resurrection* §11; cf. *Against Heresies* 4.38f., where Irenaeus characterizes gnosticism as a refusal of human infancy – that is, of the elementary distinction between the created and the uncreated – that leads not to the ascent of man but to his self-abortion.

points from the depth of the divine life, and a possibility of future developments which can never be circumscribed in advance'.[67] That is how gnosticism continues to witness to Bythos, to Abraxas, to the Unknown Father. That is how it dispenses with Jesus, and opens the way for 'other historical personalities' who can also 'receive their due and be seen in some sense as visible symbols and guarantees of faith'.[68] That is how it denies (as it has always denied) that Jesus himself is the Christ, that he ascended into heaven and is seated at the right hand of God, and that he will come again in glory to judge both the quick and the dead.

Where Christ is preached as ascended in the flesh, however, it cannot be said that the flesh is unworthy of the upward call. And if it is worthy, or has been made worthy, then it follows that the flesh can neither be despised nor indulged, but must rather be disciplined and trained for glory. 'For the grace of God has appeared for the salvation of all men, training us to renounce irreligion and worldly passions, and to live sober, upright, and godly lives in this world, awaiting our blessed hope, the appearing of the glory of our great God and Savior Jesus Christ, who gave himself for us to redeem us from all iniquity and to purify for himself a people of his own who are zealous for good deeds.'[69] This appearance of the *shekinah*, and of Jesus Christ in the *shekinah*, will do away with every obfuscation. It will make clear, even to those who do not wish it to be clear, who God is and who and what man is, and what God has done for man. It will introduce the final act of the drama of atonement, in which the will of God will be realized on earth as it is in heaven. But this means that earth, like heaven, must be winnowed and purged, that its peoples must be sifted and its powers subjected to judgement. 'For the advent of the Son comes indeed alike to all,' says Irenaeus, 'but is for the purpose of judging, and separating the believing from the unbelieving.'[70]

---

67 Troeltsch, *Writings on Theology and Religion*, 193.

68 Ibid., 201. Troeltsch, of course, thinks that Jesus is not altogether dispensable just yet, and proposes to rescue him from the dustbin of history with the tools of historical science; in that sense he advocates a very patient form of gnosticism. (See further 'Melchizedek and Modernity', in Bauckham *et al.*, 294ff.)

69 Tit. 2:11-14. This is decisive for a right understanding of the Christian *contemptus mundi* and of the ascension of the mind, which ought not to be viewed as a metaphysical enterprise but as a form of spiritual discipline that regards life in the saeculum as a preparation for glory.

70 'The Word comes preparing a fit habitation for both' (see *Against Heresies* 5.27f.; cf. Exodus 32).

IF GOD HAS CIRCUMSCRIBED in advance the course of history, if the upward call is not merely that of Nature, which in Marlowe's phrase 'doth teach us all to have aspiring minds',[71] but the call of God himself in Jesus Christ, then the drama of atonement will have indeed a final act. The moving of the restless spheres will cease, and the heavens be rolled up as a scroll, when the Son of Man appears. Then shall 'the ripest fruit of all' be picked, 'perfect bliss and sole felicity' found. Not, as Marlowe's Tamburlaine suggests, in 'the sweet fruition of an earthly crown' – that is, in the collapse of aspiration into ambition – but in the gift of eternal life through union and communion with God. The tree of life itself shall appear in the great anaphora that follows the last judgement, and partaking of it man shall be glorified. For God has circumscribed human history in Christ so that he might prise it open to a more divine history.

That more divine history is often spoken of as deification, which is the glorification of man through the indwelling of God. We have had occasion to pass judgement on right and wrong ways of speaking of deification, ways connected with the decision for or against the doctrine of ascension in the flesh. We might even say that, on one level, that is what the entire exercise has been about: affirming that deification is hominization through the commending of the whole man, body and soul, to the Spirit, as the eternal inheritance of God in Jesus Christ. Sergius Bulgakov is certainly right that the parousia of Jesus Christ, which 'the helpless language of our spatiality and temporality cannot describe or express', is, precisely as a coming in *glory*, the outpouring of the Holy Spirit that Pentecost only anticipated.[72] The glory of the God-man does not lie solely in the accomplishments of

---

71 Thus *Tamburlaine* part I, act 2, scene 6:
        Nature, that fram'd us of four elements
        Warring within our breasts for regiment,
        Doth teach us all to have aspiring minds:
        Our souls, whose faculties can comprehend
        The wondrous architecture of the world,
        And measure every wandering planet's course,
        Still climbing after knowledge infinite,
        And always moving as the restless spheres,
        Will us to wear ourselves, and never rest,
        Until we reach the ripest fruit of all,
        That perfect bliss and sole felicity,
        The sweet fruition of an earthly crown.

72 *The Bride of the Lamb*, 396; cf. 399: 'The parousia signifies the power not only of the incarnation but also of the Pentecost, of Christ in glory and glory in Christ, the appear-

his journey of descent and ascent. It does not lie merely in the authority he is given over heaven and earth. It lies also and ultimately in his own unlimited capacity to receive and transmit the Spirit.[73] What then will his parousia be but a determination of that which is fit, and that which is not fit, to share in the Spirit and so in eternal life? What will it be but the fullness of Pentecost? Even he who speaks in the tongues both of men and of angels will have difficulty describing that. Yet what he is describing will still be hominization.

Once that is acknowledged, we need not be afraid to put things the other way round and affirm that hominization is deification. Indeed, we cannot take the final step of our journey in ascension theology without doing so. But this requires us to make more explicit the way in which atonement teleology – concerned as it is with the glorification of man by the living God and the glorification of God through a living man – rests on the ontology of the incarnation as well as on the history of Jesus. It requires us to place the story of Jesus' descent and ascent where it belongs, where the creed itself places it: under the interpretive rubric of the *homoousion*. We need to be careful here, however, lest we confuse the ontology with the history (as Bulgakov has done) and so try to narrate the unnarratable, while accidentally short-circuiting our account of man's deification. Which is to say, we need to make a distinction between two different kinds of descent.[74]

---

ance in Him, with Him, and through Him of the Holy Spirit.' See below, however, for criticism of Bulgakov.

73 The Golden Legend, in treating the ascension, quotes Bede to the effect that Jesus 'opened the tavern of heaven and poured out the wine of the Holy Ghost', and speaks of the saints as those who are 'of great capacity in receiving the Holy Ghost'. It is in that connection also that we may understand his work of preparing a place, as *per* his promise in John 14, to which the Legend rightly attaches a prayer of Augustine (cf. Tractate 68.3): 'Thou arrayest us Lord to thee, and thou arrayest thee to us, when thou makest ready the place, to the end that to thee in us, and in thee to us, may be the preparation of the place and the mansion of the everlasting health.'

74 The failure to do this unites theologians as diverse as Bulgakov and Barth. The problem in Barth I have treated in several places, including 'Ascension and Atonement' (Colin Gunton, *The Theology of Reconciliation*, 77ff.) and 'Karl Barth on the Ascension' (*International Journal of Systematic Theology* 2.2, (2000), 127–50), with which cf. Andrew Burgess, *The Ascension in Karl Barth*, and my response in *IJST* 7.2, (2005), 205–8. The problem in Bulgakov likewise appears in the form of universalism, which is one legacy of Origenism, and in his unwillingness to allow that the parousia, though not *merely* an event in this world, is *also* an event in this world (*Bride of the Lamb*, 395f.; cf. 482ff.). It appears in many other forms as well, but this is not the place to embark on a critique of the concepts of godmanhood and sophianicity and the deification of creation.

The first kind can be narrated, because it is the descent in which Jesus journeys with the human race, following our own path and story in order to rewrite that story and give it a new ending. The creed speaks of this descent when it says, 'for our sake he was crucified under Pontius Pilate; he suffered, died, and was buried'. The second, however, cannot be narrated. It can only be posited and confessed in explanation of the narrative, looking back from the vantage point of the new ending. The creed speaks of it when it says, 'for us men and for our salvation he came down from heaven'. To understand this explanation, we may begin by turning once again to Isaiah, who cries out to God:

> Oh, that You would rend the heavens and come down!
> that the mountains would tremble before you!
> As when fire sets twigs ablaze
> and causes water to boil,
> come down to make your name known to your enemies
> and cause the nations to quake before you!
> For when you did awesome things that we did not expect,
> you came down, and the mountains trembled before you.
> Since ancient times no one has heard,
> no ear has perceived,
> no eye has seen any God besides you,
> who acts on behalf of those who wait for him.
> You come to the help of those who gladly do right,
> who remember your ways.
> But when we continued to sin against them, you were angry.
> How then can we be saved?
> All of us have become like one who is unclean,
> and all our righteous acts are like filthy rags;
> we all shrivel up like a leaf,
> and like the wind our sins sweep us away.
> No one calls on your name
> or strives to lay hold of you;
> for you have hidden your face from us
> and made us waste away because of our sins.
> Yet, O LORD, you are our Father.
> We are the clay, you are the potter;
> we are all the work of your hand.
> Do not be angry beyond measure, O LORD;

do not remember our sins forever.
Oh, look upon us, we pray,
for we are all your people.
Your sacred cities have become a desert;
even Zion is a desert, Jerusalem a desolation.
Our holy and glorious temple, where our fathers praised you,
has been burned with fire,
and all that we treasured lies in ruins.
After all this, O LORD, will you hold yourself back?
Will you keep silent and punish us beyond measure?

And God replies:

I revealed myself to those who did not ask for me;
I was found by those who did not seek me.
To a nation that did not call on my name,
I said, 'Here am I, here am I.'[75]

Here am I in your midst, as the answer to your prayers! That is the right context, the first and indispensable context, for understanding the statement 'he came down from heaven', which joins the narrative of Jesus to the narrative of Israel and her God.

Yet there is much more to be said by way of understanding. The appeal to God to 'come down' means, in Isaiah, to come down from that heaven of which we spoke earlier – the place from which God's rule over the world is effected through the angels, the place where, as in the terrible, magnificent vision of Isaiah 6, God's presence to and for creation is manifested and known. God comes down from this place by intervening directly on earth, by showing his own hand, as it were, as he did at Sinai.[76] In the creed, however, 'for us men and for our salvation he came down from heaven' introduces the parallel statement: 'by the power of the Holy Spirit he became incarnate of the Virgin Mary and was made man'. Taken together, these statements speak of an answer to Isaiah's prayer much more literal than anything Isaiah had in mind. They tell us that God, by way of intervention, actually became a human being; that the mighty works that

---

75 Isa. 64:1–65:1 (*NIV*)
76 Then, however, the people begged that he would not come down again, or not in that fashion; see Exod. 20:18-20.

he did of old were all for the sake of his becoming a human being;[77] that the narrative of God with man, of the righteous God with unrighteous Israel, could not go on unless God came down in this unprecedented manner, to do as man what only God could do.

To make such claims is not only to assert that the narrative that begins in Mary's womb *is* the narrative of the longed-for intervention of God. It is to posit a 'descent' in God, whereby God the Son situates himself on our side and determines himself for us in the most radical way possible. This descent cannot be narrated, however, not even by appeal to the Spirit who moves to dispose both Mary and Mary's womb for the reception of the Son. The point of the *homoousion* (which sums up a long list of credal reinforcements for the claim that the Son, both before and after this descent, is none other than God himself) is to make clear that it cannot be narrated. One can tell the story of God with Israel. One can tell the story of Jesus, conceived by Mary and born in Bethlehem, as the story of God with Israel. One cannot tell the story of God coming down to Bethlehem in order to *be* Jesus. That is to say, there is no story to be told, in gnostic fashion, of a decline or fall in God. There is no story to be told, in Origenist fashion, of the descent of the soul of Jesus or of some 'sophianic' prototype of man. There is no story to be told, in Arian fashion, of a first and best creature setting out on a mission to the lower reaches of the cosmos. There is only God himself, God the Son, who by the power of the Spirit appears in our midst as a man and yet remains, even as he does so, God in the heights as well as God in the depths.[78]

Now the purpose of all this is to make an ontological statement about the identity of the protagonist of the narrative portion of the second article of the creed: to say that the man whose story this is is not merely from God, but of God; that he is 'God of God'. And the point of that, of course, is to secure the salvation he offers as a real and true salvation. But what is salvation if not the restoration of man to his appointed end, namely, fellowship with God? As soon as we ask

---

77 Though the creed itself does not say so, we may agree with Pannenberg – while differing on important details – that 'human nature as such is ordained for the incarnation of the eternal Son in it' (*Systematic Theology*, vol. 2, 385f.).

78 Cf. Hilary, *On the Trinity* 3.16 and 10.54. That the Son of God *homo factus est* permits and requires us to speak of the Father who sends, of the Son who is sent and comes, and of the Spirit who is the means by which he comes. It requires trinitarian dogma and theology. But this is not, as some suppose, the narrative of God. The life of Jesus is the narrative of God, in the only sense that God can be narrated.

about that, however, we realize that even Isaiah's exalted vision of a 'new' heaven and earth, of a world in which the covenant shall be truly lived and its promises fulfilled, is not an adequate description of the end of man. For if God has appeared in the depths, if he has subjected himself in person to time and place, to history and narration, it is evident that things have changed for man. Man is no longer man by himself or for himself, as he had made it his ambition to be. He is once again, but this time much more fundamentally and indeed irrevocably, man with and for God. Since God has invested himself in man, man is no longer merely man, with his own creaturely interests and responsibilities under God. Man is now an internal communicant in the very life of God, for God has made himself internally communicant in the life of man.

In Eden, it is said, God used to come calling on man, on Adam and Eve, to converse with them in the cool of the evening. After their fall into sin, and the long dark descent into violence and deceit, into lovelessness and lies and forlornness, 'the very same Word of God came to call on man, reminding him of his ways, living in which he had been hidden from the Lord'. 'For just as at that time God spake to Adam at eventide, searching him out', writes Irenaeus, so 'by means of the same voice, searching out his posterity, he has visited them.' Only this visitation was different. It was no mere theophany. It was incarnation, absolute solidarity. And in this absolute solidarity was the secret of the image-bearing creature called man, the creature who has God both as his source and as his end. 'For in times long past it was *said* that man was created in the image of God, but it was not yet shown' – not until man was assimilated 'to the invisible Father through means of the visible Word'.[79]

Things have changed for man since God appeared in the depths. Man's maker has taken up his place in and with man, and as a man elevated man beyond man. God has made room in man for himself, and at the same time made room in himself for man. That is the moment of truth in every system of thought that, eschewing the

---

79 See *Against Heresies* 5.14-16. He who is *consubstantialem Patri secundum deitatem* is also *consubstantialem nobis eundem secundum humanitatem*, as Chalcedon puts it, and this double solidarity not only effects redemption but completes the creation of man, informing and qualifying everything to be said about man. Irenaeus, I note, may be added to the list of those who have suggested that the incarnation is 'due to the primal and absolute purpose of love foreshadowed in creation' (B. Horne, in Holmes and Rae, *The Person of Christ*, 113).

homoousion, nonetheless exalts man beyond his own proper creaturehood and presents him as something divine. But the purpose of all such theological simulacra is to hide from man the true facts of his exaltation. Narrating the descent of God, they narrate also the ascent of God, the return of God to himself. And that invariably means the decline and disappearance of man *qua* man. The *homoousion*, however, insists that God has invested himself in man permanently. The ascension of Jesus Christ is not the return of God to God. It is the ascension of the God-man to his rightful place, the place of glory that Adam and Eve never knew, but are yet destined to know.[80] They could not have known it without him, but they are not now without him, and neither are we. We ourselves, when in the final anaphora we are presented to the Father in perfect union with the Son, are destined to enter his glory, and to add our own glory to it, the glory that comes to those who feast from the tree of life.[81]

God has appeared in the depths as a man. As a man he has descended, and as a man he has ascended, scaling the heights of heaven. Having recovered from our fall, we too will ascend the heights of heaven, and there participate in God, which is our proper end. This participation will have to be learned, for it is not ours by nature but by adoption.[82] But we will learn it, and when we have learned it we will no longer say with the Psalmist: 'What is man that thou art mindful of him, the son of man, that thou dost care for him? Thou hast made him a little lower than the angels, and crowned him with glory and honour.' For then we will no longer be a little lower than the angels, or even on a par with the angels. We will have surpassed the angels, for we will be like God.[83] We will not live and move, as they do, *coram Deo*; we will live and move with Christ *in Deo*. We will be deified by the Spirit, knowing God by way of God.[84]

Dante hints at this in the famous closing lines of the *Paradiso*, when, transfixed by that internal reflection of the divine Light which seems,

---

80   On the 'return' of the God-man to glory in John 17, see *Ascension and Ecclesia*, 294; cf. Bulgakov (*Bride of the Lamb*, 397), who entangles that passage in narratives of divine kenosis for both the Logos and the Spirit.

81   Revelation 21f.

82   'For it is one thing to be God, another thing to be partaker of God' (*City of God* 22.30).

83   In the sense and to the extent that this is already the case, we have already surpassed the angels, who participate in eternal life only 'in the context of the relationships of the Church' (see J. Zizioulas, *Communion and Otherness*, 283; cf. 1 Cor. 6:3, Ps. 8:4f.).

84   Yet 'without losing our own proper substance' (Athanasius, *De Decretis* 3.14; cf. *Against the Arians* 1.39).

all impossible, to be painted with our human effigy, the eye of his mind is assailed by a sudden glory, by an all-encompassing flash of divine lightning, and he grasps the Mystery from the inside.[85] We should not be surprised at this, nor wonder at the fact that it is Beatrice, Bernard and Mary, not Raphael, Michael and Gabriel, whose prayers and service assist him in rising at last to his communion with 'the Glory Infinite'. That is only fitting, because the way of atonement is not just any way, but the way of the incarnation. When God comes down for our salvation, he comes down 'for us men' in a way that transforms what it means to be a man. As the seventh canto says,

> God more bounteous was himself to give,
> to make man able to uplift himself
> than if he only of himself had pardoned.[86]

The end of man, and his beginning too, are determined from his middle, the middle that is the incarnation. Of every creature created, from immortal angels to 'every brute and plant on earth', man alone is the one whose life God personally informs and 'immediately inspires', making us to love himself and kindling us 'to fresh desires'.[87] Man alone is permitted to penetrate the perichoretic spheres of the one whom Dante, in his final canto, dares himself to address, the trinitarian One 'who in thyself alone dwell'st, and thyself know'st, and self-understood, self-understanding, smilest on thine own!'[88]

In the closing lines of *Against the Heresies* Irenaeus (who may serve as our Bernard) is more explicit, if more prosaic. 'For there is one Son', he says, 'who accomplished His Father's will; and one human

---

85   *Paradiso* xxxiii, 124ff.

86   *Paradiso* vii, 115-117 (Longfellow). It is in this light that we should understand the poet's praise of Mary as *termine fisso d'etterno consiglio* (xxxiii, 3; cf. vii, 94-96). Mary is the destination of the descent of God the Son, in the non-narratable sense discussed above, and her womb the place from which the Son begins his (narratable) descent in recapitulation of our fall. As such she is also the threshold or point of access to the Son, in heaven as on earth, and to the deification which his ascent makes possible, though she remains her Son's daughter (vii, 1).

87   *Paradiso* vii, 142-144 (Binyon). Dante (145ff.) unites the doctrines of creation and of resurrection to the doctrine of the incarnation, and his Anselmian theology of atonement in canto vii lays the groundwork for his theology of deification in canto xxxiii. (On the former, see Brian Horne, 'The Cross and the Comedy', in Gunton, *The Theology of Reconciliation*, 141–57; on the latter, see Steven Botterill's *Dante and the Mystical Tradition*, noting his remarks at 228f.)

88   *Paradiso* xxxiii, 124-126 (Binyon).

race also in which the mysteries of God are wrought, "into which the angels desire to look." And they are not able to search out the wisdom of God, by means of which his handiwork, confirmed and incorporated with his Son, is brought to perfection: that his offspring, the first-begotten Word, should descend to the creature, that is, to what had been moulded, and that it should be contained by him; and, on the other hand, that the creature should contain the Word and ascend to him, passing beyond the angels, and be made after the image and likeness of God.'[89]

Has anything like this ever been said in treating of the destiny of man? Yet it is no violation of the canons of caution to speak of having an end that is higher, not merely in degree but in kind, than that of the angels, not if the *homoousion* is true. In the heavenly city, when that 'supremely cooperative, supremely ordered association of those who enjoy God and one another in God' is finally attained,[90] it will be the human race, fully incorporated into Christ in the eucharistic mystery that is the church, that shall reign over the new creation, participating from within in the three-personed Love that set everything in motion. The angels will see and rejoice, and nature itself shall be glad.

---

89   5.36.3 (altered)
90   *City of God* 19.13

Epilogue

IT IS SAID THAT EARLY CHRISTIANS, to escape detection, celebrated the ascension in a nearby cave on the Mount of Olives, before eventually moving to the spot thought to be the site of that event, on which a chapel was built toward the end of the fourth century, when the feast became more widely kept. The chapel was destroyed by the Persians in AD 614 but restored by St Modestus shortly afterwards. It was open to the sky, and by night lamps shone from its windows towards Jerusalem. Five hundred years later the chapel was again rebuilt – and roofed – by the crusaders, before passing back into Muslim hands, which added the dome. It was some three hundred years after that, in the mid-fifteenth century, when Pope Nicholas V, having finally drawn a line under the Avignon débâcle, dreamed on his deathbed of rebuilding in grand fashion the Basilica of St Peter, which also dated to the fourth century. He dared hope that the new structure would so far outstrip its Constantinian predecessor as to seem 'rather a divine than a human creation'.[1]

The task was undertaken by Julius II at the turn of the sixteenth century, and St Peter's today – in starkest contrast to the Chapel of the Ascension – bespeaks all the majesty, power and permanence that Nicholas had in mind. In sedimentary travertine and metamorphic

---

1   R. A. Scotti, *Basilica*, 21.

marble, it witnesses to the biblical promise of a city at once earthly and heavenly: a city that emerges gradually from below, shaped and contoured by temporal forces, yet somehow descends miraculously from on high, evincing its divine origin and destiny. Though seemingly chaotic in the making, that city will never pass away. For Christ himself is its cornerstone, and all its stones are living stones, quickened and transformed by the same Spirit who carried its Lord on high.

In *Basilica: The Splendor and the Scandal*, R. A. Scotti has retold with great skill the story of St Peter's own decidedly chaotic construction, not omitting the sobering fact that in answering to one great schism it helped create another. Therein lies a sad irony indeed – schism being the very opposite of the process of atonement by which the city of God is built. It is not for that reason, however, that we began our journey in Jerusalem rather than in Rome, crossing the humble portal of the Chapel of the Ascension before making, so to say, our *ad limina* in chapter five. We began with the Chapel of the Ascension both for the obvious reason that it is the Chapel of the Ascension and for the less obvious reason that (roof or no roof) it is a symbol of absence rather than presence, of provisionality rather than permanence. We began with the Chapel because it reminds us that the New Jerusalem cannot be built *by* man, but is built *for* man.

Nevertheless, we come round in the end to St Peter's, as we must. We do not come round merely because history itself has, for history still keeps some secrets, at least from us. Please God, the basilica will still be there – a church not a mosque or a museum – when Christians have done with schism. Please God, Peter's bones, which are nestled beneath it awaiting the last trumpet, will rest securely a little longer. Yet St Peter's itself is no more permanent than the Chapel of the Ascension, and its fate is likewise entangled with the vicissitudes of history. Pope Benedict, while still prefect of the Holy Office, rendered the judgement that the third secret of Fatima speaks to events now past, and no doubt it does. But prophecies too have their secrets, and the mystery of lawlessness still grows. St Peter will have his successors until the Lord comes, but his successors may not always have St Peter's.[2]

Nor do we come round because the glories of St Peter's, however

---

2   Hugh Benson was very likely right about this, though one may imagine other scenarios than that provided in the final pages of *Lord of the World* – pages that do nothing to rectify,

transitory, may serve to point us to the final redemption and affirmation of the human that God has worked in Jesus Christ. Point they do, and not (as some imagine) at the expense of any witness to absence or provisionality. For it is not Giacomo della Porta's dome, soaring upward to the heavens, that says most clearly what the basilica has to say about history under the sign of Christ. Rather it is Maderno's *confessio*, which leads downward to the crypt where Peter's bones lie. There one is reminded that the kingdom of God is glimpsed best in the company of the martyrs. Between dome and crypt, of course, there stands the papal altar to which Bernini's *baldacchino* draws the eye. That altar – or rather the eucharist celebrated on it – draws the whole of St Peter's, and all that it contains or represents, into the eschatological reality of the Christ who in his absence is nevertheless willing and able to make himself present.

Nor yet do we come round simply because Protestantism (as I learned in writing *Ascension and Ecclesia*) poses the greater threat of a false presence, a presence without absence, a presence that deifies by eliding the distinction between the divine and the human. In the end we come round to St Peter's because it houses – pre-eminently not exclusively – the true footprint of Jesus.[3] Faith requires a footprint; that is certain. It is equally certain that the footprint of Christ is a living one, impressed not merely in rock but in our humanity as a communion of confessors and confession, a communion in which the absent one is still present, in which his body is still given. But a footprint, to be recognizable, must have sufficient definition, and the definition Christ gave it was Petrine. Protestants (and some Orthodox too) like to say that it was the confession only, and not the confessor, that Christ intended by way of definition. Yet there is no confession without a confessor; and when Christ recognized Peter's confession as a gift from his Father, he committed to Peter, as to a steward, the keys to his Father's kingdom. Those who would enter with Peter must also confess with Peter, and confessing with Peter is done most truly in the humility of inward and outward harmony with Peter.

We come round, in other words, because ascension is atonement, and atonement is unity, and unity is visible, if ascension is ascension in the flesh. We come round because there is one church that is both

---

but only serve to highlight, the book's unfortunate lack of attention to the ineradicable Jewishness of salvation history.

3   The *confessio*, viewed from a certain angle, looks rather like a great footprint.

earthly and heavenly, both provisional and permanent. We come round because it is right and proper to confess with the fathers of Vatican II that this church, even *in hoc mundo*, is 'constituted and organized as a society' and is governed as such, under Christ, by 'the successors of Peter and by the bishops in communion with him'. That Peter's successors and their brothers have themselves sinned in such a way as to provoke schism does not alter that fact; neither does it justify schism.

To come round to St Peter's is not to abandon the old Jerusalem, whence the saviour departed and to which he shall return. It is, however, to admit that those who look to Jerusalem for the beginning and ending of the age must look to Rome in the meanwhile. For though they be scattered everywhere, as aliens and strangers even in their own lands, and though they worship not on this mountain or that, but wherever the Spirit leads them, there is a centre to their place of exile – which is why the New Testament sometimes refers to Rome as Babylon. That centre claims to be the Eternal City, when it is no such thing. But hard by it there is a sign, set there by the God who rules over the vicissitudes of history, of the city that is in fact eternal. That sign is the chair of Peter, without which Rome would already have ceased to be, for it is the city of God in its midst that preserves the city of man. When the city of man rises up and refuses to have the city of God in its midst, when it is as weary of the sign in Rome as of the sign in Jerusalem, then will the end of history come – the end, that is, of secular history, but the beginning of the kingdom that shall have no end.

A Summary of the Anaphoric Work of Christ

in his

offering to man & offering to God

| birth/baptism | V | passion/death |
| | *trial* | |

| descent to Hades | V | ascent to Heaven[†] |
| | *resurrection* | |

| coming in glory | V | cosmic regeneration |
| | *last judgement* | |

[†]Participation of the Church from Pentecost to Parousia

gospel mission		martyrdom
works of love		intercession
baptism		eucharist

V

epiclesis

Prayers for Ascensiontide

O my soul, magnify the Lord who in glory ascended bodily into the heavens.

Son of God, who ascended in glory, save us who sing to you; for behold the kings of the earth have assembled, they have come together.

The Divine Liturgy[1]

Father in heaven, our minds were prepared for the coming of your kingdom when you took Christ beyond our sight so that we might seek him in his glory. May we follow where he has led and find our hope in his glory, for he is Lord forever.

God our Father, make us joyful in the ascension of your Son Jesus Christ. May we follow him into the new creation, for his ascension is our joy and hope.

The Sunday Missal[2]

Grant, we beseech thee, Almighty God, that like as we do believe thy only-begotten Son our Lord Jesus Christ to have ascended into the heavens; so we may also in heart and mind thither ascend, and with him continually dwell; who liveth and reigneth with thee and the Holy Spirit, one God, world without end.[3]

O God the King of Glory, who hast exalted thine only Son Jesus Christ with great triumph unto thy kingdom in heaven: We beseech thee, leave us not comfortless; but send to us thine Holy Ghost to comfort us, and exalt us unto

1   P. Galadza, ed.; see also the Alternate Ambo Prayer on p. 550.
2   New Edition (Collins, London, 1984); based on a sermon of Leo the Great. See Bruce Harbert, 'Paradise and the Liturgy' (*New Blackfriars* 83.971, (2002), 31–40) on the challenges posed by cosmology and eschatology in the wording of such prayers, though his premises and conclusions are not my own.
3   Based on the ancient prayer in the Roman missal: *Grant, we beseech thee, almighty God, that we who believe thine only-begotten Son our Redeemer, to have ascended this day into heaven, may ourselves dwell in spirit amid heavenly things.*

the same place whither our Saviour Christ is gone before; who liveth and reigneth with thee and the Holy Ghost, one God, world without end.

Book of Common Prayer

Only-begotten Son of God! who, having conquered death, didst pass from earth to heaven: who, as Son of Man, art seated in great glory on thy throne, receiving praise from the whole angelic host! grant that we, who in the jubilant devotion of our faith, celebrate thine Ascension to the Father, may not be fettered by the chains of sin to the love of this world; and that the aid of our hearts may unceasingly be directed to the heaven, whither thou didst ascend in glory, after thy Passion.

O Lord Jesus Christ! who ascendest above the heaven of heavens to the East, after triumphing over thine own setting in the West; complete the work of our redemption, by raising us to the courts above. Thou, our Head, hast preceded us in glory; oh! draw thither, after thee, the whole body of thy Church, thy members whom thou callest to share thine honour. Leave not, we beseech thee, in the inglorious West of this world, those whom thou, the triumphant Conqueror, hast raised, by thine own Ascension, to the everlasting East.

Mozarabic Breviary [4]

Thou, O God, didst on this day raise up, together with thyself, above all Principalities and Powers, the nature of Adam, which had fallen into the deep abyss, but which was restored by thee. Because thou lovedst it, thou placedst it on thine own throne; because thou hadst pity on it, thou unitedst it to thyself; because thou hadst thus united it, thou didst suffer with it; because thou, the impassible, didst thus suffer, thou gave it to share in thy glory.

In Assumptione Domini, ad Vesperas [5]

---

4   Dom Prosper Guéranger, *The Liturgical Year*, vol. 3, 193 and 234f. The whole section from 130–265 (especially 167ff.) should be consulted. For additional biblical and patristic texts useful in the formation of ascension prayers, cf. Davies, *He Ascended into Heaven*, and Gerrit Dawson, *Jesus Ascended* – though Guéranger's riches make such works, including the present one, almost redundant.

5   Guéranger, *Liturgical Year*, vol. 3, 190. The following sentence is noteworthy: '*The Angels cried out: "Who is this beautiful Man? nay, not Man only, but God and Man, having the Nature of both!"*'

Prepare thus, Lord, what you are preparing; for you are preparing us for yourself, and yourself for us, inasmuch as you are preparing a place both for yourself in us, and for us in you.

Augustine, *Tractates on John* 68.3

WE BLESS YOU, *God our Father, for displaying your love both in the depths below and in the heights above. He whom the Virgin bore, who for our sake descended to the dead, has risen and ascended into heaven! Descending and ascending, he commended to your Holy Spirit our lost humanity, that we too might enter your glory.*

QUICKEN US, *O God, by the same Spirit, that we may lift from the depths our anthem of praise and receive from the heights your heavenly benediction. Keep us faithful to Jesus, our great priest and king, until he comes to make all things new. Grant that we, with him, may inhabit forever the mansions of your love.*

AMEN.[6]

---

6   I have based this on Irenaeus, *Against Heresies* 3.19.3. A eucharistic context is presupposed, in which the eucharist too is understood as a sign in the depths and in the heights: 'Lord, receive our offering as we celebrate the ascension of Christ your Son. May his gifts help us rise with him to the joys of heaven.' Or more richly in the Latin: *Sacrificium, Domine, pro Filii tui supplices venerabili nunc ascensione deferimus: praesta, quaesumus, ut his commerciis sacrosanctis ad caelestia consurgamus.*

Bibliography*

Appleyard, Bryan. *Understanding the Present*. An Alternative History of Science. Macmillan, New York, 2004.

Barth, Karl. *Church Dogmatics*. Ed. G. W. Bromiley and T. F. Torrance. T&T Clark, Edinburgh, 1956–75.

——— *Theology and Church*. Shorter Writings 1920–1928. Trans. L. P. Smith. SCM, London, 1962.

Bartsch, H. W., ed. *Kerygma and Myth*. A Theological Debate. Vol. 1. Trans. R. H. Fuller. SPCK, London, 1953.

Bauckham, Richard, Daniel Driver, Trevor Hart and Nathan MacDonald, eds. *The Epistle to the Hebrews and Christian Theology*. Eerdmans, Grand Rapids, 2009.

Benson, Robert Hugh. *Lord of the World*. BiblioBazaar, 2007.

Black, Conrad, Kathy Clark, Douglas Farrow and others. *Canadian Converts*. The Path to Rome. Justin Press, Ottawa, 2009.

Bonhoeffer, Dietrich. *Letters and Papers from Prison*. SCM, London, 1953.

——— *Christ the Center*. Trans. Edwin H. Robertson. Harper & Row, New York, 1978.

Botterill, Steven. *Dante and the Mystical Tradition*. Bernard of Clairvaux in the Commedia. Cambridge University Press, Cambridge, 1994.

Bowden, John. *Jesus*. The Unanswered Questions. SCM, London, 1988.

Buckley, Michael J. *At the Origins of Modern Atheism*. Yale University Press, Newhaven and London, 1987.

Bulgakov, Sergius. *The Bride of the Lamb*. Trans. Boris Jakim. Eerdmans, Grand Rapids, 2002.

---

* This bibliography includes only books cited from the modern period; for other and older literature, please consult the index and the notes.

Burgess, Andrew. *The Ascension in Karl Barth*. Ashgate, Aldershot, 2004.

Busch, Eberhard. *Karl Barth*. His Life from Letters and Autobiographical Texts. Trans. J. Bowden. Fortress, Philadelphia, 1976.

Canlis, Julie. *Calvin's Ladder*. A Spiritual Theology of Ascent and Ascension. Eerdmans, Grand Rapids, 2010.

Carnell, Corbin Scott. *Bright Shadow of Reality*. C. S. Lewis and the Feeling Intellect. Eerdmans, Grand Rapids, 1974.

Carpenter, Humphrey and Christopher Tolkien, eds. *The Letters of J. R. R. Tolkien*. Houghton Mifflin, Boston, 1981.

Cere, Daniel and Douglas Farrow, eds. *Divorcing Marriage*. Unveiling the Dangers in Canada's New Social Experiment. McGill-Queen's University Press, Montreal, 2004.

Chadwick, Owen. *The Reformation*. The Pelikan History of the Church. Vol. 3. Penguin, Hammondsworth, 1964.

Clarkson, John F., J. H. Edwards, W. J. Kelly, and J. Welch. *The Church Teaches*. Documents of the Church in English Translation. Herder, St Louis, 1955.

Collins, Jeffrey R. *The Allegiance of Thomas Hobbes*. Oxford University Press, Oxford, 2005.

Connelly, Matthew. *Fatal Misconception*. The Struggle to Control World Population. Belknap, Cambridge, MA, 2008.

Davies, J. G. *He Ascended into Heaven*. Lutterworth, London, 1958.

Dawson, Gerrit Scott. *Jesus Ascended*. The Meaning of Christ's Continuing Incarnation. Continuum, London, 2004.

Delors, Jacques. *Learning: The Treasure Within*. Report to UNESCO of the International Commission on Education for the Twenty-first Century. Odile Jacob, Paris, 1996.

Denzinger, H. *The Sources of Catholic Dogma*. Trans. R. J. Defarrari from the 30th edn of Enchiridion Symbolorum. Loreto, Fitzwilliam, NH, 2002.

Dru, Alexander, ed. *The Journals of Søren Kierkegaard*. Oxford University Press, London, 1938.

Dumbrell, William J. *The End of the Beginning*. Revelation 21–22 and the Old Testament. Baker, Grand Rapids, 1985.

Edwards, Mark Julian. *Origen against Plato*. Ashgate, Aldershot, 2002.

Ericson, Edward E. Jr, and Daniel J. Mahoney, eds. *The Solzhenitsyn Reader*. New and Essential Writings 1947–2005. ISI Books, Wilmington, 2006.

Farrow, Douglas. *Ascension and Ecclesia*. On the Significance of the Doctrine of the Ascension for Ecclesiology and Christian Cosmology. T&T Clark, Edinburgh, 1999.

——— *Nation of Bastards*. Essays on the End of Marriage. BPS, Toronto, 2007.

Farrow, Douglas, ed. *Recognizing Religion in a Secular Society*. Essays on Pluralism, Religion and Public Policy. McGill-Queen's University Press, Montreal, 2004.

Frye, Northrup. *The Great Code*. The Bible and Literature. Academic Press, Toronto, 1982.

Gage, Warren Austin. *The Gospel of Genesis*. Studies in Protology and Eschatology. Eisenbrauns, Winona Lake, 1984.

Gairdner, William D. *The Trouble with Democracy*. A Citizen Speaks Out. Stoddart, Toronto, 2001.

Galadza, Peter, ed. *The Divine Liturgy*. An Anthology for Worship. Metropolitan Andrey Sheptytsky Institute of Eastern Christian Studies, Ottawa, 2004.

Guéranger, Dom Prosper. *The Liturgical Year*. Pascal Time. Vol. 3. James Duffy, Dublin, 1871.

Gunton, Colin E., ed. *The Theology of Reconciliation*. T&T Clark, London, 2003.

Hall, Douglas John. *Professing the Faith*. Christian Theology in a North American Context. Fortress, Minneapolis, 1993.

Hart, David Bentley. *The Beauty of the Infinite*. The Aesthetics of Christian Truth. Eerdmans, Grand Rapids, 2003.

Hegel, G. W. F. *The Philosophy of History*. Trans. J. Sibree. Dover, New York, 1956.

Heine, Heinrich. *Religion and Philosophy in Germany*. A Fragment. Trans. John Snodgrass. Beacon Press, Boston, 1959.

Heron, Alasdair I. C. *A Century of Protestant Theology*. Westminster, Philadelphia, 1980.

——— *Table and Tradition*. Toward an Ecumenical Understanding of the Eucharist. Handsel, Edinburgh, 1983.

Hitler, Adolf. *Mein Kampf*. Trans. Ralph Manheim. Pimlico, London, 1992.

Holmes, Stephen and Murray Rae, eds. *The Person of Christ*. T&T Clark, London, 2005.

Hong, Howard V. and Edna H. Hong. *Søren Kierkegaard's Journals and Papers*. Vol. 2. Indiana University Press, Indiana, 1976.

Jaki, Stanley L. *The Saviour of Science*. Scottish Academic Press, Edinburgh, 1990.

Joyce, James. *A Portrait of the Artist as a Young Man*. Penguin, London, 1992.

Kant, Immanuel. *Religion within the Limits of Reason Alone*. Trans. T. M. Greene and H. H. Hudson. Harper Torchbooks, New York, 1960.

Kelly, J. N. D. *Early Christian Doctrines*. 5th edn. A&C. Black, London, 1977.

Khalidi, Tarif, ed. and trans. *The Muslim Jesus*. Sayings and Stories in Islamic Literature. Harvard, Cambridge, MA, 2001.

Kierkegaard, Søren. *Attack upon 'Christendom'*. Trans. Walter Lowrie. Beacon, Boston, 1956.

——— *Practice in Christianity*. Ed. and trans. H. V. Hong and E. H. Hong. Princeton University Press, Princeton, 1991.

Lindars, Barnabus. *The Theology of Hebrews*. Cambridge University Press, Cambridge, 1991.

McCabe, Herbert. *God Still Matters*. Continuum, London, 2002.

McCormack, Bruce L. *Karl Barth's Critically Realistic Dialectical Theology. Its Genesis and Development 1909–1936*. Clarendon Press, Oxford, 1995.

MacCullough, Diarmaid. *Thomas Cranmer*. A Life. Yale University Press, New Haven, 1996.

MacKintosh, Hugh R. *Types of Modern Theology*. Schleiermacher to Barth. Nisbet, London, 1937.

Marion, Jean-Luc. *God Without Being*. Hors-Texte. Trans. T. A. Carlson. University of Chicago, Chicago, 1991.

——— *Prolegomena to Charity*. Trans. Stephen Lewis. Fordham, New York, 2002.

Marius, Richard. *Thomas More*. A Biography. Harvard, Cambridge, MA, 1984.

Marrevee, W. H. Th*e Ascension of Christ in the Works of St. Augustine*. University of Ottawa Press, Ottawa, 1967.

Mill, John Stuart. *Utilitarianism, On Liberty*, and *Considerations on Representative Government*. Ed. H. B. Acton. Dent, London, 1972.

Murphy-O'Connor, Jerome. *The Holy Land*. An Archaeological Guide from Earliest Times to 1700. 2nd edn. Oxford University Press, Oxford, 1986.

Nietzsche, Friedrich. *Twilight of the Idols & The Anti-Christ*. Trans. R. J. Hollingdale. Penguin, London, 1968.

O'Donovan, Oliver and Joan O'Donovan, eds. *From Irenaeus to Grotius*. A Sourcebook in Christian Political Thought, 100–1625. Eerdmans, Grand Rapids, 1999.

Pannenberg, Wolfhart. *Systematic Theology*. Vol. 2. Trans. G. W. Bromiley. Eerdmans, Grand Rapids, 1994.

Parsons, Mikeal C. *The Departure of Jesus in Luke-Acts*. Journal for the Study of the New Testament Supplement Series 21, JSOT Press, Sheffield, 1987.

Pearson, John. *An Exposition of the Creed*. Ed. James Nichols. William Tegg, London, 1857 [1659].

Pelikan, Jaroslav. *The Christian Tradition*. A History of the Development of Doctrine, 5 vols. University of Chicago Press, Chicago and London, 1971–89.

Perry, Tim. *Mary for Evangelicals*. Toward an Understanding of the Mother of our Lord. IVP Academic, Downers Grove, 2006.

Petry, Ray C., ed. *A History of Christianity*. Readings in the History of the Church. Vol. 1. Baker, Grand Rapids, 1981.

Playoust, Catherine Anne. Lifted up from the Earth: The Ascension of Jesus and the Heavenly Ascents of Early Christians. UMI, Ann Arbor, 2007 (Harvard Divinity School dissertation, 2006).

Raeder, Linda C. *John Stuart Mill and the Religion of Humanity*. University of Missouri Press, Columbia and London, 2002.

Ratzinger, Joseph. *Eschatology*. Death and Eternal Life. Trans. Michael Waldstein. Ed. Aidan Nichols. Catholic University of America Press, Washington, 1988.

Rummel, Erika, ed. *The Erasmus Reader*. University of Toronto Press, Toronto, 1990.

Sahas, Daniel J. *John of Damascus on Islam*. The 'Heresy of the Ishmaelites'. Brill, Leiden, 1972.

Satinover, Jeffrey. *Homosexuality and the Politics of Truth*. Baker, Grand Rapids, 1996.

——— *The Quantum Brain*. The Search for Freedom and the Next Generation of Man. Wiley, New York, 2002.

Schleiermacher, Friedrich. *On Religion*. Speeches to its Cultured Despisers. Trans. John Oman. Harper & Row, New York, 1958.

Scotti, R. A. *Basilica*. The Splendor and the Scandal: Building St Peter's. Viking, New York, 2006.

Seitz, Christopher, ed. *Nicene Christianity*. The Future for a New Ecumenism. Brazos, Grand Rapids, 2001.

Sleeman, Matthew. *Geography and the Ascension Narrative in Acts*. Society for New Testament Studies Monograph Series 146. Cambridge University Press, Cambridge, 2009.

Sri, Edward. *Queen Mother*. A Biblical Theology of Mary's Queenship. Emmaus Road, Steubenville, 2005.

Strauss, David F. *The Life of Jesus, Critically Examined*. Trans. George Eliot. 2nd edn. SCM, London, 1973.

——— *The Life of Jesus for the People*. 2nd edn. 2 vols. Williams & Norgate, London, 1879.

Tait, Arthur, J. *The Heavenly Session of our Lord*. An Introduction to the History of the Doctrine. Robert Scott, London, 1912.

Teilhard de Chardin, Pierre. *Le Milieu Divin*. Ed. Bernard Wall. Collins / Fontana, London, 1960.

——— *The Future of Man*. Trans. Norman Denny. Collins, London, 1964.

———— *Hymn of the Universe*. Trans. Gerald Vann. Collins, London, 1965.

———— *The Phenomenon of Man*. Trans. B. Wall. Collins/Fontana, London, 1970.

Toon, Peter. *The Ascension of our Lord*. Nelson, Nashville, 1984.

Torrance, T. F. *Theology in Reconciliation*. Essays towards Evangelical and Catholic Unity in East and West. Geoffrey Chapman, London, 1975.

Trigg, Joseph Wilson. *Origen*. SCM, London, 1983.

Troeltsch, Ernst. *Writings on Theology and Religion*. Trans. and ed. R. Morgan and M. Pye. Westminster/John Knox, Louisville, 1990.

Walker, William Garnett. The Doctrine of the Ascension of Christ in Reformed Theology, UMI, Ann Arbor, 1991 (Vanderbilt University dissertation, 1968).

Walls, Jerry, ed. *The Oxford Handbook of Eschatology*. Oxford University Press, Oxford, 2007.

Webster, John. *Barth's Ethics of Reconciliation*. Cambridge University Press, Cambridge, 1995.

Webster, John, Kathryn Tanner and Iain Torrance, eds. *The Oxford Handbook of Systematic Theology*. Oxford University Press, Oxford, 2007.

Weikart, Richard. *From Darwin to Hitler*. Evolutionary Ethics, Eugenics, and Racism in Germany. Palgrave Macmillan, New York, 2004.

Wright, N. T. *The New Testament and the People of God*. Christian Origins and the Question of God. Vol. 1. Fortress, Minneapolis, 1992.

———— *Surprised by Hope*. Rethinking Heaven, the Resurrection, and the Mission of the Church. HarperCollins, New York, 2008.

Wybrew, Hugh. *The Orthodox Liturgy*. The Development of the Eucharistic Liturgy in the Byzantine Rite. SPCK, London, 1989.

Zizioulas, John D. *Being as Communion*. Studies in Personhood and the Church. St Vladimir's, Crestwood, 1985.

———— *Communion and Otherness*. Further Studies in Personhood and the Church. Ed. Paul McPartlan. T&T Clark, London, 2006.

List of Images

Front cover: Apocalyptic Christ
Cimabue, San Francesco, Assisi, c. AD 1290
Scala/Art Resource, NY

p. xi: Saint Clement and the Graoully
Vie de saint Clément, c. AD 1380
Bibliothèque Nationale de France (MS 5227 fol. 18r)

p. 1: The Ascension
Sacramentary, Mont-Saint-Michel, c. AD 1060
The Pierpont Morgan Library, NY (MS M.641 fol. 75v)

p. 11: The Tabernacle (Exodus 26)
Christoph Weigel, *Biblia ectypa* [Pictorial Bible], AD 1695
Pitts Theology Library, Candler School of Theology,
Emory University, Atlanta

p. 15: Chapel of the Ascension, Jerusalem
Photo: Odone Augusto Moojen

p. 33: The Ascension
Giotto di Bondone, Scrovegni Chapel, Padua, AD 1305
Scala/Art Resource, NY

p. 51: Protestant bishop blessing Nazi banners
Reichsbischof Ludwig Müller weiht in der Gustav-Adolf-Kirche in Berlin-Charlottenburg die Fahnen der angetretenen SA-Verbände, AD 1934
Süddeutsche Zeitung Photo/Scherl
The Bridgeman Art Library

p. 63: Elijah on the Fire-Cart
Giotto di Bondone, detail from The Ascension,
Scrovegni Chapel, Padua, AD 1305
Scala/Art Resource, NY

p. 89: Pope John Paul II at Mass in Downsview, Ontario,
15 September 1984
Canadian Press Photo/Luciano Mellace

p. 121: The New Jerusalem
Bamberg Apocalypse, c. AD 1020
Source: Wikimedia Commons

p. 153: Confessio, St Peter's Basilica
Photo: Franco Cosimo Panini Editore
Source: http://saintpetersbasilica.org/Confessio/Confessio.htm

p. 158: The Ascension (detail)
Mont-Saint-Michel sacramentary
The Pierpont Morgan Library, NY

p. 161: The central doors of the twelfth-century Basilica St Denis
(photographs by the author). Their roundels, read in descending
order on the left and ascending order on the right, tell the story
of recapitulation from the kiss of Judas in the garden to the
priestly blessing bestowed by Jesus at his return to the Father. The
tympanum (not shown) presents the last judgement. Anyone who
has walked through the doors of St Denis into its glorious interior
has some small inkling, at least, of the world to come – to which,
as the inscription overhead reminds us, 'Christ is the true door'.

Rear cover: *The Woman Clothed with the Sun Fleeth from the
Persecution of the Dragon*
Reproduction of a painting (c. AD 1797) by Benjamin West
Source: The Athenaeum (www.the-athenaeum.org)

The author would like to thank everyone who assisted with these images.

Index

* For guidance on the main topics and themes of this work, please see pp. ix–x.

Aaron 3, 11, 64, 140
Abel 13, 125
abortion 87, 110
Abraham 3
academia 80–1
Act of Supremacy 114
Adam 34–5, 45, 85, 142, 149–50, 160
adoptionism 52
Ambrose 91, 114–16, 129n. 20
Anselm 93n. 7, 114
anti-Semitism 101–2
Antichrist 95–6, 98, 100n. 29, 111–13, 118, 137, 142
Antiochus Epiphanes 4–5
Apocalypse, author of 13, 99n. 24, 112–13
Apollinarianism 82
Appleyard, Bryan 44n. 37
Aquinas, Thomas 25, 41, 69, 74, 76, 99n. 26, 117n.74, 134n. 46
Archangels 64, 90, 124n. 7, 136, 151
Arianism 29, 100n. 29, 148
Aristotle 17n. 7, 77

Ascension Day xi–xii, 11, 40–1, 90, 139
Athanasius 150n. 84
atonement 4, 52, 93n. 7, 120, 122–42, 155
Augustine of Hippo 21–4, 26, 38–9, 41–2, 58, 80n. 41, 86n. 54, 92–4, 105n. 41, 106n. 43, 124n. 8, 132–4, 137–8, 145n. 73, 150n. 82, 152, 161

Babel 3, 98n. 21
baptism 7, 9, 71, 75, 78, 82–3, 122, 128, 158
Barmen Declaration 102, 119
Barth, Karl 54, 61, 81–2, 101–3, 119, 145n. 74
Basil of Caesarea 43n. 35
Basilica of St Denis 170
Basilica of St Peter 153–6
Basilides 18, 19n. 11
Bathsheba 3
Beatrice 151
Bede, the Venerable 65–6, 145n. 73
Bellarmine, Robert 99n. 26

Benedict XVI 75, 81, 90–1, 98–9, 134–5, 154
Benson, Hugh 111n. 59, 154n. 2
Bentham, Jeremy 104n. 38
Berengarius 74
Bernard of Clairvaux 41, 151
Bernini, Gian Lorenzo 155
Black Rubric 73n. 21
Blonie Park 90–1
Blowers, Paul 130n. 25
Bonhoeffer, Dietrich 59–61, 134n. 46
Boniface, St 103
Boniface VIII 114
Botteril, Steven 151n. 87
Bowden, John 60n. 26
Buckley, Michael 44n. 36
Bulgakov, Sergius 144–5, 150n. 80
Bultmann, Rudolf 16–17, 37, 43
Burgess, Andrew 145n. 74
Busch, Eberhard 120n. 82

Cain 2, 110
Calvin, John 25, 27n. 31, 41, 49n. 52, 73, 78–9
Canlis, Julie 79n. 36
Capuchins 69
Carnell, Corbin Scott 132n. 31
Catherine of Aragon 113
Catholicism 54, 56, 59, 70–1, 79, 122n. 1, 139
Chagall, Marc 102n.32
Chapel of the Ascension 16, 31, 34, 37, 52, 153–4
Charles I, King of England 117
Charles V, Holy Roman Emperor 115
Christendom 58–9, 111n. 59, 114
Chrysostom, John 40
Church of England 113

Cicero 106n. 43
Clarkson, J. F. 40n. 23
Clement V 115
Clement of Alexandria 17n. 7
Collins, Jeffrey 117n. 77
communion of saints 85n. 53, 125n. 11, 136
communism 56, 100, 104
compassion 105
Connelly, Matthew 110
consubstantiation 75
contraceptive mentality 87, 108
church councils
    Chalcedon 149n. 79
    Constantinople II 23, 40n. 23
    Lateran IV 43n. 35, 73n. 19
    Toledo XI 40n. 23
    Trent 69, 73n. 21, 77n. 31
    Vatican II 69, 83n. 49, 156
Counter-Reformation 69
creatio ex nihilo 43n. 35, 46, 77, 84n. 51
Cynewulf 125n. 10
Cyril of Jerusalem 98, 113, 122

Daniel 4
Dante Alighieri 43, 124–5, 132, 134n. 44, 150–1
Darwin, Charles 57
Darwinism 55–6, 101
David 3, 5–8, 11, 64
Davies, J. G. 7n. 15, 8n. 19, 40n. 22, 41n. 24, 160n. 7
Dawson, Gerrit. 160n. 7
deification 34–6, 55–6, 109, 111, 122, 144–5, 150–1, 155
Delors, Jacques 106n. 44
democracy 104–5, 107, 109n. 52, 111, 118–19
Denzinger, Henry 74n. 24

Derrida, Jacques 58n. 19
devil 92–5, 97n. 20, 105, 118, 122, 125–6, 129n. 22, 137
dignity 100, 105, 107–8
Dix, William 64, 71
docetism 29, 52–3
Dumbrell, William 6n. 10

Eden 2–3, 6, 9, 34, 45, 149
Edwards, Mark 31n. 42
Einstein, Albert 43
Elijah 2, 65, 124, 137, 170
Elisha 64–5
Enoch 2, 45, 84n. 52, 124
epektasis 130n. 125
Erasmus, Desiderius 23
Erastianism 70, 113–14
eschaton 5, 80, 86
ethics 80, 82, 86
eucharistic adoration 78n. 33
eucharistic controversies 24–5, 64–82
eugenics 110
euthanasia 110
Eve 34, 82–5, 150
evil 18–19, 92, 96–8, 110, 112, 135n. 47
exclusivism 58

fall 3, 18–20, 34–6, 60, 105n. 40, 108, 122, 123n. 2, 142, 148–50, 160
Faraday, Michael 43
fascism 56, 100–4, 110
Fatima 140n. 63, 154
feminism 16, 53, 87
Fichte, Johann 104
Fifth Monarchists 118
Fisher, John 75
Fox, Matthew 56

Francis I, King of France 115
freedom of religion and of conscience 119n. 79
Frye, Northrup 9n. 23
fundamentalism 60

Gabriel 99n. 25, 151
Gage, William 3n. 3
Gairdner, William 105
Galilei, Galileo 25
Gelasian sacramentary 41n. 28
German Christian movement 101–2
Gibbons, Linda 141n. 6
gnosticism 17–20, 31, 34, 36–8, 44, 52, 54–6, 60n. 26, 78, 87, 109, 142–3, 148
Golden Legend xiv, 125, 145n. 73
grace 25, 66, 69n. 12, 76, 79, 81, 83–6, 122–3, 131, 134n. 46, 143
Gregory of Nazianzus 39–40
Gregory of Nyssa 130n. 25
Griffiths, Paul 134n. 46
Guéranger, Prosper xiv, 160

Haeckel, Ernst 103
Hall, Douglas 38n. 13
Harbert, Bruce 159n. 5
harm principle 105–6
harrowing of hell and heaven 125
Hart, David 130n. 25
heaven 123–9, 135, 137n. 55, 140, 147, 149, 152
heavenly session 2, 6–11, 22, 27n. 34, 40, 45–7, 53–4, 76–7, 90–3, 111, 113, 122–3, 136, 140
Hebrews, author of 11–13, 92, 123–4
Hegel, G. W. F. 28, 52, 54, 57–8, 60, 66, 81, 100–3, 123

Heidegger, Martin 58n. 19
Heine, Heinrich 103, 111
Henry VIII 113–19
Hick, John 60n. 26
Hilary of Poitiers 148
Hill, Edmund 39n. 19
Hitler, Adolf 101, 103, 110, 112, 115
Hobbes, Thomas 115, 117
homoousion 145, 148, 150, 152
homophobia 108
Horne, Brian 149n. 79, 151n. 87
human rights 108–10
Huxley, Aldous 111

imago dei 37, 105, 122, 129, 149, 152
Irenaeus xiii–xiv, 34–8, 40, 42, 67–9, 72–4, 78, 79n. 36, 83, 94, 98, 105n. 40, 111, 117–18, 123–5, 129–37, 142–3, 149, 151–2, 161
Isaiah 3, 46–7, 101n. 32, 124, 146–7, 149
Islam 52–3, 99–100

Jacobus de Voragine 124n. 8
Jaki, Stanley 44n. 35
Jantzen, Grace 16n. 3
Jerome 124n. 8
Jerusalem 3–4, 7, 9, 46–7, 68, 95–6, 126, 140, 153–4, 156
Jews 52, 99–103, 137, 142, 155n. 2
John, St 9–11, 80, 85, 95, 137n. 54
John of Damascus 40n. 22, 45, 100n. 29
John Paul II 71, 87n. 58, 90, 98n. 21, 108–9, 111–12, 120
John Philoponos 43n. 35
Josephus 5n. 9, 8
Joshua 8, 11
Joyce, James 74n. 24

Judaism 58, 66
Judas 52n. 2, 170
judgement 2–3, 9, 12, 54, 77n. 30, 80, 94–7, 126, 133–7, 140, 143–4, 154
Julian of Norwich 122–3
Julius II 153
Junge, Traudl 112n. 62
justice 19, 91–9, 104, 106, 108, 111, 117–18, 124n. 8, 134n. 46

Kant, Immanuel 26–7, 30, 101, 104
Kelly, J. N. D. 20n. 15, 29n. 38
Khalidi, Tarif 53n. 6
Kierkegaard, Søren 57–61
kingdom of God 5, 38n. 14, 17, 90, 129, 136n. 52, 155

Leo XIII 108, 119
Leo the Great 159n. 5
Leo the Martyr 75n. 26
Lessing, G. E. 27, 101
Leviathan 106–7
Lewis, C. S. 132
liberalism 53, 56
liberty 60, 68–9, 86, 96–7, 104–6, 108–12, 115, 130, 132
Lindars, Barnabus 139n. 59
Linge, Heinz 112n. 62
Locke, John 119
Lot 3n. 1
Luke, St 6–8, 12, 16
Luther, Martin 25, 59n. 22, 72n. 17, 101–3

McCabe, Herbert 77–8
McCormack, Bruce 102n. 34
MacCulloch, Diarmaid 114n. 67
MacKintosh, Hugh 28n. 37
Maderno, Carlo 155

Marcionism 45–6, 101
Marion, Jean-Luc 79
Marius, Richard 114n. 66,
116n. 73, 117
Mark, St 8
Marlowe, Christopher 144
Marrevee, William 22n. 20
marriage 87, 107–8, 113
martyrs 13, 55n. 10, 70, 91–5,
99–100, 114–20, 126, 140–1, 155
Mary 52, 54, 64, 82–7, 122,
129n. 20, 136n. 53, 147–8, 151
assumption 83–5
immaculate conception 83
Matthew, St 8, 11–12
Maximus Confessor 23–4,
128n. 19, 130n. 25
Maximus of Turin 40
Maxwell, J. C. 43
Melchizedek 6, 12–13
Mill, J. S. 104–6, 109, 119n. 79
millennium 134–7
Modestus of Jerusalem 153
Montanism 38n. 14
Montgomery, James 140
du Moulin, Pierre 117
More, Thomas 115–20
Moses 3, 8, 10–11, 60, 64
Muhammad 2
Müller, Ludwig 169
Murphy-O'Connor, Jerome
34
mystery of lawlessness 95–118,
125, 142, 154

National Socialism 102, 119–20
natural law 94, 107–8, 110
Nehemiah 4
Neuhaus, Richard xiii
Newton, Isaac 26, 43–4

Nicene creed xi, 40n. 23, 129n. 21,
145–9
Nicholas V 153
Nietzsche, Friedrich 99n. 27, 112
Noah 2–3

O'Donovan, Oliver and Joan
114n. 67
Origen 17–23, 26, 28–31, 37–8, 42,
54, 58, 130n. 25, 134–5
original sin 83
Orthodoxy 54, 59, 82, 128n. 19,
155

Pannenberg, Wolfhart 148n. 77
paradise 12, 125
parousia 2, 26, 29n. 38, 39, 48, 61,
66, 71, 76, 79–80, 86, 90, 94, 98,
113, 127–8, 134–41, 143–5, 158
Parsons, Mikeal 7n.15
Paul, St 10–11, 21, 41–2, 54, 95,
97n. 20, 100, 104, 116, 123, 130,
135–7
Paul VI 108n. 49
Pearson, John 43–5, 48, 138n. 56
Pelagianism 41, 69, 81–3, 87
Pentecost xii, 7, 11, 35, 133, 144–5
Perry, Tim 84n. 52, 86n. 54
Peter, St 7, 64, 85, 127, 133, 154–6
Petry, Ray 24n. 25
Philip IV, King of France 114–15
Philo of Alexandria 42
Pius XII 84n. 52
Planned Parenthood 109n. 55
Plato 17n. 7, 31
Platonism 17–20, 31
Playoust, Catherine 17n. 6
Plotinus 17, 27, 42
pluralism 16, 110–11
della Porta, Giacomo 155

positive law 104, 117n. 74
Prisca 142
Proclus 42
progress 16n. 3, 27–8, 34, 55, 58, 98, 130–3
Protestant principle 86n. 54, 111n. 59
Protestantism xiii, 27, 56, 59, 70, 71n. 15, 80–7, 101–2, 111n. 59, 155
Pseudo-Dionysius 24, 42–3
Ptolemaic universe 43–4
purgatory 134–7
Pythagoras 17n. 7

Raeder, Linda 106n. 44
Ratzinger, Joseph 134n. 46, 140
Rawls, John 119
real presence 70–80, 102
recapitulation 82n. 48, 111, 151n. 86, 129, 150–1, 170
Reformation 25, 59, 67, 81, 101, 111n. 59, 113–14
Religion of Humanity 30–1, 57
Religionsgeschichte and *Religionswissenschaft* 81
Reno, R. R. 87n. 58
reproductive rights 108
retroactivity 83–4
Rogation Days xii, xiv, 125
Rome 117, 153–6
Roschini, Gabriel 85n. 52
Rousseau, Jean-Jacques 119
Rublev's icon (The Old Testament Trinity) 82

Saccas, Ammonias 17, 19n. 10, 27
sacramental nominalism 76–8, 81, 102–3
sacramentalism 24, 47n. 46, 48, 55

Sahas, Daniel 100n. 29
Sartre, Jean-Paul 58n. 19
Satinover, Jeffrey 43n. 33, 104n. 37
schism 113–14, 118, 120, 154–6,
Schleiermacher, Friedrich 26–30, 52, 57, 60, 132–3
scientism 44
Scotti, R. A. 153–5
secularity 105
semiotics 80
Sinai 3, 10, 13n. 35, 46, 60, 138, 147
slavery 108
Sleeman, Matthew 7n. 15, 47n. 46
Soja, Ed 47n. 46
solus christus 85
Solzhenitsyn, Aleksandr 100n. 30
Sophia 87, 145n. 74, 148
space and spatiality 36, 44–7, 74, 90, 134n. 44, 136–7
Sri, Edward 83n. 49
Stanislaus of Szczepanów 91
state 58, 106–8, 110, 113–15, 117n. 74, 118
Strauss, D. F. 29, 103
substance–accident distinction 25n. 27, 74–8

Teilhard de Chardin, Pierre 54–6
temple 3–4, 9–11, 113, 136
Tertullian 38n. 14, 133, 142n. 66
theologia crucis 57
Theotokos 83. n. 49, 84–6
Thor 103–4
Time and temporality 5, 19–22, 28n. 36, 37, 42–9, 55, 65, 75–8, 85n. 53, 130–1, 133–9, 144
tolerance 105, 110–11
Tolkien, J. R. R. 95
Torrance, J. B. 82
Torrance, T. F. 43, 82

totus christus 55n. 9
transhumanism 56
transignification 81
transubstantiation 55, 66, 69–81
Trigg, Joseph 20n. 15
Trinity 91, 95, 129n. 21, 132n. 35, 148n. 78
Troeltsch, Ernst 53n. 5, 60, 143
Tyburn 117–18

ubiquity 24–5, 28, 54–5
United Nations 109n. 52, 106n. 44
universal commonwealth 111n. 59
Universal Declaration of Human Rights 113n. 65
universalism 56, 99 145n. 74
utopianism 80, 100, 106n. 44, 117

Valentinian II 116
Valentinus 18

Vermigli, Peter Martyr 81–2
Vespasian 5n. 9
visio dei 24, 150–2

Wagner, Richard 59n. 20
Waldron, Jeremy 104n. 38
Wars of Religion 115n. 70
West, Benjamin. 13n. 37
Wright, N. T. 5n. 9, 7n. 13, 45n. 42
Wybrew, Hugh 128n. 19

Yom Kippur 4, 12, 123, 139–40

Zeno 17n. 7
Zion 3–7, 12, 46, 138
Zizioulas, J. 59n. 21, 75n. 28, 150n. 83
Zwingli, Ulrich 82